LIQUID-METAL FLOWS AND MAGNETOHYDRODYNAMICS

The Third International Seminar in the
MHD Flows and Turbulence Series

Edited by
H. Branover
Ben-Gurion University of the Negev
Beer-Sheva, Israel

P.S. Lykoudis
Purdue University
West Lafayette, Indiana

A. Yakhot
Ben-Gurion University of the Negev
Beer-Sheva, Israel

Volume 84
PROGRESS IN
ASTRONAUTICS AND AERONAUTICS

Martin Summerfield, Series Editor-in-Chief
Princeton Combustion Research Laboratories, Inc.
Princeton, New Jersey

Technical papers from the Proceedings of the Third Beer-Sheva
International Seminar on Magnetohydrodynamic Flows and Turbulence,
Ben-Gurion University of the Negev, Beer-Sheva, Israel, March 23–27,
1981, and subsequently revised for this volume.

Published by the American Institute of Aeronautics and Astronautics
1633 Broadway, New York, New York 10019

American Institute of Aeronautics and Astronautics, Inc.
New York, New York

Library of Congress Cataloging in Publication Data

Beer-Sheva Seminar on MHD-flows and Turbulence
 (3rd: 1981: Ben-Gurion University of the Negev)
 Liquid-metal flows and magnetohydrodynamics.

 (Progress in astronautics and aeronautics; v. 84)
 Includes bibliographical references.
 1. Fluid dynamics—Congresses. 2. Liquid metals—Congresses.
 3. Magnetohydrodynamics—Congresses. I. Branover, H.
 II. Lykoudis, P. S. (Paul S.), 1926-   . III. Yakhot, A.
 IV. American Institute of Aeronautics and Astronautics, Inc.
 V. Title. VI. Series.
 TL507.P75 vol. 84 [TA357] 629.1s [538'.6] 83-2610
 ISBN 0.915928-70-1

Copyright © 1983 by
American Institute of Aeronautics and Astronautics, Inc.

All rights reserved. No part of this book may be reproduced in any form or by any means, electronic or mechanical, including photocopying, recording, or by any information storage and retrieval system, without permission in writing from the publisher.

Progress in Astronautics and Aeronautics

Series Editor-in-Chief

Martin Summerfield
Princeton Combustion Research Laboratories, Inc.

Series Associate Editors

Burton I. Edelson
*National Aeronautics
and Space Administration*

Leroy S. Fletcher
Texas A&M University

Allen E. Fuhs
Naval Postgraduate School

Rudolf X. Meyer
The Aerospace Corporation

J. Leith Potter
Vanderbilt University

Norma J. Brennan
Director, Editorial Department
AIAA

Camille S. Koorey
Series Managing Editor
AIAA

Table of Contents

Preface ix

List of Series Volumes xiv

Chapter I   Fundamental Studies in MHD and Turbulence 1

   Three-Dimensional Laminar MHD Flows in Rectangular Ducts
   with Thin Conducting Walls and Strong Transverse
   Nonuniform Magnetic Fields 3
     J.S. Walker, *University of Illinois, Urbana, Ill.*

   Why, How, and When MHD Turbulence
   Becomes Two Dimensional 20
     R. Moreau, *Université de Grenoble, Grenoble, France*

   Relaminarization—Magnetohydrodynamic and Otherwise 30
     R. Narasimha, *Indian Institute of Science, Bangalore, India*

   Experimental Studies on Liquid-Metal Two-Phase MHD Flow
   in Traveling and dc Magnetic Fields 53
     Y. Fujii-e, *Nagoya University, Nagoya, Japan* and K. Miyazaki, S. Inoue,
     and N. Yamaoka, *Osaka University, Osaka, Japan*

   Some Peculiarities of Turbulence Suppression by Magnetic Fields
   and Their Use for Magnetic Flow Control in Liquid Metals 83
     H. Branover, *Ben-Gurion University of the Negev, Beer-Sheva, Israel*
     and S. Claesson, *University of Uppsala, Uppsala, Sweden*

   Geostrophic Turbulence, Stable Baroclinic Eddies, and Self-Exciting
   Dynamos in Rotating Fluid Systems: A Summary 90
     R. Hide, *Geophysical Fluid Dynamics Laboratory, Bracknell, England*

   Velocity and Particle Size Measurements by a Laser Technique 94
     Y.M. Timnat, *Technion—Israel Institute of Technology, Haifa, Israel*

   Interactions of Turbulent Eddies with a Free Surface 98
     D. Naot, *Technion—Israel Institute of Technology, Haifa, Israel*
     and W. Rodi, *University of Karlsruhe, Karlsruhe,*
     Federal Republic of Germany

   Diagnostics of Flowfields 113
     S. Lederman, P. Brown, and T. Posillico, *Polytechnic Institute*
     of New York, Farmingdale, N.Y.

Chapter II MHD Generation and Electromagnetic Flowmeters 125

Conceptual Design of a Coal-Fired Retrofit Liquid-Metal MHD Power System 127
E.S. Pierson, H. Herman, and M. Petrick, *Argonne National Laboratory, Argonne, Ill.*

Solar-Powered Liquid-Metal MHD Performance and Cost Studies 138
E.S. Pierson and H. Herman, *Argonne National Laboratory, Argonne, Ill.*

An Operating Model of Two-Phase Heavy-Metal MHD Power Systems 160
H. Branover, A. El-Boher, and A. Yakhot, *Ben-Gurion University of the Negev, Beer-Sheva, Israel* and S. Claesson, *University of Uppsala, Uppsala, Sweden*

An Analytical Model of a Two-Phase Liquid-Metal Magnetohydrodynamic Generator 176
A. Yakhot and H. Branover, *Ben-Gurion University of the Negev, Beer-Sheva, Israel*

Critique of MHD Electrical Power Generation 193
W.D. Jackson, *Energy Consultant, Inc., Washington, D.C.*

Formulation of the Slip Loss in a Two-Phase Liquid-Metal Magnetohydrodynamic Generator 216
G. Fabris, *Argonne National Laboratory, Argonne, Ill.*

A Review of Recent Developments in Electromagnetic Flow Measurement 225
R.C. Baker, *Cranfield Institute of Technology, Cranfield, England*

Chapter III Electromagnetic Pumps, Flow Couplers, Fission and Fusion Applications 261

High Interaction Parameter Studies in a Large NaK Loop 263
I.R. McNab, C.C. Alexion, A.R. Keeton, and P.A. Ciarelli, *Westinghouse Electric Corporation, Pittsburgh, Pa.*

High-Efficiency Direct-Current Electromagnetic Pumps and Flow Couplers for Pool-Type LMFBRs 266
I.R. McNab and C.C. Alexion, *Westinghouse Electric Corporation, Pittsburgh, Pa.* and R.K. Winkleblack, *Electric Power Research Institute, Palo Alto, Calif.*

A Quasi-One-Dimensional Analysis of an Electromagnetic Pump Including End Effects 287
W.F. Hughes, *Carnegie-Mellon University, Pittsburgh, Pa.*
and I.R. McNab, *Westinghouse Electric Corporation, Pittsburgh, Pa.*

A Finite-Element Analysis of Two-Dimensional MHD Flow 313
N.S. Winowich and W.F. Hughes, *Carnegie-Mellon University, Pittsburgh, Pa.*

Fusion Application of an Imploding Shell Initially Formed by Falling Liquid Metal 323
Y. Itoh, T. Kanagawa, N. Yamaoka, and K. Miyazaki, *Osaka University, Osaka, Japan* and Y. Fujii-e, *Nagoya University, Nagoya, Japan*

Geometrical Integrity of a Metallic Cylindrical Shell Magnetically Imploded on an Axial Magnetic Field 339
Y. Itoh, T. Kanagawa, S. Umezawa, and K. Miyazaki, *Osaka University, Osaka, Japan* and Y. Fujii-e, *Nagoya University, Nagoya, Japan*

Chapter IV Metallurgical Magnetohydrodynamics 357

Electromagnetic Stirring in the Coreless Induction Furnace 359
D.J. Moore and J.C.R. Hunt, *University of Cambridge, Cambridge, England*

Single-Phase Electromagnetic Stirring in Coreless Induction Furnaces 374
Y. Fautrelle, *Université de Grenoble, Grenoble, France*

Liquid-Solid Separation in a Molten Metal by a Stationary Electromagnetic Field 387
Ch. Vives and R. Ricou, *Centre Universitaire d'Avignon, Avignon, France*

Pressure and Velocity Distribution around an Obstacle Immersed in Liquid Metal Subjected to Electromagnetic Forces 402
Ph. Marty and A. Alemany, *Université de Grenoble, Grenoble, France*
and R. Ricou and Ch. Vives, *Centre Universitaire d'Avignon, Avignon, France*

Liquid-Metal Columns Confined by External Parallel Conductors and Surface Tension Part I: Two-Dimensional Theory 414
J.A. Shercliff, *University of Cambridge, Cambridge, England*

Liquid-Metal Columns Confined by External Parallel Conductors and Surface Tension Part II: Experiment 422
J. Etay and M. Garnier, *Université de Grenoble, Grenoble, France*

Electromagnetic Devices for Molten Metal Confinement 433
M. Garnier, *Université de Grenoble, Grenoble, France*

Magnetic Levitation of Liquid Metals. 442
J. Mestel, *University of Cambridge, Cambridge, England*

**New Electromagnetic Measuring Methods
in Continuous Casting** 445
F.R. Block, *Technical University, Aachen,
Federal Republic of Germany*

Author Index for Volume 84 454

Preface

Liquid-metal flows, especially when influenced by external magnetic fields, manifest a number of very unusual phenomena which appear extremely interesting to anybody used to conventional fluid mechanics. Indeed, M-shaped velocity profiles in uniform straight ducts, many-fold amplified or many-fold reduced friction, depending on the direction of the magnetic field, strongly anisotropic and almost two-dimensional turbulence, and uncommon heat-transfer properties are frequently found. This is just a very much abbreviated list of phenomena, most of which have no analogs in other fluid flow situations.

But liquid-metal magnetohydrodynamic flows still would not attract so much attention if it were not for a number of important practical applications in metallurgy and foundries, in nuclear reactors and electrical power generation systems.

The number of books on magnetohydrodynamics (MHD) in general is very limited. A small but very valuable book, *Magnetohydrodynamics* by T.G. Cowling, appeared back in 1957 (Interscience Publishers, New York). The well-known *Textbook of Magnetohydrodynamics*, written by J.A. Shercliff, was published in 1965 (The Commonwealth and International Library of Science, Technology, Engineering, and Liberal Studies, London). Among several other books on MHD, the most application oriented are: *Engineering Magnetohydrodynamics* by A. Sherman and G. Sutton (McGraw-Hill, New York, 1965) and *An Introduction to Magneto-Fluid-Mechanics* by V.C.A. Ferraro and C. Plumpton (Second Edition, Oxford University Press, London, 1966). The twenty volumes of the *Proceedings of the Annual Conferences on Engineering Aspects of MHD Power Generation* as well as the *Proceedings of World Congresses on MHD Power Generation* should also be mentioned.

The number of books on liquid-metal MHD is even more limited. Actually, the only monograph in this field is *Magnetohydrodynamic Flow in Ducts* by H. Branover, published in 1978. There are also two volumes of the *Proceedings of the Beer-Sheva International Seminars on MHD Flows and Turbulence*, published in 1976 and

1980, respectively, containing papers related mainly to liquid-metal MHD problems.

A number of books on both physical aspects and applications of liquid-metal MHD flows have been published in the Soviet Union. However, they have unfortunately never been translated into English. The only systematic review of Soviet work in this field published in English is "Liquid Metal Magnetohydrodynamics" by O.A. Lielausis in *Atomic Energy Review* [1975, Vol. 13(3), pp. 527-581].

In view of this situation, it is more than timely to publish this collection of papers representing the present status of studies of liquid-metal MHD flows, including some related problems of MHD of a general kind and different applications of these flows. The papers contained herein were presented at the Third Beer-Sheva International Seminar on MHD Flows and Turbulence, held in March 1981. They can be divided into four groups.

Chapter I deals with fundamental problems of single- and two-phase liquid-metal MHD flows and turbulence, with and without magnetic fields. The paper of *Walker* presents new results related to one of the long-established fields of MHD duct flow theory. It should be mentioned that Walker has for a long time been one of the most active contributors to this field, along with Ludford in the U.S.A., Shercliff and Hunt in England, and Kulikovskii, Lyubimov, and Regiere in the Soviet Union. In the present paper, Walker is concerned with flow in ducts with walls parallel to the magnetic field, when the walls are thin and finitely conducting and the field or the cross section is nonuniform. The paper establishes the existence of a thick inviscid layer on the walls parallel to the magnetic field outside the usual thin viscous layer. This layer is characterized mainly by electric current circulation in the streamwise direction. The important implication is that end losses at the edge of the magnetic field are essentially higher than previously expected.

The paper of *Moreau* presents a review of all the presently available information on the anisotropy of turbulence in liquid-metal MHD flows and on the extreme case of two-dimensional MHD turbulence. Data accumulated in Riga, Purdue, Beer-Sheva, and Grenoble are analyzed, and it is established that weakly dissipative quasi-two-dimensional turbulence exists when the Hartmann-to-Reynolds number ratio is high and the insulating walls of the duct are perpendicular to the magnetic field. The paper suggests that MHD flows of this type offer a good opportunity for realization of experiments with two-dimensional turbulence. Another review written by *Narasimha* is dedicated to the relaminarization phenomenon. The author treats this physically very interesting and practically important phenomenon from the most general point of view, and turbulent flow relaminarization by magnetic fields appears just as one of the various relaminarization cases.

The next two papers deal with two-phase liquid-metal and gas flows under the influence of magnetic fields. The paper of *Fujii-e, Miyazaki, Inoue,* and *Yamaoka* presents experimental results obtained with sodium, potassium, and mercury flows in both steady and traveling magnetic fields. It should be mentioned that there are very few research projects performed in the field of two-phase MHD flows and this paper covers a great deal of the presently available knowledge in this field.

The next paper, written by *Branover* and *Claesson,* presents experimental results on pressure drop in an MHD duct with a high-aspect ratio cross section. Because of different conditions of turbulence suppression, the pressure drop differs almost 10 times, depending on whether the magnetic field is perpendicular to the short or long side of the cross section. This offers a possibility for a noncontact flow rate control in liquid-metal flows.

The extensive review by *Hide* deals with the very special motions that can be produced in rotating laboratory systems and other fluid-mechanical phenomena that are relevant to Earth and planetary science. This paper is only partially connected with present-day magnetohydrodynamics but may eventually impact the field.

This chapter includes three additional papers without specific reference to MIID, dealing with theories of turbulent flows and experimental techniques for measuring instantaneous velocities and droplet size in multiphase flows and with diagnostics of flowfields.

Chapter II is related to MHD power generation and electromagnetic flowmeters. The concept of two-phase liquid-metal and gas (or vapor) MHD power generation systems is dealt with in rather great detail in a number of papers. According to this concept, a liquid-metal "armature" is propelled in the MHD channel by a second, lighter phase. The specific thermodynamic cycle characteristic for these systems is inherently more efficient than conventional cycles performed between the same temperature limits, and this makes the system very attractive.

Pierson, Herman, and *Petrick* are concerned with the conceptual design of a large-scale system using liquid copper and coal combustion gases. Another paper by *Pierson* and *Herman* presents assessments of schemes for high-temperature systems using highly concentrating solar collectors as a heat source.

Branover, El-Boher, Yakhot, and *Claesson,* describe small complete models of two-phase liquid-metal (mercury) MHD systems, which have been built and tested in Beer-Sheva, Israel and

Uppsala, Sweden. The theory of the flow in two-phase MHD generators is developed in the paper of *Yakhot* and *Branover*. The equations are averaged, and inhomogeneity of the void fraction over the cross section is taken into account by correlation coefficients. Fair agreement with experimental data of the Argonne National Laboratory is demonstrated. Different aspects of the conceptually difficult topic of energy losses due to slip between the liquid-metal and the gaseous phase are analyzed in the paper by *Fabris*.

The history and present status of high-temperature plasma MHD systems is given by *Jackson*. Finally, a very detailed and fundamental review of electromagnetic flowmeters is presented by *Baker*. Various alternative types of flowmeters are categorized, the theory of each is outlined, and the problems still unsolved are described.

Chapter III deals with electromagnetic pumps, flow couplers, fission, and fusion applications. Liquid alkali metals are the accepted coolants for fast breeder fission reactors, and therefore electromagnetic pumping is considered as a serious alternative to mechanical pumping. The paper by *McNab, Alexion, Keeton,* and *Ciarelli* presents results obtained from a large NaK loop. The MHD channel was tested in a pumping regime with electrical currents up to 18 kA. In another paper presented by *McNab, Alexion,* and *Winkleblack* a flow coupler is theoretically analyzed. In a flow coupler the pumping of the primary coolant takes place exactly as in a dc pump, but the necessary current is supplied by an adjoining liquid-metal dc generator in the secondary coolant circuit. This presents a number of advantages, some of which are related to the safety of the reactor.

Hughes and *McNab* describe a simple one-dimensional model used to study the effect of design parameters on overall performance of electromagnetic pumps. The paper of *Winowich* and *Hughes* presents an analysis of two-dimensional MHD duct flows using finite-element methods. Two papers by *Itoh, Kanagawa, Yamaoka, Miyazaki,* and *Fujii-e* and by *Itoh, Kanagawa, Umezawa, Miyazaki,* and *Fujii-e* deal with liquid-metal MHD flow aspects related to fusion technology.

Chapter IV represents metallurgical magnetohydrodynamics. This group of papers demonstrates the essential progress that has been achieved in applications of MHD to metallurgy during the last several years. Both theoretical and actual industrial applications are presented.

Moore and *Hunt* describe the results of an experimental study of liquid-metal motions that occur in a coreless induction furnace. Mean flow and turbulence were measured. The paper by *Fautrelle* deals with the theory of the flow of an induction furnace. The problem of separation of undesirable ingredients from the liquid metal is dealt with in two complementary papers by *Vives* and *Ricou* and by *Marty, Alemany, Ricou,* and *Vives*; theory and experiment are both presented.

Papers by *Shercliff* and by *Etay* and *Garnier* deal with theoretical and experimental aspects, respectively, of the possibility of shaping a falling liquid-metal column by means of vertical conductors bearing high-frequency ac current. The use of magnetic fields for keeping the liquid metal away from the walls of devices is discussed in the papers by *Garnier* and by *Mestel*.

Finally, direct industrial applications are discussed by *Block*. This paper deals with new techniques for determining the level in the mold in continuous casting of steel.

An overview of the papers gathered in this volume gives one a definite feeling of the maturity of liquid-metal MHD and of its serious accomplishments from the practical point of view. Indeed, it turns out that this relatively new field has not only satisfactory theoretical solutions to its credit and a large amount of experimental information in relation to a great variety of most interesting phenomena, but also a number of advantageous applications in power generation, flow measurement, and metallurgy, which have already reached the preindustrial or even the industrial stage. The editors must admit that this was rather unexpected even to themselves, and became convinced that this volume would stimulate more engineers to use MHD methods in their respective fields and more researchers to enter the field of MHD studies.

The editors would like to acknowledge the invaluable assistance of Ms. Brenda Hio and Ms. Camille S. Koorey, Managing Editor of the Series, and the irreplaceable and skillful guidance by Dr. Martin Summerfield, the Series Editor-in-Chief. The whole book was typed by Mrs. Lili Lang. She deserves very special thanks for exactness, dedication, and patience. Finally, all of the contributors are to be thanked sincerely for their cooperation and care in the preparation of their papers.

The Editors
May 1982

Progress in Astronautics and Aeronautics

Volume Titles | Volume Editors

*1. **Solid Propellant Rocket Research.** 1960
Martin Summerfield
Princeton University

*2. **Liquid Rockets and Propellants.** 1960
Loren E. Bollinger
The Ohio State University
Martin Goldsmith
The Rand Corporation
Alexis W. Lemmon Jr.
Battelle Memorial Institute

*3. **Energy Conversion for Space Power.** 1961
Nathan W. Snyder
Institute for Defense Analyses

*4. **Space Power Systems.** 1961
Nathan W. Snyder
Institute for Defense Analyses

*5. **Electrostatic Propulsion.** 1961
David B. Langmuir
Space Technology Laboratories, Inc.
Ernst Stuhlinger
NASA George C. Marshall Space Flight Center
J. M. Sellen Jr.
Space Technology Laboratories, Inc.

*6. **Detonation and Two-Phase Flow.** 1962
S. S. Penner
California Institute of Technology
F. A. Williams
Harvard University

*7. **Hypersonic Flow Research.** 1962
Frederick R. Riddell
AVCO Corporation

*8. **Guidance and Control.** 1962
Robert E. Roberson
Consultant
James S. Farrior
Lockheed Missiles and Space Company

*9. **Electric Propulsion Development.** 1963
Ernst Stuhlinger
NASA George C. Marshall Space Flight Center

---
*Now out of print.

*10. Technology of Lunar
Exploration. 1963

Clifford I. Cummings and
Harold R. Lawrence
Jet Propulsion Laboratory

*11. Power Systems for Space
Flight. 1963

Morris A. Zipkin and
Russell N. Edwards
General Electric Company

*12. Ionization in High-
Temperature Gases. 1963

Kurt E. Shuler, Editor
National Bureau of Standards
John B. Fenn, Associate Editor
Princeton University

*13. Guidance and Control—II.
1964

Robert C. Langford
General Precision Inc.
Charles J. Mundo
Institute of Naval Studies

*14. Celestial Mechanics and
Astrodynamics. 1964

Victor G. Szebehely
Yale University Observatory

*15. Heterogeneous Combustion.
1964

Hans G. Wolfhard
Institute for Defense Analyses
Irvin Glassman
Princeton University
Leon Green Jr.
Air Force Systems Command

*16. Space Power Systems
Engineering. 1966

George C. Szego
Institute for Defense Analyses
J. Edward Taylor
TRW Inc.

*17. Methods in Astrodynamics
and Celestial Mechanics. 1966

Raynor L. Duncombe
U. S. Naval Observatory
Victor G. Szebehely
Yale University Observatory

*18. Thermophysics and
Temperature Control of
Spacecraft and Entry
Vehicles. 1966

Gerhard B. Heller
*NASA George C. Marshall Space
Flight Center*

*19. Communication Satellite
Systems Technology. 1966

Richard B. Marsten
Radio Corporation of America

*20. Thermophysics of Spacecraft and Planetary Bodies: Radiation Properties of Solids and the Electromagnetic Radiation Environment in Space. 1967

Gerhard B. Heller
NASA George C. Marshall Space Flight Center

*21. Thermal Design Principles of Spacecraft and Entry Bodies. 1969

Jerry T. Bevans
TRW Systems

*22. Stratospheric Circulation. 1969

Willis L. Webb
Atmospheric Sciences Laboratory, White Sands, and University of Texas at El Paso

*23. Thermophysics: Applications to Thermal Design of Spacecraft. 1970

Jerry T. Bevans
TRW Systems

24. Heat Transfer and Spacecraft Thermal Control. 1971

John W. Lucas
Jet Propulsion Laboratory

25. Communications Satellites for the 70's: Technology. 1971

Nathaniel E. Feldman
The Rand Corporation
Charles M. Kelly
The Aerospace Corporation

26. Communications Satellites for the 70's: Systems. 1971

Nathaniel E. Feldman
The Rand Corporation
Charles M. Kelly
The Aerospace Corporation

27. Thermospheric Circulation. 1972

Willis L. Webb
Atmospheric Sciences Laboratory, White Sands, and University of Texas at El Paso

28. Thermal Characteristics of the Moon. 1972

John W. Lucas
Jet Propulsion Laboratory

29. Fundamentals of Spacecraft Thermal Design. 1972

John W. Lucas
Jet Propulsion Laboratory

30. Solar Activity Observations and Predictions. 1972 — Patrick S. McIntosh and Murray Dryer
   Environmental Research Laboratories, National Oceanic and Atmospheric Administration

31. Thermal Control and Radiation. 1973 — Chang-Lin Tien
   University of California, Berkeley

32. Communications Satellite Systems. 1974 — P. L. Bargellini
   COMSAT Laboratories

33. Communications Satellite Technology. 1974 — P. L. Bargellini
   COMSAT Laboratories

34. Instrumentation for Airbreathing Propulsion. 1974 — Allen E. Fuhs
   Naval Postgraduate School
   Marshall Kingery
   Arnold Engineering Development Center

35. Thermophysics and Spacecraft Thermal Control. 1974 — Robert G. Hering
   University of Iowa

36. Thermal Pollution Analysis. 1975 — Joseph A. Schetz
   Virginia Polytechnic Institute

37. Aeroacoustics: Jet and Combustion Noise; Duct Acoustics. 1975 — Henry T. Nagamatsu, Editor
   General Electric Research and Development Center
   Jack V. O'Keefe, Associate Editor
   The Boeing Company
   Ira R. Schwartz, Associate Editor
   NASA Ames Research Center

38. Aeroacoustics: Fan, STOL, and Boundary Layer Noise; Sonic Boom; Aeroacoustics Instrumentation. 1975 — Henry T. Nagamatsu, Editor
   General Electric Research and Development Center
   Jack V. O'Keefe, Associate Editor
   The Boeing Company
   Ira R. Schwartz, Associate Editor
   NASA Ames Research Center

39. Heat Transfer with Thermal Control Applications. 1975 — M. Michael Yovanovich
   University of Waterloo

40. **Aerodynamics of Base Combustion.** 1976

S. N. B. Murthy, Editor
Purdue University
J. R. Osborn, Associate Editor
Purdue University
A. W. Barrows and J. R. Ward, Associate Editors
Ballistics Research Laboratories

41. **Communication Satellite Developments: Systems.** 1976

Gilbert E. LaVean
Defense Communications Engineering Center
William G. Schmidt
CML Satellite Corporation

42. **Communication Satellite Developments: Technology.** 1976

William G. Schmidt
CML Satellite Corporation
Gilbert E. LaVean
Defense Communications Engineering Center

43. **Aeroacoustics: Jet Noise, Combustion and Core Engine Noise.** 1976

Ira R. Schwartz, Editor
NASA Ames Research Center
Henry T. Nagamatsu, Associate Editor
General Electric Research and Development Center
Warren C. Strahle, Associate Editor
Georgia Institute of Technology

44. **Aeroacoustics: Fan Noise and Control; Duct Acoustics; Rotor Noise.** 1976

Ira R. Schwartz, Editor
NASA Ames Research Center
Henry T. Nagamatsu, Associate Editor
General Electric Research and Development Center
Warren C. Strahle, Associate Editor
Georgia Institute of Technology

45. **Aeroacoustics: STOL Noise; Airframe and Airfoil Noise.** 1976

Ira R. Schwartz, Editor
NASA Ames Research Center
Henry T. Nagamatsu, Associate Editor
General Electric Research and Development Center
Warren C. Strahle, Associate Editor
Georgia Institute of Technology

46. **Aeroacoustics: Acoustic Wave Propagation; Aircraft Noise Prediction; Aeroacoustic Instrumentation.** 1976

Ira R. Schwartz, Editor
NASA Ames Research Center
Henry T. Nagamatsu, Associate Editor
General Electric Research and Development Center
Warren C. Strahle, Associate Editor
Georgia Institute of Technology

47. **Spacecraft Charging by Magnetospheric Plasmas.** 1976

Alan Rosen
TRW Inc.

48. **Scientific Investigations on the Skylab Satellite.** 1976

Marion I. Kent and Ernst Stuhlinger
NASA George C. Marshall Space Flight Center
Shi-Tsan Wu
The University of Alabama

49. **Radiative Transfer and Thermal Control.** 1976

Allie M. Smith
ARO Inc.

50. **Exploration of the Outer Solar System.** 1977

Eugene W. Greenstadt
TRW Inc.
Murray Dryer
National Oceanic and Atmospheric Administration
Devrie S. Intriligator
University of Southern California

51. **Rarefied Gas Dynamics, Parts I and II (two volumes).** 1977

J. Leith Potter
ARO Inc.

52. **Materials Sciences in Space with Application to Space Processing.** 1977

Leo Steg
General Electric Company

53. Experimental Diagnostics in Gas Phase Combustion Systems. 1977

Ben T. Zinn, Editor
Georgia Institute of Technology
Craig T. Bowman, Associate Editor
Stanford University
Daniel L. Hartley, Associate Editor
Sandia Laboratories
Edward W. Price, Associate Editor
Georgia Institute of Technology
James G. Skifstad, Associate Editor
Purdue University

54. Satellite Communications: Future Systems. 1977

David Jarett
TRW Inc.

55. Satellite Communications: Advanced Technologies. 1977

David Jarett
TRW Inc.

56. Thermophysics of Spacecraft and Outer Planet Entry Probes. 1977

Allie M. Smith
ARO Inc.

57. Space-Based Manufacturing from Nonterrestrial Materials. 1977

Gerard K. O'Neill, Editor
Princeton University
Brian O'Leary, Assistant Editor
Princeton University

58. Turbulent Combustion. 1978

Lawrence A. Kennedy
State University of New York at Buffalo

59. Aerodynamic Heating and Thermal Protection Systems. 1978

Leroy S. Fletcher
University of Virginia

60. Heat Transfer and Thermal Control Systems. 1978

Leroy S. Fletcher
University of Virginia

61. Radiation Energy Conversion in Space. 1978

Kenneth W. Billman
NASA Ames Research Center

62. Alternative Hydrocarbon Fuels: Combustion and Chemical Kinetics. 1978

Craig T. Bowman
Stanford University
Jørgen Birkeland
Department of Energy

63. Experimental Diagnostics in Combustion of Solids. 1978
Thomas L. Boggs
Naval Weapons Center
Ben T. Zinn
Georgia Institute of Technology

64. Outer Planet Entry Heating and Thermal Protection. 1979
Raymond Viskanta
Purdue University

65. Thermophysics and Thermal Control. 1979
Raymond Viskanta
Purdue University

66. Interior Ballistics of Guns. 1979
Herman Krier
University of Illinois at Urbana-Champaign
Martin Summerfield
New York University

67. Remote Sensing of Earth from Space: Role of "Smart Sensors." 1979
Roger A. Breckenridge
NASA Langley Research Center

68. Injection and Mixing in Turbulent Flow. 1980
Joseph A. Schetz
Virginia Polytechnic Institute and State University

69. Entry Heating and Thermal Protection. 1980
Walter B. Olstad
NASA Headquarters

70. Heat Transfer, Thermal Control, and Heat Pipes. 1980
Walter B. Olstad
NASA Headquarters

71. Space Systems and Their Interactions with Earth's Space Environment. 1980
Henry B. Garrett and Charles P. Pike
Hanscom Air Force Base

72. Viscous Flow Drag Reduction. 1980
Gary R. Hough
Vought Advanced Technology Center

73. Combustion Experiments in a Zero-Gravity Laboratory. 1981
Thomas H. Cochran
NASA Lewis Research Center

74. Rarefied Gas Dynamics, Parts I and II (two volumes). 1981
Sam S. Fisher
University of Virginia at Charlottesville

75. **Gasdynamics of Detonations and Explosions.** 1981

J. R. Bowen
University of Wisconsin at Madison
N. Manson
Université de Poitiers
A. K. Oppenheim
University of California at Berkeley
R. I. Soloukhin
Institute of Heat and Mass Transfer, BSSR Academy of Sciences

76. **Combustion in Reactive Systems.** 1981

J. R. Bowen
University of Wisconsin at Madison
N. Manson
Université de Poitiers
A. K. Oppenheim
University of California at Berkeley
R. I. Soloukhin
Institute of Heat and Mass Transfer, BSSR Academy of Sciences

77. **Aerothermodynamics and Planetary Entry.** 1981

A. L. Crosbie
University of Missouri-Rolla

78. **Heat Transfer and Thermal Control.** 1981

A. L. Crosbie
University of Missouri-Rolla

79. **Electric Propulsion and Its Applications to Space Missions.** 1981

Robert C. Finke
NASA Lewis Research Center

80. **Aero-Optical Phenomena.** 1982

Keith G. Gilbert
and Leonard J. Otten
Air Force Weapons Laboratory

81. **Transonic Aerodynamics.** 1982

David Nixon
Nielsen Engineering & Research, Inc.

82. **Thermophysics of Atmospheric Entry.** 1982

T. E. Horton
The University of Mississippi

83. **Spacecraft Radiative Transfer and Temperature Control.** 1982

T. E. Horton
The University of Mississippi

84. **Liquid-Metal Flows and Magnetohydrodynamics.** 1983

H. Branover
Ben-Gurion University of the Negev
P. S. Lykoudis
Purdue University
A. Yakhot
Ben-Gurion University of the Negev

(Other volumes are planned.)

Chapter I   Fundamental Studies in MHD and Turbulence

Three-Dimensional Laminar MHD Flows in Rectangular Ducts with Thin Conducting Walls and Strong Transverse Nonuniform Magnetic Fields

John S. Walker*
University of Illinois, Urbana, Ill.

Abstract

This paper treats the steady laminar flow of a liquid metal in a rectangular duct with thin, electrically conducting walls and with a plane, nonuniform magnetic field applied parallel to one pair of walls and normal to the flow direction. Induced magnetic fields are neglected, while the applied magnetic field is assumed to be sufficiently strong that inertial effects are negligible everywhere and that viscous effects are confined to boundary layers. A new type of boundary layer occurs in these flows, but does not occur in ducts with either insulating or highly conducting walls. This new type of layer has recently been treated for variable-area ducts with uniform magnetic fields, and the purpose of the present paper is to begin to extend that treatment to constant-area and variable-area ducts with nonuniform magnetic fields.

I. Introduction

A number of important applications of magnetohydrodynamics involve ducts with thin, electrically conducting walls. Insulating materials are incompatible with hot alkaline me-

---

Paper presented at Third Beer-Sheva International Seminar on Magnetohydrodynamic Flows and Turbulence, Ben-Gurion University of the Negev, Beer-Sheva, Israel, March 23-27, 1981. Copyright © American Institute of Aeronautics and Astronautics, Inc., 1982. All rights reserved.
*Professor, Department of Theoretical and Applied Mechanics.

tals and with high neutron fluxes, so that the walls of an electromagnetic feed pump for a liquid-metal fast breeder reactor and the walls of a liquid-lithium blanket for a magnetically confined fusion reactor must be metal. On the other hand, pressure drops and energy losses increase with increases of the thicknesses of all walls except the pump's electrodes, so that the thinnest metal walls compatible with strength requirements are desirable.

Walker and Ludford (1975) presented solutions for variable-area circular ducts with thin conducting walls and with uniform transverse magnetic fields, while Holroyd and Walker (1978) presented solutions for constant-area and variable-area circular ducts with thin conducting walls and with nonuniform transverse magnetic fields. These results indicated that the flows in ducts with thin conducting walls did not involve any basic phenomena which did not also occur in ducts with either insulating or highly conducting walls. However, Walker (1981) discovered that a new type of boundary layer occurs in rectangular ducts with thin conducting walls, but does not occur in rectangular ducts with any combination of insulating or highly conducting walls. The new layers do not occur in constant-area rectangular ducts with thin conducting walls and with uniform magnetic fields, which explains why they were not discovered during the previous analyses of fully developed flows in these ducts. Walker (1981) presented solutions for variable-area ducts with uniform transverse magnetic fields. The present paper presents some preliminary results for constant-area and variable-area rectangular ducts with thin conducting walls and with nonuniform transverse magnetic fields. The forms of the solutions are presented for the core and for the new outer side layers adjacent to the walls which are parallel to the magnetic field.

II. Governing Equations and Boundary Conditions

The dimensionless equations governing the steady flow of an electrically conducting, incompressible fluid in the presence of a magnetic field are

$$N^{-1}(\bar{v}\cdot\nabla)\bar{v} = -\nabla p + \bar{j} \times \bar{b} + M^{-2}\nabla^2 \bar{v} \tag{1a}$$

$$\bar{j} = -\nabla\phi + \bar{v} \times \bar{b}, \qquad \nabla\cdot\bar{j} = 0 \tag{1b,c}$$

$$\nabla\cdot\bar{v} = 0, \qquad \nabla\cdot\bar{b} = 0, \qquad \nabla \times \bar{b} = R_m \bar{j} \tag{1d,e,f}$$

(see Shercliff 1965). Here $\bar{v}$, p, $\bar{j}$, $\bar{b}$, and ϕ are the velocity, pressure, electric current density, magnetic field, and electric potential function, respectively, which are normalized with respect to V, $\sigma VB^2 L$, σVB, B, and VBL, respectively, where V, B, and L are the characteristic velocity and magnetic field strength and length, respectively, while σ is the fluid's electrical conductivity. In addition the dimensionless parameters

$$N = \sigma B^2 L/\rho V, \qquad M = BL(\sigma/\eta)^{1/2}, \qquad R_m = \mu\sigma VL$$

are the interaction parameter, Hartmann number, and magnetic Reynolds number, respectively, where ρ, η, and μ are the fluid's density, viscosity, and magnetic permeability, respectively. The fluid's physical properties are assumed to be constant.

The present analysis treats the flow in a rectangular duct with one pair of parallel walls (sides) at $z = \pm 1$ and one pair of symmetrically converging or diverging or parallel walls (top and bottom) at $y = \pm F(x)$, where the duct's centerline coincides with the x axis and L is half the distance between the sides (see Fig. 1). The average axial velocity at the $x = 0$ cross section is chosen as V, so that the dimensionless solution must satisfy a volume-flux condition

$$\int_{-1}^{1} \int_{-F(x)}^{F(x)} v_x dy dz = 4F(0) \qquad (2)$$

The duct considered here is made from a single piece of metal so that all four walls have the same electrical conductivity σ_w and the same thickness t, and the walls are electrically connected at the corners. In general, the electric current density and electric potential function in the fluid are coupled to these quantities in the duct's walls and the surrounding medium. However, if the thickness of the walls $t \ll L$ and if the surrounding medium is an electrical insulator, then a thin conducting wall boundary condition on the fluid variables can be derived. A thin conducting wall condition was first derived by Shercliff (1956) for fully developed flow in a constant-area duct, while a general condition for three-dimensional flows was recently derived by Walker (1981). For the walls of the present duct, the general thin

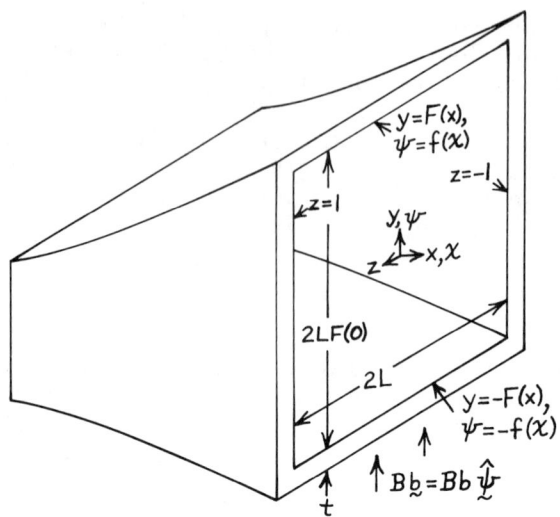

Fig. 1 The rectangular duct.

conducting wall condition reduces to

$$\bar{j} \cdot \bar{n} = c(\partial^2 \phi/\partial s^2 + \partial^2 \phi/\partial z^2), \quad \text{at} \quad y = \pm F(x) \quad (3a)$$

$$j_z = \mp c(\partial^2 \phi/\partial x^2 + \partial^2 \phi/\partial y^2), \quad \text{at} \quad z = \pm 1 \quad (3b)$$

where $\bar{n}$ is a unit vector which is normal to the top or bottom and which points into the fluid, $c = \sigma_w t/\sigma L$ the wall conductance ratio, and s the distance measured along the top or bottom in any z = const plane. The boundary conditions on the velocity are

$$\bar{v} = 0, \quad \text{at} \quad y = \pm F(x), \quad z = \pm 1 \quad (4a,b)$$

In general the magnetic field $\bar{b}$ in the fluid is coupled to $\bar{b}$ in the walls, surrounding medium and external magnet, so that the fluid boundary value problem is not yet decoupled from the external problems, even with the thin conducting wall assumption. However, if $R_m \ll 1$, then the right-hand side of Eq. (1f) is negligible and $\bar{b}$ is a potential field which is independent of the flow and which depends only on the geometry, electrical circuit, etc., of the external magnet. Henceforth, $\bar{b}$ will be taken as a known, applied magnetic field, thus decoupling the fluid and external problems. Only plane magnetic fields which are parallel to the duct's sides are considered here, i.e., $b_z = 0$ and $\partial \bar{b}/\partial z = 0$. Since the nonzero components of the magnetic field, $b_x(x,y)$ and $b_y(x,y)$, satisfy Eqs. (1e) and (1f) with the right-hand side of Eq. (1f) replaced by zero, a potential function $\psi(x,y)$ and a stream function $\chi(x,y)$ can be introduced for $\bar{b}$, such that

$$b_x = \partial\psi/\partial x = -\partial\chi/\partial y$$

$$b_y = \partial\psi/\partial y = \partial\chi/\partial x$$

where χ and ψ are conjugate harmonic functions which depend only on the external magnet's characteristics (Ludford and Walker 1980). In any z = const plane, the χ = const lines coincide with the magnetic field lines, while the ψ = const lines are perpendicular to $\bar{b}$ at each point. A strong magnetic field suppresses variations along the magnetic field lines, so that an intrinsic orthogonal curvilinear coordinate system (χ, ψ, z) with χ and ψ varying normal to and along magnetic field lines, respectively, is advantageous. The scale factors for the curvilinear system are

$$h_\chi = h_\psi = b^{-1}, \qquad h_z = 1$$

so that the length dS of a differential line element is given by

$$(dS)^2 = b^{-2}(d\chi)^2 + b^{-2}(d\psi)^2 + (dz)^2$$

where $b(\chi,\psi) = |\bar{b}|$ is the magnetic field strength at each point and is taken as a known scalar function as far as the flow problem is concerned. The magnetic field is $\bar{b} = b\bar{\psi}$, where $\bar{\psi}$ is a unit vector in the ψ direction, i.e., tangent to the χ = const, z = const line, at each point. Only essentially transverse magnetic fields which are symmetric about the $y = 0$ plane are considered here, so that b_x and b_y are odd and even functions of y, respectively. For symmetric fields and for $\chi = \psi = 0$ at $x = y = 0$, the $\psi = 0$ surface coincides with the $y = 0$ plane, the χ axis coincides with the x axis and the ψ axis coincides with the magnetic field line that passes through the origin. In a region where the magnetic field is uniform, the χ and ψ coordinates coincide with the x and y coordinates, respectively, while in a region where the magnetic field is nonuniform but still essentially transverse, the χ and ψ coordinates are still essentially axial and transverse, respectively, like the x and y coordinates, respectively. The remaining governing equations (1a-1d), the volume flux condition [Eq. (2)], and the boundary conditions [Eqs. (3) and (4)] are now rewritten in terms of the intrinsic orthogonal curvilinear coordinates [see, for example, Morse and Feshbach (1953)], while the fluid variables $\bar{v}$, p, $\bar{j}$, and ϕ are now functions of χ, ψ, and z. The locations of the inside surfaces of the top and bottom are now given by $\psi = \pm f(\chi)$, while

$$\bar{n} = (f'\bar{\chi} \mp \bar{\psi})/[1 + (f')^2]^{1/2}, \quad \text{at } \psi = \pm f(\chi)$$

where $\bar{\chi}$ is a unit vector in the χ direction at each point.

The magnetic field is assumed to be sufficiently strong that inertial effects are negligible everywhere and that viscous effects are confined to thin boundary layers adjacent to the walls and to thin free shear layers parallel to the magnetic field. It turns out that the flow considered here involves sheet jets adjacent to the sides, i.e., large $O(M^{1/2})$ velocities in the $O(M^{-1/2})$ thickness boundary layers adjacent to the sides, so that the inertialess assumption requires that $N \gg M^{3/2}$, where $M \gg 1$ (Walker et al. 1971). If the electrical conductivities of the fluid and walls are comparable, then the thin conducting wall assumption, $t \ll L$, implies that $c \ll 1$. If $c \ll M^{-1}$, then the flow is the same as the flow in a duct with insulating walls, which has been discussed by Ludford and Walker (1980) and by Holroyd (1976), so only values of c in the range $M^{-1} \ll c \ll 1$ are considered here. Under these assumptions the interior of the duct can be divided into subregions and certain terms in the governing equations (1a-1d) can be neglected in each subregion. The subregions (shown in Fig. 2) are the core (c), the Hart-

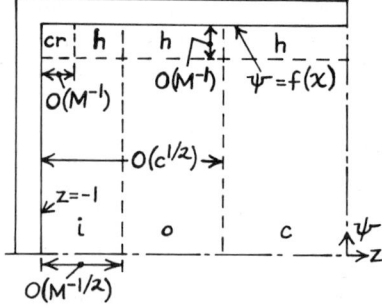

Fig. 2 One quarter ($\psi > 0$, $z < 0$) of a χ = const section of the duct showing the subregions of the flow for $M^{-1} \ll c \ll 1$: c) core, h) Hartmann layer, o) outer side layer, i) inner side layer, and cr) corner region.

mann layers of $O(M^{-1})$ thickness adjacent to the top and bottom (h), the outer side layers of $O(c^{1/2})$ thickness adjacent to the sides (o), the inner side layers of $O(M^{-1/2})$ thickness which lie between the outer side layers and the sides (i), and the corner regions with $O(M^{-1}) \times O(M^{-1})$ dimensions at the four corners (cr). All of these subregions are standard features of large Hartmann number flows in rectangular ducts [see, for example, Roberts (1967)] except the outer side layers which only occur in ducts with thin conducting walls.

The variables in the Hartmann layers are determined locally by the tangential velocity in the adjacent core or side layer evaluated at the wall. The Hartmann layer solution satisfies the boundary conditions of Eq. (4a) and matches the core or side layer solution provided the latter satisfies

$$f'v_\chi \mp v_\psi = 0, \qquad \text{at } \psi = \pm f(\chi) \tag{5}$$

where $O(M^{-1}\overline{v})$ terms are neglected (Walker et al. 1972). The jumps in the normal component of the electric current density and in the electrical potential functions across the Hartmann layers are $O(M^{-1}\overline{v})$ and $O(M^{-2}\overline{v})$, respectively. A corner region is needed to properly represent a singularity in the inner side layer solution at each corner. The solutions in the other regions can be obtained to first order without considering the corner region solution in detail (Temperley and Todd 1976).

Each variable in each subregion is written as an asymptotic expansion in two small parameters where the parameters are two independent combinations of c and M and where the coefficients are functions of the possibly rescaled coordinates but are independent of c and M. The order of magnitude of the leading term in each expansion is justified by the compatibility of the results, i.e., the solutions in adjacent subregions match and together satisfy the volume flux condition [Eq. (2)], which scales the otherwise homogeneous boundary value problem. Variables with the subscripts c, o, and i henceforth denote the leading terms in the asymptotic expansions for variables in the core, outer side layers, and inner side layers, respectively. Because of symmetry only the side layers at z = -1 are considered, while

$$v_\psi = j_\psi = 0, \qquad \text{at } \psi = 0 \tag{6a,b}$$

III. Core Solution

In the core the inertial term on the left-hand side and the viscous term on the right-hand side of Eq. (1a) are both negligible, all derivatives are $O(1)$ and all variables are $O(\gamma_c)$, where γ_c is a combination of M and c which is determined by matching and by Eq. (2). The ψ component of Eq. (1a) now indicates that p_c is a function of χ and z, i.e., that the core pressure is constant along magnetic field lines, while the other two components of Eq. (1a) are

$$j_{\chi c} = b^{-1} \frac{\partial p_c}{\partial z}, \quad j_{zc} = -\frac{\partial p_c}{\partial \chi} \quad (7a,b)$$

To obtain an expression for $j_{\psi c}$, Eqs. (7) are introduced into Eq. (1c), the resulting equation is integrated with respect to ψ and the symmetry condition [Eq. (6b)] is used to evaluate the integration function. The result is

$$j_{\psi c} = -b I_1 \frac{\partial p_c}{\partial z} \quad (8a)$$

where

$$I_1(\chi,\psi) = \int_0^\psi \frac{\partial [b(\chi,\tau)]^{-2}}{\partial \chi} d\tau \quad (8b)$$

The thin conducting wall condition [Eq. (3a)] becomes

$$f' j_{\chi c} \mp j_{\psi c} = 0, \quad \text{at} \quad \psi = \pm f(\chi)$$

where $O(c\gamma_c, M^{-1}\gamma_c)$ terms are neglected, and this condition, together with Eqs. (7a) and (8), gives

$$b\alpha \frac{\partial p_c}{\partial z} = 0 \quad (9a)$$

where

$$\alpha(\chi) = \frac{d}{d\chi} \int_0^{f(\chi)} [b(\chi,\psi)]^{-2} d\psi \quad (9b)$$

The integral in Eq. (9b) is the same as the integral derived by Holroyd and Walker (1978) to define the characteristic surfaces in ducts with insulating walls, since the differential length along a magnetic field line is $dS = b^{-1} d\psi$.

If $\alpha = 0$, then the MHD flow is "free," corresponding to a geostrophically free rotating flow (Howard 1969). In this case Eq. (9a) places no restrictions on p_c. It turns out that the Hartmann condition [Eq. (5)] does place a restriction on the core variables for $\alpha = 0$, but the $O(\gamma_c)$ solution still includes an integration function of χ and z, which has no corresponding restrictions or governing equation. The $O(c\gamma_c)$ thin conducting wall condition [Eq. (3a)] must be used to complete the analysis for the $O(\gamma_c)$ core solution. One special case of free flows with $\alpha = 0$ is a constant-area duct in a uniform magnetic field, i.e., $\bar{b} = \bar{y}$ and $F' = 0$. This special case has been treated by Hunt and Ludford (1968) for ducts with insulating top and bottom and by Walker (1982) for ducts with thin conducting top and bottom, while all free flows are qualitatively similar to the flows for this special case. For free flows, $\gamma_c = 1$, the volume flux of $4F(0)$ which is required by Eq. (2) is carried by the $O(1)$ core velocity, there are no outer side layers, and the inner side layers simply match the core velocity and satisfy the boundary condition [Eq. (4b)]. Since the objective of the present paper is to describe the outer side layers, free flows with $\alpha = 0$ are not considered any further here.

With the assumption that $\alpha \neq 0$, p_c is a function of χ alone, while

$$j_{\chi c} = j_{\psi c} = 0, \qquad j_{zc} = -dp_c/d\chi \qquad (10)$$

so that the core pressure is constant on, and the core electric current must flow transversely along, the $\chi = $ const sections of the duct. This MHD flow is "guided," correspondingly to a geostrophically guided rotating flow (Howard 1969). For $\alpha \neq 0$, γ_c is not yet determined. The ψ component of Eq. (1b) indicates that ϕ_c is a function of χ and z, i.e., that the core electric potential is constant along magnetic field lines for these guided flows, while the other two components of Eq. (1b), together with Eq. (10), are

$$v_{\chi c} = b^{-1}\frac{\partial \phi_c}{\partial z} - b^{-1}\frac{dp_c}{d\chi}, \qquad v_{zc} = -\frac{\partial \phi_c}{\partial \chi} \qquad (11a,b)$$

To obtain an expression for $v_{\psi c}$, Eqs. (11) are introduced into Eq. (1d), the resulting equation is integrated with respect to ψ, and the symmetry condition [Eq. (6a)] is used to evaluate the integration function. The result is

$$v_{\psi c} = b \int_0^{\psi} \frac{\partial}{\partial \chi} \{ \frac{dp_c}{d\chi} (\chi) [b(\chi,\tau)]^{-2} \} d\tau - bI_1 \frac{\partial \phi_c}{\partial z} \quad (12)$$

To obtain an expression for ϕ_c, Eqs. (11a) and (12) are substituted into the Hartmann condition [Eq. (5)] which neglects $O(M^{-1}\gamma_c)$ terms, the resulting equation is integrated with respect to z, and the symmetry condition (ϕ = 0 at z = 0) is used to determine the integration function. The result is

$$\phi_c = \frac{z}{\alpha} \frac{d}{d\chi} \{ \frac{dp_c}{d\chi} (\chi) \int_0^{f(\chi)} [b(\chi,\psi)]^{-2} d\psi \} \quad (13)$$

and the three components of the core velocity are determined in terms of $dp_c/d\chi$ by substituting Eq. (13) into Eqs. (11) and (12).

This guided flow core solution [Eqs. (10-13)] involves a transverse electric current j_{zc} which must be accepted by the side layers at z = ±1. If the sides were perfect conductors, then an O(1) core current could pass directly through the side layers and into the sides, so that an O(1) axial core velocity would be possible and at least part of the O(1) volume flux could be carried by the core (Walker et al. 1971). However, the sides of the present duct are thin conductors which cannot accept an O(1) electric current for c << 1. It turns out that the order of magnitude of the transverse electric current density which can be matched by the outer side layers in the present duct is $O(M^{-1/2})$ for M^{-1} << c << $M^{-1/2}$ or O(c) for $M^{-1/2}$ << c << 1, so that $\gamma_c = M^{-1/2}$ or c for these two ranges of values of c, respectively (Walker 1981). Therefore, the core velocity in the present duct is too small to carry any of the O(1) volume flux required by Eq. (2), and the MHD flow is "blocked," corresponding to geostrophically blocked rotating flows (Howard 1969). The small $O(\gamma_c)$ core solution is given by Eqs. (10-13) once $p_c(\chi)$ is determined by matching the core and outer side layer pressures. Thus the core flow represents a perturbation in an essentially stagnant core region and this perturbation flow is driven by the inner and outer side layers which must carry all of the volume flux.

IV. Outer Side Layer Solution

For the outer side layer at $z = -1$, the coordinate normal to the side is stretched by substituting $z = -1 + c^{1/2}\xi$, so that the derivatives with respect to χ, ψ, and ξ are all $O(1)$. If the outer side layer pressure is $O(\gamma_0)$, where γ_0 is an as yet undetermined combination of M and c, then $v_{\chi 0}$ and $v_{\psi 0}$ are $O(c^{-1}\gamma_0)$; v_{z0}, ϕ_0, $j_{\chi 0}$, and $j_{\psi 0}$ are $O(c^{-1/2}\gamma_0)$; and j_{z0} is $O(\gamma_0)$. Matching the core and the outer side layer solutions indicates that $\gamma_0 = \gamma_c$ and gives the boundary conditions

$$\phi_0 \to 0, \qquad p_0 \to p_c(\chi), \qquad \text{as} \qquad \xi \to \infty \qquad (14a,b)$$

Matching the other variables shows that $j_{z0} \to j_{zc}$, while the other outer side layer variables vanish as $\xi \to \infty$, but these conditions are automatically satisfied if the governing equations (1a-1d) and the conditions of Eqs. (14) are satisfied. Two steps remain in order to determine γ_0 and γ_c, as well as the order of magnitude of each inner side layer variable: matching the inner and outer side layer solutions and satisfying the volume flux condition [Eq. (2)]. As pointed out in the previous section, the result is that $\gamma_0 = \gamma_c = M^{-1/2}$ or c for $M^{-1} \ll c \ll M^{-1/2}$ or $M^{-1/2} \ll c \ll 1$, respectively (Walker 1981). Therefore, the essentially axial outer side layer velocity $v_{\chi 0}$ is large, namely $O(c^{-1}M^{-1/2})$, or $O(1)$ for $c \ll M^{-1/2}$ or $c \gg M^{-1/2}$, respectively, but the volume flux carried by the outer side layers is small in both cases, namely $O(c^{-1/2}M^{-1/2})$ or $O(c^{1/2})$, respectively. All of the $O(1)$ volume flux which is required by Eq. (2) is carried by the large $O(M^{1/2})$ essentially axial inner side velocity $v_{\chi i}$, while both the outer side layer and core solutions represent perturbation flows in essentially stagnant regions, which are driven by the electric current circulation required by the inner side layers. While the outer side layers make a negligible contribution to the volume flux, arguments and analysis to follow show that they are important parts of the electric current circuit and account for significant pressure variations.

For the outer side layers, the viscous term in Eq. (1a) is small, namely $O(c^{-2}M^{-2})$, compared to the pressure and electromagnetic body force terms, so that both the viscous and inertial terms are negligible. The governing equations (1a-1d) are essentially the same for both the core and the outer side layers, and the fundamental difference between these subregions arises from the fact that the right-hand side of the thin conducting wall condition [Eq. (3a)] is negligible for the core, but is comparable to the left-hand side for the outer side layers. For most inner regions in matched asymptotic expansions schemes the coordinate stretching is chosen so that a term in the governing equations, which is negligible compared to lower order terms for the outer region, is comparable to other terms for the inner region. This is true for the inner side layers in the present duct since an $O(M^{1/2})$ coordinate stretching normal to the magnetic field makes the viscous term in Eq. (1a) comparable to the pressure and electromagnetic body force terms. The outer side layers are unusual because the governing equations are unchanged by the coordinate stretching and the stretching scale arises from a change in a boundary condition. This situation is analogous to that for rotating flows with walls parallel to the axis of rotation (sides) and with small Ekman number E. These rotating flows involve inner viscous side layers of $O(E^{1/3})$ thickness and outer inviscid side layers of $O(E^{1/4})$ thickness, while the fundamental difference between the latter and the core arises because the Ekman pumping due to the Ekman layers, which are adjacent to the top and bottom, is negligible in the core, but is comparable to the velocity in the outer side layers (Howard 1969). The control of the $E^{1/4}$ layers by the Ekman pumping in rotating flows is analogous to the control of the $c^{1/2}$ layers by the thin conducting wall electric current "pumping," given by the right-hand side of Eq. (3a) here. In both cases, changes in the top and bottom boundary conditions produce more than local effects because variations along magnetic field lines or along lines parallel to the axis of rotation are suppressed by strong magnetic fields or by high speeds of rotation, respectively.

For the outer side layer at $z = -1$, the ψ component of Eq. (1a) indicates that p_o is a function of χ and ξ, i.e., that the outer side layer pressure is also constant along magnetic field lines, while the other two components of Eq. (1a) are

$$j_{\chi o} = b^{-1} \frac{\partial p_o}{\partial \xi}, \quad j_{zo} = -\frac{\partial p_o}{\partial \chi} \quad (15a,b)$$

After Eqs. (15) are introduced into Eq. (1c), this equation is integrated with respect to ψ and the symmetry condition [Eq. (6b)] is used to determine the integration function. The result is

$$j_{\psi o} = -bI_1 \frac{\partial p_o}{\partial \xi} \quad (16)$$

where $I_1(\chi,\psi)$ is defined by Eq. (8b). After Eqs. (15) and (16) are introduced into Eq. (1b), the ψ component of this equation can be integrated with respect to ψ to obtain

$$\phi_0 = \Phi + \frac{\partial p_0}{\partial \xi} \int_0^\psi I_1(\chi,\tau) d\tau \qquad (17)$$

where $\Phi(\chi,\xi)$ is an integration function which equals the outer side layer electric potential in the $\psi = 0$ plane of symmetry. The other two components of Eq. (1b) are now

$$v_{\chi 0} = b^{-1} \frac{\partial \phi_0}{\partial \xi}, \qquad v_{z0} = -\frac{\partial \phi_0}{\partial \chi} - b^{-2} \frac{\partial p_0}{\partial \xi} \qquad (18a,b)$$

After Eqs. (18) are introduced into Eq. (1d), the latter is integrated with respect to ψ and the symmetry condition [Eq. (6a)] is used to determine the integration function. The result is

$$v_{\psi 0} = bI_2 \frac{\partial^2 p_0}{\partial \xi^2} - b \int_0^\psi \frac{\partial \phi_0}{\partial \xi}(\chi,\tau,\xi) \frac{\partial}{\partial \chi} [b(\chi,\tau)]^{-2} d\tau \qquad (19a)$$

where

$$I_2(\chi,\psi) = \int_0^\psi [b(\chi,\tau)]^{-4} d\tau \qquad (19b)$$

When Eq. (17) is substituted into Eqs. (18) and (19a) and the terms which are independent of τ are moved outside the integrals in Eq. (19a), the results are expressions for the three components of the outer side layer velocity in terms of ξ derivatives of $p_0(\chi,\xi)$ and $\Phi(\chi,\xi)$ and in terms of integrals of the known function $b(\chi,\psi)$. When the results from Eqs. (18a) and (19a) are introduced into the Hartmann conditions [Eq. (5)], which neglects terms that are $O(c^{-1/2}M^{-1})$ compared to the retained terms, the resulting equation can be integrated with respect to ξ and the boundary conditions [Eqs. (14)] can be used to evaluate the integration functions. The result is

$$\Phi = \alpha^{-1} \frac{\partial p_0}{\partial \xi} \{I_2(\chi,f) - f' [b(\chi,f)]^{-2} \int_0^f I_1(\chi,\tau) d\tau$$

$$-\int_0^f \int_0^\psi I_1(\chi,\tau)d\tau \frac{\partial[b(\chi,\psi)]^{-2}}{\partial \chi} d\psi\} \qquad (20)$$

The thin conducting wall condition [Eq. (3a)] for the outer side layer is

$$[1 + (f')^2]^{-1/2} (f'j_{\chi 0} \mp j_{\psi 0}) = \frac{\partial^2 \phi_0}{\partial \xi^2}, \quad \text{at } \psi = \pm f \qquad (21)$$

where terms which are O(c) compared to the retained terms are neglected. When Eqs. (15a), (16), (17), and (20) are introduced into Eq. (21), the result is a differential equation involving derivatives of p_0 with respect to ξ. Since χ is a passive parameter, this equation can be treated as an ordinary differential equation, and the solution which satisfies the boundary condition [Eq. (14b)] is

$$p_0 = p_c + A \exp\{-\xi |\alpha|[b(\chi,f)]^{1/2}/[1 + (f')^2]^{1/4} q^{1/2}\} \qquad (22a)$$

where

$$q(\chi) = I_2(\chi,f) + \int_0^f \int_\psi^f I_1(\chi,\tau)d\tau \frac{\partial[b(\chi,\psi)]^{-2}}{\partial \chi} d\psi \qquad (22b)$$

and $A(\chi)$ is an integration function, which together with $p_c(\chi)$ is determined by matching the inner and outer side layer variables after the inner side layer solution has been found (Walker 1981). The outer side layer variables can now be expressed in terms of $A(\chi)$ and $p_c(\chi)$: Eq. (22) is substituted into Eqs. (15-17), (18b), (19), and (20), then the solution [Eq. (20)] is substituted into Eq. (17), and finally the solution [Eq. (17)] is substituted into Eqs. (18) and (19).

V. Conclusion

The core and outer side layer solutions appear to be local solutions in that the forms of these solutions have been found independently in each χ = const section of the duct. However, these forms involve the two functions $p_c(\chi)$ and $A(\chi)$ which cannot be determined until the inner side layer solution is known. The inner side layer boundary value problem reduces to a pair of integro-differential equations governing two unknown functions of χ and ψ with derivatives with respect to χ and integrals with singular kernels with respect to ψ. These equations cannot be solved independently for each χ value, so that the local nature of the core and outer side layer is lost in the inner side layer and the functions p_c and A cannot be determined without specifying the geometry and magnetic field everywhere, as well as boundary conditions at the upstream and downstream ends of the duct. Walker (1981) has presented complete solutions for straight expansion and contractions with uniform magnetic fields, $\overline{b} = \overline{y}$, and some of the characteristics of those solutions can be generalized for all of the flows considered here.

The structure of an inner side layer is completely determined by the geometry, the magnetic field, the end conditions, and the fact that it must carry half the volume flux required by Eq. (2). This structure involves a specific electric current flowing from the top or bottom, through the inner side layer in χ = const sections and into the outer side layer as $j_{zo}(\chi,0)$. Consideration of the electric currents in the walls leads to the conclusion that this electric current also determines the value of

$$\int_0^\infty [f'j_{\chi o}(\chi, \pm f, \xi) \mp j_{\psi o}(\chi, \pm f, \xi)]d\xi$$

These two conditions on the outer side layer electric current determine p_c and A, thus completing the core and outer side layer solutions. Physically, the inner side layer must have a specific circulation of electric current in χ = const sections in order to carry the required volume flux; the circuit for this electric current is completed in χ = const sections through the outer side layer and the top or bottom; the structure of the outer side layer which can accommodate this required electric current circulation involves a net axial electric current which is not the same at different χ = const sections; excesses and deficiencies in the axial outer side layer electric current can be taken care of only by a transverse core current j_{zc} flowing between the two outer side layers and this determines the structure of the core.

The net essentially axial outer side layer electric current $I_{\chi o}$ and the transverse core current density j_{zc} complete an electric current circuit in ψ = const surfaces, while this current circuit is driven by the electric current circuit in χ = const sections through the top or bottom, inner side layer and outer side layer. The current circulation in ψ = const surfaces accounts for the important pressure variations: j_{zc} accounts for the pressure drop or rise along the duct and

$$I_{\chi 0} = -2\gamma_0 A(\chi) \int_0^f [b(\chi,\psi)]^{-2} d\psi$$

accounts for a pressure jump, $\Delta p_0 = A$, across the outer side layers. This pressure jump has practical implications for experiments. In many experiments it is difficult to measure pressure along the top and bottom because of the close proximity of the magnet's pole faces, so pressure is only measured along the sides. However, the side pressure may be quite different from the pressure throughout most of the duct because of the pressure jump across the side layers. Indeed, there are cases where the side pressure is dropping while the core pressure is rising in the flow direction (Walker 1981).

Acknowledgment

This work was supported by the National Science Foundation under Grant ENG-7820146.

References

Holroyd R.J. (1976) "MHD duct flows in non-uniform magnetic fields," PhD dissertation, Univ. of Cambridge, Cambridge, England.

Holroyd R.J. and Walker J.S. (1978) "A theoretical study of the effects of wall conductivity, non-uniform magnetic fields and variable-area ducts on liquid-metal flows at high Hartmann numbers," J. Fluid Mech., Vol. 84, pp. 471-495.

Howard L.N. (1969) Rotating Flow, Lectures at Royal Institute of Technology of Stockholm, Royal Institute of Technology, Stockholm.

Hunt J.C.R. and Ludford G.S.S. (1968) "Three-dimensional MHD duct flows with strong transverse magnetic fields, Part I: Obstacles in a constant area channel," J. Fluid Mech., Vol. 33, pp. 693-714.

Ludford G.S.S. and Walker J.S. (1980) "Current status of MHD duct flow," MHD-Flows and Turbulence II, Proc. 2nd Bat-Sheva Int. Seminar, Beer-Sheva, Israel Universities Press, Jerusalem, pp. 83-95.

Morse P.M. and Feshbach H. (1953) Methods of Theoretical Physics, McGraw-Hill Book Co., New York, Vol. I, p. 115.

Roberts P.H. (1967) An Introduction to Magnetohydrodynamics, American Elsevier, New York, pp. 186-190.

Shercliff J.A. (1956) "The flow of conducting fluids in circular pipes under transverse magnetic fields," J. Fluid Mech., Vol. 1, pp. 644-666.

Shercliff J.A. (1965) A Textbook of Magnetohydrodynamics, Pergamon Press, London, p. 24.

Temperley D.J. and Todd L. (1976) "Some remarks on a class of plane, linear boundary value problems," J. Inst. Math. Its Appl., Vol. 18, pp. 309-324.

Walker J.S. (1981) "Magnetohydrodynamic flows in rectangular ducts with thin conducting walls, Part I: Constant-area and variable-area ducts with strong uniform magnetic fields," J. Méc., Vol. 20, pp. 79-112.

Walker J.S. (1983) "Magnetohydrdynamic flows in rectangular ducts with thin conducting walls, Part II: Pumps with finite length electrodes and with strong uniform magnetic fields," in preparation.

Walker J.S. and Ludford G.S.S. (1975) "MHD flow in circular expansions with thin conducting walls," Int. J. Eng. Sci., Vol. 13, pp. 261-269.

Walker J.S. Ludford G.S.S., and Hunt J.C.R. (1971) "Three-dimensional MHD duct flows with strong transverse magnetic fields, Part 2: Variable-area rectangular ducts with conducting sides," J. Fluid Mech., Vol. 46, pp. 657-684.

Walker J.S., Ludford G.S.S., and Hunt J.C.R. (1972) "Three-dimensional MHD duct flows with strong transverse magnetic fields," Part 3: Variable-area rectangular ducts with insulating walls," J. Fluid Mech., Vol. 56, pp. 121-141.

Why, How, and When MHD Turbulence Becomes Two Dimensional

René Moreau*
Université de Grenoble, Grenoble, France

Abstract

A review is first proposed of the very impressive work done during the last 20 years on MHD turbulence and MHD turbulent flows. It shows a clear experimental evidence of the existence of two-dimensional turbulence, persisting for very large distances with very weak ohmic and viscous dissipation and exhibiting the classical k^{-3} spectral law. The presence of insulating walls perpendicular to the uniform magnetic field appears to be a necessary ingredient. To explain this behavior two mechanisms are proposed. One is the electromagnetic diffusion along magnetic streamlines, in which Alfén waves degenerate and which transforms three-dimensional structures into two-dimensional ones. The other is the Hartmann effect generalized to account for the inertia and vorticity of the turbulence, which leads to some quantization of the energy-containing zone in the Fourier space. Finally, it is concluded that eddies of length scale ℓ perpendicular to the magnetic field included between two limits have their dynamics described by ordinary two-dimensional Navier-Stokes equations.

I. Introduction

MHD turbulence and MHD turbulent flows have been the subject of very impressive work, essentially experimental,

---

Paper presented at Third Beer-Sheva International Seminar on Magnetohydrodynamic Flows and Turbulence, Ben-Gurion University of the Negev, Beer-Sheva, Israel, March 23-27, 1981. Copyright © American Institute of Aeronautics and Astronautics, Inc., 1982. All rights reserved.
*Professor.

during the last 20 years, primarily in four places: Riga, Purdue, Beer-Sheva, and Grenoble. A review of the main ideas arising from this very important work is first presented (Secs. II and III). Then the three basic questions will be addressed: Why, How, and When? In doing this, several original ideas will be introduced, most of which have been proposed by Joel Sommeria (1980) in his recent thesis.

II. Experimental Evidence of the Existence of Two-Dimensional Turbulence

The main conclusion which arises from the first experimental results on turbulent duct flows is the existence of some critical value of the parameter M/Re, above which friction measurements agree with laminar law (Shercliff 1953, 1956). Of course, the critical value depends on the geometry of the duct, namely the aspect ratio of the rectangular ducts. But the interesting point is that in supercritical conditions turbulence is still present, although at a lower level than at the same Reynolds number without the magnetic field, and still persists with a very slow decay at very long distances however high the magnetic field can be. This persistence, already observed in previous work (Branover et al. 1965), has been well documented and later pinpointed at Riga and also Purdue (Lielausis 1975).

When hot films were introduced into these turbulent flows, strange spectra such as k^{-3} were recognized. All of these observations followed the development of the theory on ordinary two-dimensional turbulence (Batchelor 1969) in which the quantity following the cascade process is no longer kinetic energy but enstrophy, thus providing a very weak viscous dissipation. At that time the explanation put forward to interpret the persistence of MHD turbulence was that, because such turbulence becomes two-dimensional according to Maxwell equations and Ohm's law, the induced current is exactly zero and therefore the Joulean and viscous dissipations are both extremely weak.

However, the situation was far from clear, since the theories of that time [namely the linear theory proposed by Moffatt (1967), but also those trying to introduce nonlinear effects (Moreau 1968)] could not explain why the velocity component in the field direction was smaller than the two others. These theories which assumed that turbulence was homogeneous could explain only a tendency to some partial two-dimensionality in the sense $\partial/\partial z \ll \partial/\partial x, \partial/\partial y$ (z in the direction of $\overline{B}$).

The progress of the experimental work in Riga during the 1970s has not reduced this contradiction. On the contrary, some very clear direct confirmations of the fairly good two-dimensionality have been obtained, in the full sense $w \ll u,v$, and $\partial/\partial z \ll \partial/\partial x, \partial/\partial y$ (Platnieks and Freibergs 1972; Kolesnikov and Tsinober 1974).

III. Necessary Ingredients for Two-Dimensionality

The controversy reached its peak in 1977 and 1978 when measurements made in Grenoble (Alemany and Moreau 1977) began to be published. In this special arrangement (a grid falling in a vertical column of mercury placed inside a big solenoid) the parallel velocity component could be measured and was found to be of the same order with or without the magnetic field. It was also observed that the rate of dissipation was higher with than without the magnetic field, and that the k^{-3} energy spectra were taking the place of the $k^{-5/3}$ spectra. Clearly this experiment and those in in ducts under a transverse field could not be explained by the same and unique theory. And we must consider that the presence of walls perpendicular to the uniform magnetic field is one of the necessary ingredients to get two-dimensional turbulence in the full sense.

Now, comparing the experiment made by Gel'fgat et al. (1971) with insulating walls to that made by Platnieks and Freibergs (1972) with highly conducting walls perpendicular to the magnetic field, and showing a very quick dissipation, demonstrates that these perpendicular walls should be insulating to insure the persistence of turbulence at large distances. This comparison is of great interest, since it suggests that the Hartmann effect plays some crucial role in the establishment of the two-dimensional structure and in whether it damps out or not.

Some other experiments also designated a third necessary ingredient to cause nondissipative two-dimensional turbulence: the presence of some entry effect. When no entry is present, as in the annular duct used by Gel'fgat et al. (1971), no turbulence exists in a high magnetic field. Branover and Gershon (1979) clearly demonstrated that the origin of the remaining turbulence in duct flows is not the result of some regular input from the instability of M-shaped mean velocity profiles balancing the regular dissipation, but only the influence of some entry or of some initial production.

So we must consider that two-dimensional turbulence with large scales exists and has already been observed in MHD duct flows under the three following conditions:

1) $M/Re \gg (M/Re)_{crit}$.

2) The walls perpendicular to the magnetic field are insulating.

3) Some turbulence promoter is present.

Of course many questions still have to be answered to understand this conclusion, mainly the following: Why and through which mechanism do initially three-dimensional disturbances transform into two-dimensional turbulence? How does the Hartmann effect limit this two-dimensionality and influence the dynamics of this turbulence? Is such a quasi-two-dimensional turbulence well described by ordinary two-dimensional Navier-Stokes equations, or is it a specific MHD phenomenon?

IV. A Mechanism of Transformation, Three- to Two-Dimensionality

At the scale of the laboratory, the magnetic Reynolds number is so weak that Alfén waves degenerate into some diffusion in the direction of the magnetic field B (z direction). This mechanism yields quite important consequences for quasi-two-dimensional velocity fields. Any disturbance of length scale ℓ and velocity scale U in the plane perpendicular to $\vec{B}$ is diffused with the diffusivity $\sigma B^2 \ell^2/\rho$. During its turnover time $\tau_{tu} = \ell/U$, good correlations are established over distances of the order of $\ell \sqrt{\sigma B^2 \ell /\rho U}$ in the z direction. Then it is clear that the typical length scale $\ell_\parallel$ in the direction of the magnetic field becomes very large compared with ℓ,

$$\frac{\ell_\parallel}{\ell} \simeq \sqrt{N} \quad \text{if} \quad N = \frac{\sigma B^2 \ell}{\rho U} \gg 1$$

Now, if a denotes the gap between the walls perpendicular to $\vec{B}$, it is clear that the time necessary to establish good correlations over all a is

$$\tau_a \simeq \frac{\rho a^2}{\sigma B^2 \ell^2} \simeq \frac{1}{N} \frac{a^2}{\ell^2} \tau_{tu}$$

It appears that in many laboratory experiments τ_a is much smaller than τ_{tu}. This certainly explains why the observed properties of MHD turbulence are rather independent of the details of the promoter (regular or irregular grid, entry effect, etc.).

V. Influence of the Insulating Walls

Clearly the classical Hartmann theory cannot answer the question about insulating walls because of the presence of inertia and vorticity in the outer flow. It is then necessary to develop a second-order Hartmann theory, using a double perturbation based on the two small parameters $1/N$ and $1/M$ ($M = B\ell \sqrt{\sigma/\rho\nu}$).

The presence of inertia is necessarily forcing some Ekman suction at the ends of each quasi-two-dimensional eddy, but the layer in which this phenomenon takes place is the Hartmann layer (if $N \gg 1$). Then some radial velocity component is driven, necessarily of the order of U/N. Because of continuity the velocity component in the direction of the magnetic field cannot be zero and must be of the order of U/NM. The interesting point, easy to understand without making the detailed calculations, is that the parallel velocity component $u_\parallel$ in outer flow is extremely small compared to the perpendicular one,

$$u_\parallel/u = O(1/NM)$$

Since the component E_z of the electric field must vanish at the wall, it is elementary to deduce its value at the edge of the Hartmann layer from the equation

$$\frac{\partial^2 \phi}{\partial z^2} = B\omega_z$$

expressing the continuity of the current [ϕ stands for the electric potential, and the z component of vorticity ω_z is known in the Hartmann layer as $\omega_z = \Omega_z(1 - e^{-Mz/\ell})$],

$$E_z = \Omega_z \sqrt{\rho\nu/\sigma}$$

This result is of great interest, at first because it gives a direct diagnostic technique of the vorticity in the outer flow (but at the edge of the Hartmann layer), but also because it can be turned into a very striking condition on the velocity by eliminating E_z taking the curl of Ohm's law twice. One gets

$$\frac{\partial \Omega_z}{\partial z} = - \frac{\ell}{M} \Delta \Omega_z$$

This demonstrates that the normal derivative at the wall, of the order of $1/M\ell$, is much smaller than the same z derivative in the turbulent core, of the order of $1/\ell_{||} = 1/\ell \sqrt{N}$. This property has quite a drastic consequence: the wavelength λ in the z direction cannot vary continuously, but must be an integer fraction of the gap between the two walls a. In the Fourier space this means that the energy-containing zone is not all of the domain between the cones in which the Joule effect is predominant. The wave vectors must have their extremity(ies) on a set of planes $k_z = \pi/a, 2\pi/a, \ldots, n\pi/a$. Thus the energy-containing zone appears to be quantized (Fig. 1).

As a consequence, for scales larger than some critical value corresponding to $k^* = (\pi/a)\sqrt{N}$, there is absolutely no other possibility than being exactly two-dimensional.

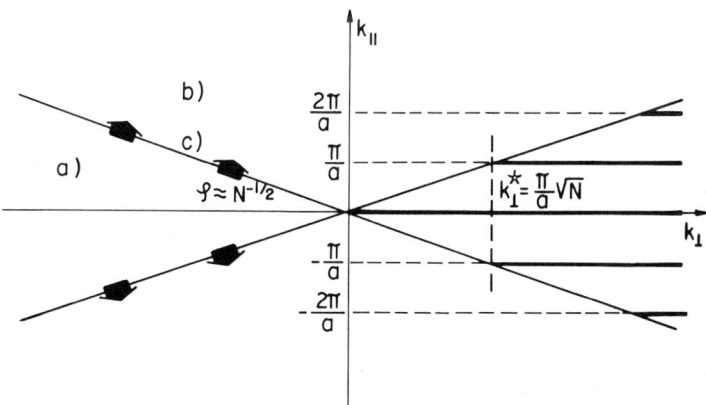

Fig. 1 Left side: localization of a) energy-containing, b) dissipating, and c) transferring zones in homogeneous turbulence after Alemany et al. (1979); right side: quantization of the energy-containing zone when perpendicular walls are insulating.

VI. Closure of Electric Currents

The ideas just developed explain why and how MHD turbulence becomes <u>kinematically</u> two-dimensional. But we still have to examine whether its dynamics agree with ordinary Navier-Stokes equations, a point essentially controlled by the condition of closure of electric currents. This condition may be written

$$\oint_C \left(\int_0^a \bar{J} dz \right) d\bar{s} = 0$$

for any arbitrary closed circuit C (s is the abcissa along C) in a plane perpendicular to $\bar{B}$. Clearly it demands that the vector integral taken from one wall to the other should be a pure gradient. Writing it as a sum of three terms, the current in each Hartmann layer [expressed from Hartmann theory in terms of the core velocity V (0) or V (a)], and the current in the core expressed from the motion equation, gives the following equation (nondimensional notations)

$$<\frac{D\bar{V}}{Dt}> = - <\nabla P> + \frac{1}{Re} <\nabla \bar{V}> - \frac{M}{\lambda Re} [\bar{V}(0) + \bar{V}(a)]$$

where P is a total pressure, and where the brackets < > stand for the average value between $z = 0$ and $z = a$. The following notations are also used: $\nabla = (\partial/\partial x, \partial/\partial y, 0)$, $D/Dt = \partial/\partial t + \bar{V} \nabla$, and $\lambda = a/\ell$.

Notice that the ratio N/M, which appears quite obviously between the current in the core (with inertia present) and the current in Hartmann layers, explains the relevance of the parameter M/Re for turbulent as well as laminar duct flows. But the most interesting result arises when two-dimensionality is fairly well achieved; then the brackets may be suppressed and $\bar{V}(0) = \bar{V}(a)$. It essentially consists in the linear electromagnetic braking term $-2M/\lambda Re\, \bar{V}$, which appears able to damp out the turbulence in a time $\tau_H = \lambda Re/2M\tau_{tu} = (a/B)\sqrt{\rho/\sigma\nu}$ independent of the scale of the eddies. One can conclude that scales such as this (Hartmann time being much larger that their turnover time) are well described by the ordinary two-dimensional Navier-Stokes equation. However, for very large scales MHD two-dimensional turbulence does not coincide with ordinary two-dimensional turbulence.

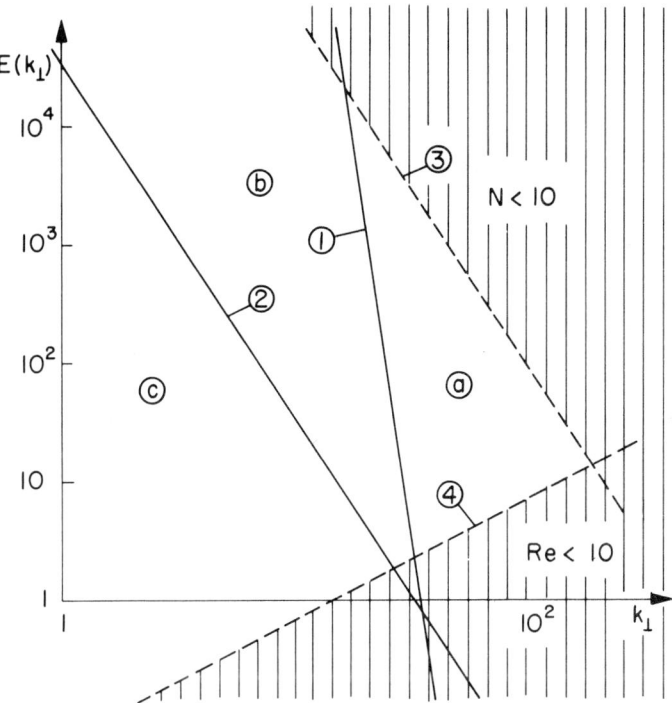

Fig. 2 Three typical kinds of MHD turbulence between insulating walls perpendicular to a strong uniform magnetic field: 1) condition for kinematically two-dimensional turbulence, 2) condition for negligible Hartmann braking effect, 3) $N = 10$, 4) $Re = 10$, a) equilibrium between angular transfer and Joule effect, b) dynamically two-dimensional eddies, c) two-dimensional eddies with Hartmann effect predominant.

VII. Conclusion

Two conditions of prime importance have been established: 1) the condition for two-dimensionality $[k < (\pi/a) \cdot (\sigma B^2 a/\rho U)^{1/3}]$, and 2) the condition for negligible Hartmann braking $[k \gg (B/Ua)(\sigma\nu/\rho)^{1/2}]$. Since the typical velocity of the considered turbulent scales $U \simeq \sqrt{k\,E(k)}$ appears in these conditions, the diagram (k,E), where $E(k)$ is the energy spectrum of this two-dimensional turbulence, seems to offer the best possibility to characterize some given experiment (Fig. 2).

It is interesting to note that when one observes a k^{-3} spectrum, the Hartmann effect is either negligible for each k value, or not negligible, since condition 2 in Fig. 2 also has a slope of -3. On the contrary, a $k^{-5/3}$ energy spectrum would cut the border of the Hartmann braking zone and would necessarily support an increasing dissipation as one goes to larger and larger scales. It would be extremely interesting to perform such an experiment, with a source of turbulent energy at some scale, demonstrating the presence of a direct cascade of entropy in k^{-3} to large k, and at the same time the presence of an inverse cascade of energy in $k^{-5/3}$ to small k. In such a situation the turbulence should adapt itself to transfer the input of energy to the larger scales which would in turn be able to dissipate it by viscous friction in the Hartmann layers.

References

Alemany A., Moreau R., Sulem P.L., and Frisch U. (1979) "Influence of an external magnetic field on homogeneous MHD turbulence," J. Mécanique, Vol. 18, No. 2, p. 277.

Batchelor G.K. (1969) "Computation of the energy spectrum in homogeneous two-dimensional turbulence," Phys. Fluids, Suppl., Vol. 12, Pt. II, p. 233.

Branover H.H., Slyusarev N.M., and Scherbinin E.V. (1965) "Some results of measuring turbulent pulsations of velocity in a mercury flow in the presence of a transverse magnetic field," Magn. Gidrodin., No. 1, p. 33.

Branover H.H. and Gershon P. (1979) "Experimental investigation of the origin of residual disturbances in turbulent MHD flows after laminarization," J. Fluid Mech., Vol. 94, Pt. 4, p. 629.

Gel'fgat Yu., Kit L.G., and Tsinober A.B. (1971) "Experimental investigation of turbulent-to-laminar transition in a magnetohydrodynamic flow," Dokl. Akad. Nauk. SSSR, Vol. 199, p. 560.

Kolesnikov Yu. B. and Tsinober A.B. (1974) "An experimental study of two-dimensional turbulence behind an array," Mekh. Gidk. i Gaza, No. 4, p. 146.

Leith C.E. and Kraichnan R.H. (1972) "Predictability of turbulent flows," J. Atmos. Sci., Vol. 29, p. 1041.

Lielausis O. (1975) "Liquid metal magnetohydrodynamics," At. Energy Rev., Vol. 13, No. 3, p. 527.

Moffatt H.K. (1967) "On the suppression of turbulence by a uniform magnetic field," J. Fluid Mech., Vol. 28, Pt. 3, p. 571.

Moreau R. (1968) "On magnetohydrodynamic turbulence," Proc. Symp. on Turbulence of Fluids and Plasmas, Polyt. Inst. of Brooklyn, New York, p. 359.

Moreau R. and Alemany A. (1977) "Lecture Notes," Physics 75, Proc. Symp. on Turbulence, August, Berlin, German, pp. 369-384.

Platnieks I.A. and Freibergs J. (1972) "Turbulence and some questions of stability in flows with M-shaped velocity profiles," Magn. Gidrodin., No. 2, p. 29.

Reed, Cl.B. and Lykoudis P.S. (1978) "The effect of a transverse magnetic field on shear turbulence," J. Fluid Mech., Vol. 89, Pt. 1, p. 147.

Shercliff J.A. (1953) "Steady motion of conducting fluids in pipes under transverse magnetic fields," J. Fluid Mech., Vol. 89, Pt. 1, p. 147.

Shercliff J.A. (1956) "The flow of conducting fluids in circular pipes under transverse magnetic fields," J. Fluid Mech., Vol. 1, p. 644.

Somméria J. (1980) "Tendance à la bidimensionalité de la turbulence MHD," PhD dissertation, Univ. Scientifique et Médicale de Grenoble, France.

Votsish A.D. and Kolesnikov Yu. B. (1976a) "Spatial correlation and vorticity in two-dimensional homogeneous turbulence," Magn. Gidrodin., Vol. 12, No. 3, p. 25.

Votsish A.D. and Kolesnikov Yu. B. (1976b) "Transition from three- to two-dimensional turbulence in a magnetic field," Magn. Gidrodin., Vol. 12, No. 3, p. 141.

Relaminarization—Magnetohydrodynamic and Otherwise

Roddam Narasimha*
Indian Institute of Science, Bangalore, India

Abstract

A brief review is presented of the different mechanisms by which a turbulent flow becomes effectively laminar. Laminarization is possible by dissipation involving a molecular transport parameter such as viscosity, conversion, or absorption of turbulent energy into a different form of energy (e.g., gravitational), and domination of turbulent transport by an externally imposed agency. A distinction is drawn between "hard" laminarization, in which turbulent intensities tend to zero sooner or later, and "soft" laminarization, in which there may be appreciable residual turbulence but in which turbulent transport plays a negligible role in mean flow dynamics. In MHD duct flows, it is argued that when the field is aligned with the flow laminarization is dissipative, the larger eddies being destroyed through the electrical resistivity and the smaller ones through viscosity. In the presence of strong normal fields the Hartmann effect becomes important — laminarization here is chiefly due to the domination of the magnetic force over the Reynolds stress gradient across the boundary layer. An analogy is drawn in the latter case with the effects of strong favorable pressure gradients in non-MHD flow, and it is suggested that the critical nondimensional group is of the form [(Reynolds number)/(Hartmann number)^n], where n is appreciably less than unity and probably around 1/2. Recent measurements in duct flow lend some support to these arguments.

Paper presented at Third Beer-Sheva International Seminar on Magnetohydrodynamic Flows and Turbulence, Ben-Gurion University of the Negev, Beer-Sheva, Israel, March 23-27, 1981. Copyright © American Institute of Aeronautics and Astronautics, Inc., 1982. All rights reserved.
*Professor, Department of Aerospace Engineering.

I. Introduction

It has been well known for a long time that the imposition of a magnetic field on certain types of turbulent shear flow in an electrically conducting fluid tends to suppress the turbulence. [There are also situations in which the field so alters the mean flow as to render it unstable and hence promotes turbulence. An early example was given by Lehnert (1955); a more recent one, due to the formation of unstable M-shaped velocity profiles in duct flow, has been discussed by Hunt (1965) and Branover (1978).] The classical experiments of Hartmann and Lazarus (1937) in channels and pipes were already suggestive of the phenomenon. The more extensive measurements of Murgatroyd (1953) showed that the pressure loss in pipes tended to laminar values as the magnetic field (applied normal to the flow) was increased. Since then many very interesting studies of the phenomenon have been made, in particular by Branover, Lykoudis, and their co-workers with magnetic field normal to the flow; some studies have also been made in flows with an aligned field. An excellent review of the present state of the subject has been recently presented by Branover (1978). MHD experiments are notoriously difficult to perform successfully; the experimental information that has been carefully collected by dedicated groups over the years is therefore of great value.

The same difficulties, however, are also responsible for making interpretation of data difficult, and it appears that there are many basic questions still open. This is most strikingly illustrated by the continuing uncertainty regarding the precise parameters that govern laminarization (whose meaning will be discussed at some length below). It has generally been assumed that the relevant parameter, certainly in duct flows, is the ratio of the Reynolds number (Re) to the Hartmann number (Ha). The chief reason for this notion appears to be that it is this combination of parameters which governs the stability and the pressure loss coefficient of laminar duct flows. However, experiments do not provide a single well-defined critical value for Re/Ha. Although Murgatroyd (1953) quoted a critical value of 225 for this ratio from his experiments on a channel, the recent work of Hua and Lykoudis (1974) suggests a significantly different value of nearly 330. Tsinober's (1975) excellent review of MHD turbulence notes that the critical value of the parameter increases with increase in Reynolds number. Branover (1978, p. 155) finds, from a detailed analysis of the data, that a more accurate criterion is Re/Ha^n, where n is a little larger than unity. On the other hand, the recent experiments of Hua and Lykoudis (1974) in a channel show that what they call the turbulence damping factor correlates with the parameter $Ha^2/Re^{0.75}$. These differences demonstrate that there are factors in the basic dynamics that are not yet well understood.

There are a variety of other questions as well connected with the structure of residual turbulence in these laminarizing flows, that have also not be satisfactorily resolved, for example, the tendency toward highly anisotropic and eventually two-dimensional turbulence.

In parallel, and apparently in little contact with MHD work, many studies have been made in the last decade or two of flows that laminarize without the aid of a magnetic field. The number of agencies that appear to laminarize a flow keeps increasing. In a recent extensive analysis Narasimha and Sreenivasan (1979, hereafter referred to as I) studied the effects of viscous dissipation, stabilizing buoyancy forces, favorable streamwise pressure gradients, flow curvature, rotation, suction, heating, etc. The same survey also considered MHD duct flows and showed how they fitted into an overall scheme devised by the authors. It is the objective of the present paper to draw parallels (whenever possible) between MHD and non-MHD situations, and to show how work in one type of flow may help to shed light on other types.

But before we proceed further, it is necessary to understand what one might mean by laminarization. (The phenomenon we are discussing has at various times been called inverse or reverse or antitransition, reversion, relaminarization, etc. We will generally call it laminarization, although the other words will also be of use.) For some, laminarization can mean only the total disappearance of turbulence, or at least a tendency toward zero turbulence as an asymptotic state. This concept is clearly what is important to engineers who might be interested, for example, in estimating the loading due to random pressure fluctuations on a surface in a flying vehicle or the disturbance levels likely to be encountered in a wind tunnel. If the intensities in these cases are not sufficiently low, it would be justifiable to hesitate in calling the flow laminar. On the other hand, there are many situations, both in MHD and non-MHD flows, where intensities may not tend to zero but many flow parameters, including such important ones as skin friction and heat-transfer coefficients, attain laminar values. Again, to engineers interested in estimating the drag of a surface in a flying vehicle or the heat transfer in a rocket nozzle, the possible presence of turbulent fluctuations in the flow is of secondary importance compared to the fact that the momentum and heat transport can be estimated as in laminar flows.

It is important to make a distinction between these two types of situations; indeed, it may be worthwhile to call the first type, in which turbulence eventually vanishes, "hard" laminarization and the second type, in which only turbulent transport is rendered negligible, "soft" laminarization. (Those who favor terminology that is more common among mathematicians may prefer to call these laminarization in the "strong" and "weak" sense, respectively.) Some will say that "soft" laminarization is not very interesting; but I must disagree with this view, as in fact it is this phenomenon on which there has been a great deal of argument and from which we have learned more; it is the "hard" type that is relatively easier to understand.

The next section presents a brief review of certain features of laminarization in non-MHD flows, in particular, those aspects that have some relevance to MHD flows, which are covered in the succeeding section.

II. Laminarization in non-MHD Flows

There are three basic archetypes in laminarizaing flows (Narasimha 1977), and we discuss these in turn before examining MHD flows.

The first of these is typified by a flow in which the Reynolds number decreases downstream. This can happen in duct flows if the effective size of the duct increases, continuing as either a single duct or a set of branches. (An interesting example of the latter is the human lung, where the flow must laminarize in the bronchioles, starting from what sometimes can be a turbulent state in the windpipe.) A pipe whose diameter increases from $2a_1$ to $2a_2$ experiences a drop in the Reynolds number by the factor $(a_2/a_1)^2$; in a two-dimensional channel whose height remains constant (at $2a$, say) downstream, but breadth increases from b_1 to b_2, the Reynolds number decreases by a factor of b_2/b_1. If the upstream Reynolds number Re_1 is sufficiently high and the downstream Reynolds number Re_2 is sufficiently low, we may expect that an initial turbulent flow should revert to a laminar state. Laufer (1962) found that this indeed happens. Later measurements (Sibulkin 1962, Badri Narayanan 1968) have revealed some interesting features of this type of reversion (I):

1) The rate of approach to the final laminar state is very slow. Thus in a channel with Re_2 equal to about half the critical value ($\simeq$ 1500, based on average velocity) at a distance x = 115a from the beginning of the broader section, the rms value of the longitudinal velocity fluctuation, $\hat{u}$, drops by a factor of only 2.

2) The turbulence decays exponentially with distance with an e-folding distance that is approximately proportional to $(Re_{cr} - Re_2)^{-3}$.

3) This decaying turbulence is in an approximate state of equilibrium as the dissipation remains very nearly equal to (but slightly higher than) the production of turbulent energy. For example, the spectrum is similar at all stations, consistent with the observed slow decay.

4) The decaying turbulence also becomes highly anisotropic in the final stage. Table 1 shows the measured ratio of the intensity of the longitudinal component $\hat{u}$ to the normal component $\hat{v}$. It is seen that the ratio is nearly 15 at x = 220a in a flow at Re_2 = 625! The explanation offered in I is that, while there is production of longitudinal turbulent energy due to interaction of the Reynolds shear stress with the mean velocity profile, there is no corresponding production for the other two components of turbulent energy. To those working with MHD flows and accustomed to thinking of the magnetic field as providing a preferred direction in the flow, the observed anisotropy may come as a surprise, but we need to remind ourselves that any shear flow has preferred directions; for example, a two-dimensional channel has two preferred directions, respectively, along the main flow itself and the vorticity.

5) Finally, the Reynolds shear stress goes down even faster than the turbulent intensities, because of a decrease in the correlation coefficient. Over a distance of about 16a, at Re_2 = 865 (more than half of the critical Reynolds number), the maximum correlation coefficient goes down from 0.32 to 0.14.

This kind of laminarization is basically dissipative; a molecular transport parameter such as viscosity converts the turbulent energy into heat.

In the second type of reversion, turbulence is suppressed chiefly by conversion or absorption into another form of energy (different from heat). For example, a turbulent flow vertically upward into a stabilizing density gradient becomes laminar if the gradient is sufficiently high; the turbulent kinetic energy is in this case transformed to gravitational potential energy. This phenomenon is commonly observed in inversions in the atmosphere and was studied by Richardson in 1920. It appears to be chiefly governed by the Richardson number. Although a precise value for the critical Richardson number cannot be identified easily, it would appear that, whatever its value, it is not very sensitive to the Reynolds number. Townsend (1958), analyzing measurement in a wind tunnel on a boundary layer with a momentum thickness Reynolds number of only about 600 (Nicholl 1970), concluded that the Richardson number just before collapse of turbulence was less than 0.1. Businger

Table 1 Turbulence intensities in laminarizing channel flow

Re_2	20	At $x/a =$ 100	220
		$\hat{u}/U_0(20) =$	
625	9.2	5.3	1.2
865	10.0	7.7	4.7
980	8.5	8.8	7.5
1250	7.8	7.5	7.5
		$\hat{v}/U_0(20) =$	
625	1.18	0.32	0.02
865	1.78	1.2	0.53
980	2.22	1.72	1.23
1250	2.74	2.35	1.94
		$\hat{u}/\hat{v} =$	
625	7.79	16.56	60
865	5.62	6.42	8.86
980	3.82	5.12	6.09
1250	2.85	3.19	3.86

Notes: $\hat{u}$, $\hat{v}$ = rms value of longitudinal and normal fluctuating velocity components. $U_0(20)$ = centerline velocity at $x/a = 20$; x is measured from the end of the divergence in the channel; 2a = height of channel. Based on Badri Narayanan (1968).

and Arya (1974) estimated a critical Richardson number of 0.21 from measurements in the atmosphere. Considering that the Richardson number varies widely across a boundary layer, making it difficult to identify a precise critical value, and that the Reynolds numbers in the atmosphere are several orders of magnitude higher than in a wind tunnel, it would seem that the effect of the Reynolds number cannot be very strong, except possibly at very low Reynolds numbers.

A characteristic feature of this type of reversion is its rapidity. A photograph of a turbulent jet issuing vertically into a stabilizing gradient, shown by Viswanath et al. (1977), demonstrates that laminarization can be complete in the order of a few jet widths.

It is the third type however that is most intriguing and has led to very intensive work over the last 15 years. A typical situation here is an initially turbulent boundary layer subjected to a large favorable pressure gradient beginning from some streamwise station x_0 (say). Normally one expects that the effect of such a favorable pressure gradient would be to increase the skin friction coefficient and reduce the boundary-layer shape factor. Furthermore, as the wall friction velocity U_* will also increase, the viscous sublayer is expected to thin down. However, observation in such flows shows that although initially the flow behaves as expected, after a point the skin friction coefficient drops steeply and the shape factor increases. Figure 1, prepared using the experimental data of Blackwelder and Kovasznay (1972), illustrates the kind of changes that occur. As may be seen, one effect of the pressure gradient is that the momentum thickness Reynolds number also drops. Eventually the mean velocity profiles observed are almost exactly what would be expected in laminar flow.

It has often been assumed that the laminarization occurring in this kind of flow is also governed by some kind of Reynolds number, although there has been no agreement on the precise length and velocity scales that enter into this Reynolds number. One of the most widely used parameters, namely

$$K = \nu |dp/dx|/U^3 = \nu(dU/dx)/U^2$$

(where dp/dx is the imposed pressure gradient and ν the kinematic viscosity), carries no boundary-layer parameter at all; it is the inverse of a Reynolds number based on the freestream velocity $U = U(x)$ and a length scale characteristic of its streamwise variation. Many years ago Preston (1957) argued that there was a critical Reynolds number, which he estimated to be $Re_\theta = 420$, below which turbulent flow cannot be sustained; he obtained this critical Reynolds number by postulating that in a fully turbulent boundary layer an inertial sublayer with a logarithmic velocity profile must exist. More recently Bradshaw (1969) has proposed an extension of the idea, the criterion now being an "eddy Reynolds number," involving the friction velocity and the dissipative length as the appropriate scales.

All of this implicitly assumes that this kind of laminarization is also dissipative, but it must be noted that a severely accelerated boundary layer exhibits many features totally different from those associated with dissipative reversion, which we discussed above. In the first place the "laminarization" due to acceleration is fairly rapid. For example in the experiments of Ramjee (1968) (see I), at a distance of about 20 initial boundary-layer thicknesses from x_0, the velocity profile has changed from a characteristically turbulent one to the Blasius distribution. Second, the turbulence intensities, especially in the outer layer,

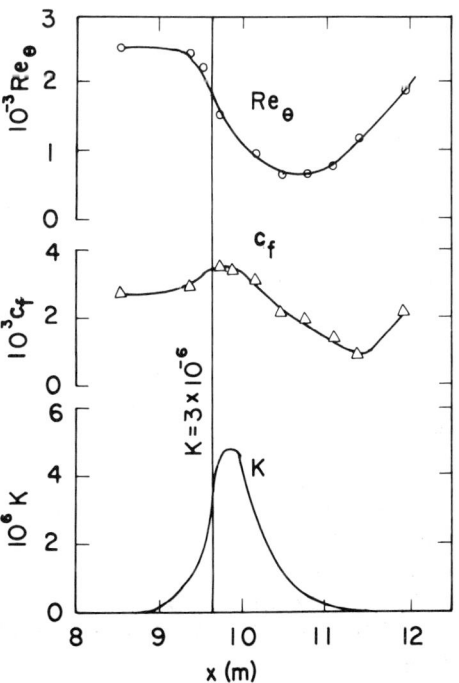

Fig. 1 Boundary-layer characteristics during laminarization in a strong favorable pressure gradient: experimental data of Blackwelder and Kovasznay (1972). The station at which K attains the value 3×10^{-6}, often suggested as the critical value for laminarization, is also marked; note that the Reynolds number R_θ here is about 1800.

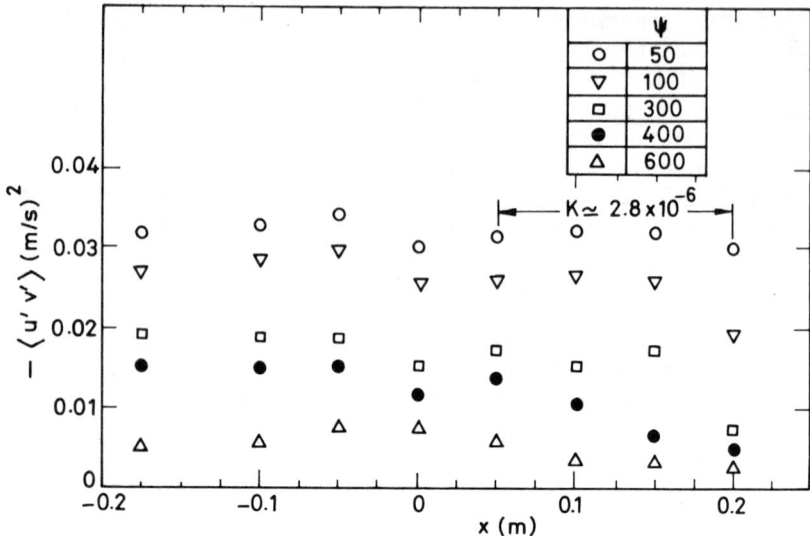

Fig. 2 Variation of Reynolds shear stress along streamlines in a boundary layer subjected to strong favorable pressure gradient, from the data of Rajagopalan (1974). The region where $K = 2.8 \times 10^{-6}$ is also marked. The stream function ψ is in arbitrary units; $\psi \simeq 700$ at the edge of the boundary layer.

hardly vary down the flow. Figure 2 shows the Reynolds shear stress $\langle u'v' \rangle$ along streamlines in such an accelerating boundary layer; it is seen that the absolute value of this stress hardly changes through the severe acceleration.

Furthermore, turbulent energy balance measurements (Fig. 3) in the flow in such a boundary layer (I) show that the dissipation always remains smaller than production; therefore the mechanism of laminarization cannot be attributed to dissipation.

There is a widely prevalent (but mistaken) view that whenever a boundary layer is subjected to a critical pressure gradient corresponding to $K \simeq 3 \times 10^{-6}$ the momentum thickness Reynolds number R_θ "inevitably" drops to its own critical value of 300-400. The error in this view can be seen in many different ways. First of all, this clearly does not happen in Blackwelder and Kovasznay's experiment, where $\overline{R_\theta} \simeq 1800$ at $K = 3 \times 10^{-6}$ (see Fig. 1). But surely this is not surprising; one can have any value of K one wants without even having a boundary layer! K does not contain any boundary-layer parameters.

It is true that in many other experiments R_θ does fall to around 300-400 at $K = 3 \times 10^{-6}$, but that is because the initial Reynolds number (i.e., at x_0 where the pressure gradient starts) was also low in those experiments. The reason that about the same range of low R_θ is encountered in many experiments is more a property of academic wind tunnels (most of which have a characteristic dimension of 1 ft, speed of 100 ft/s) than of relaminarization!

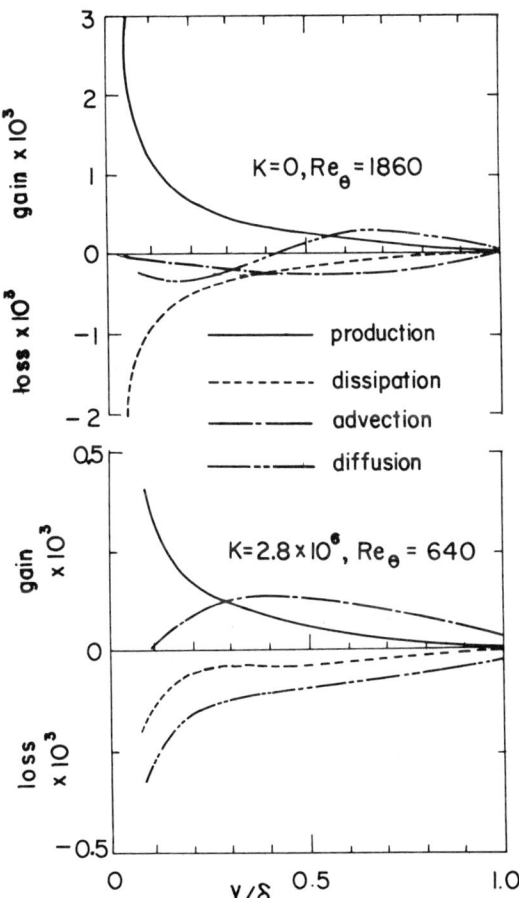

Fig. 3 Turbulent energy balance measurements in laminarizing boundary layer, at two stations where $K = 0$ and 2.8×10^{-6}, respectively. At the latter station, laminarization was complete in the sense that the predictions of quasilaminar theory agree with measurement. Each term in the energy equation is nondimensionalized using local freestream velocity U and boundary-layer thickness δ as scales. Note the enlarged scale in lower diagram ($K = 2.8 \times 10^{-6}$). Data from Rajagopalan (1974).

It is useful to examine this in still another way. The momentum integral equation is

$$\frac{\nu}{U}\frac{dR_\theta}{dx} = 1/2\, c_f - (H+1)\, K\, R_\theta \qquad (1)$$

In strong pressure gradients, the second term on the right tends to be much larger than the first; e.g., for the conditions quoted above for the Blackwelder-Kovasznay experiment, the ratio is about 7. Therefore Eq. (1) may be approximated by

$$\frac{\nu}{U}\frac{dR_\theta}{dx} = -(H+1)\, K\, R_\theta$$

with the solution

$$R_\theta(x) = R_\theta(x_0)\, \exp[-\int (H+1)\, KU\, dx/\nu]$$
$$= R_\theta(x_0)\, \exp[-\int (H+1)\, d(\ln U)] \qquad (2)$$

It is easy to see from here that if K were constant, $R_\theta(x)$ tends to go with $R_\theta(x_0)$; there is <u>no</u> tendency for R_θ to drop to a critical value of 300 or 400 <u>at</u> any given value of K. Full and explicit integration of Eq. (1) in any given flow verifies this conclusion. It is of course true that a possible solution of Eq. (1), which may be attained asymptotically in a sink flow of constant K, is that

$$R_\theta = 1/2\, \frac{c_f}{(H+1)K}$$

and remains constant (with c_f and H). But, as Fig. 1 demonstrates, there can be laminarizing flows where this does not happen and R_θ remains well above its alleged critical value at large K.

There is also the possibility that because the mean velocity u increases downstream there may be negative turbulence production. The total production is given by

$$-\langle u'v'\rangle \frac{\partial u}{\partial y} - (\hat{u}^2 - \hat{v}^2)\frac{\partial u}{\partial x}$$

where u is the mean velocity. The last term here is negative if $\hat{u} > \hat{v}$ and $\partial u/\partial x > 0$. Consider the ratio of the second term to the first,

$$\chi = \frac{(\hat{u}^2 - \hat{v}^2)\partial u/\partial x}{-\langle u'v'\rangle \partial u/\partial y}$$

It was shown by Back et al. (1964) that the maximum value of this parameter in the boundary layer is $44K/c_f$; in the many flows where reversion has been observed at $K \simeq 3 \times 10^{-6}$, $c_f \simeq 0.003$, χ could not therefore have been more than 4% even where the shear production in the boundary layer is largest, showing that any negative production, even if present, would be negligible.

The changes observed in such a boundary layer cannot therefore be attributed to either dissipation or negative turbulence production. On the other hand, an order of magnitude estimate shows that in all cases where reversion has been observed the pressure gradient is very much larger than the stress gradient across the boundary layer; i.e., the parameter

$$\Lambda = - \frac{dp}{dx} \frac{\delta}{\tau_0} \tag{3}$$

where δ is the boundary-layer thickness, tends to take very large values. For example in the data of Blackwelder and Kovasznay (1972) illustrated in Fig. 1, this number reaches a maximum value of 130.

As the Reynolds stress does not change rapidly, and therefore does not keep in step with the freestream dynamic pressure, the momentum transport due to turbulence becomes irrelevant for the dynamics of the mean flow. In fact, we have shown (Narasimha and Sreenivasan 1973) that a very large part of this flow can be described as consisting of two layers, an outer one which is stress-free but rotational and an inner laminar subboundary layer. A "quasilaminar" theory, constructed using the method of matched asymptotic expansions in the limit $\Lambda \to \infty$, shows very good agreement with many measured parameters beginning almost at x_0 and for c_f from around the station where it reaches a maximum. From the standpoint of this theory the decrease of boundary-layer thickness is seen to be a simple consequence of the conservation of vorticity in the two-dimensional inviscid flow of the outer layer. The maintenance of an effectively laminar inner layer, in spite of the highly disturbed state of the flow riding on top of it, is the result of the strong stabilizing influence of the favorable pressure gradient. A detailed calculation by Narasimha and Sreenivasan (1973) has shown that almost up to the point of retransition to turbulence the critical instability Reynolds number is higher than the actual Reynolds number of the laminar subboundary layer. Soon after this critical Reynolds number is exceeded, however, the flow goes back to its original turbulent state. [A stability calculation made by Launder (1963) gave negative results, presumably because he did not consider just the inner layer, as we do, but the whole boundary layer.]

What remains to be explained is why the turbulent stresses remain frozen in this flow. A clue to the likely mechanism can be obtained by considering the following data from the experiment of Blackwelder and Kovasznay already quoted. Here, a turbulent boundary layer with an initial thickness of about 125 mm was subjected to a freestream acceleration from 5.5 to 11.5 m/s over a streamwise distance of about a meter, giving a transit time for a fluid particle in the freestream of order 0.15 s. At $y/\delta \simeq 0.5$, the turbulent velocity at the initial station is of order $\hat{u} \simeq 0.04U \simeq 0.15$ m/s (as we have already noted, the absolute value of $\hat{u}$ does not vary strongly down the flow). Taking the relevant eddy length ℓ to be about $2\delta \simeq 250$ mm (see Sreenivasan 1974), we obtain an eddy time scale of order 1.3 s, which is an order of magnitude higher than the transit time of a fluid particle across the region of freestream acceleration. Under these conditions the eddies undergo rapid distortion; viscous and nonlinear effects are not significant, the dominant effect being vortex stretching.†

A complete rapid-distortion theory is available in isotropic homogeneous turbulence (Batchelor and Proudman 1954), but the validity of this theory for the boundary layer is open to question. However, approximate theories for both nonisotropic and nonhomogeneous turbulence have been formulated (Sreenivasan and Narasimha 1974, 1978). The conclusion from these theories is that, in the kind of two-dimensional strain that is imposed on turbulence in the laminarization experiments we are considering, $\hat{u}$ tends to go down and $\hat{v}$ to go up; both the product $\hat{u}\hat{v}$ and the Reynolds stress $-\langle u'v'\rangle$ tend to remain nearly constant for a total strain

---

†Interestingly, the distortion in such accelerated boundary layers is much more rapid than in the many grid turbulence experiments specially designed to test rapid distortion (Sreenivasan 1974). The condition for the validity of the rapid distortion limit, laid down by Batchelor and Proudman (1954), is that the ratio of time scales estimated above should be much less than $(R_\ell = \hat{u}\ell/\nu)^{1/2}$, which would be about 450 in the present case. However, it is now being realized (Hunt 1973, Narasimha and Sreenivasan 1974) that this condition is probably too stringent; when (as in the present studies) we are chiefly interested in the energy-containing eddies, it should be sufficient to require that the transit time be much less than the characteristic eddy time scale.

of as much as $U(x)/U(x_0) \simeq 5$, which covers the boundary-layer experiments in question. It appears, therefore, that the observed stress-freezing is a result of rapid distortion of shear turbulence.

Close to the wall, turbulence does decay by viscous action, but it is unnecessary to discuss it here in detail again (see I). But it should be clear that, in contrast to laminarization by dissipation or absorption, the phenomena that occur in a boundary layer subjected to severe free-stream acceleration are chiefly the result of the domination of pressure gradient over Reynolds stress gradient, and the stabilization of the laminar subboundary layer; we have here an excellent example of "soft" reversion.

III. MHD Flows

The experimental results in MHD duct flows have been discussed thoroughly by Branover (1978), who also points out the divergence in the available data. Our purpose in this section is to highlight and amplify some of the suggestions contained in I, and to show in particular that there are interesting parallels between MHD and non-MHD flows.

In MHD flows we need to consider two additional effects beyond those present in non-MHD situations, namely that each unit volume of fluid experiences a body force $\bar{J} \times \bar{B}$ and an energy loss (due to ohmic dissipation) of j^2/σ. (Here $\bar{j}$ is the current density, $\bar{B}$ the magnetic induction, and σ the electrical conductivity.) The role of these two effects in laminarization depends in particular on the orientation of the magnetic field $\bar{B}$ relative to the main flow.

Let us first examine the situation in which the imposed field is aligned with the direction of the main flow. (We consider only incompressible flow at low magnetic Reynolds number.) Many experiments in this configuration have shown that there is a tendency toward suppression of turbulence, for example, Globe (1961) and Freim and Heiser (1968) in pipes and Sajben and Fay (1967) in a jet. The experiment of Freim and Heiser suggests that reversion occurs if the ratio Re/Ha $\leqslant$ 30. This ratio, given by

$$Re/Ha = (U/B)/(\rho/\nu\sigma)^{1/2} \qquad (3)$$

where U and B are scales of velocity and magnetic field, does not involve any length scale characteristic of the flow. As earlier explanations of the relevance of this non-dimensional group (interpreting it, e.g., as "the square root of the product of the viscous and magnetic forces divided by the inertial forces") seem difficult to understand, it may be worthwhile to have a simple physical interpretation. This interpretation is based on a postulated physical mechanism underlying the laminarization (I), which we shall briefly discuss below.

The effect of a magnetic field on turbulence is a little complicated. It may be shown by a detailed examination of the vorticity equation (Moreau 1969) that the field tends to suppress vorticity normal to it, with a characteristic decay time of order $\rho/\sigma B^2$. Vorticity along the field is not directly suppressed, although we may expect that because of redistribution of energy among different directions even the component along $\overline{B}$ will eventually be affected. We do know from ordinary turbulence that this redistribution can be slow. Furthermore, the dissipation is concentrated in a cone in wave number space around the direction of the applied field $\overline{B}$. To begin with, however, let us concentrate on the total turbulent energy. A measure of the ratio of the magnetic to inertial forces is the Stuart number or interaction parameter given by

$$N = Ha^2/Re = \sigma B^2 \ell/\rho u' \qquad (4)$$

for eddies of size ℓ and velocity u'. This parameter is large for the larger eddies which therefore tend to be damped out by ohmic dissipation, as has been pointed out by Moffatt (1967) and Moreau (1969). The largest eddy that escapes ohmic dissipation is of the size $U\tau \simeq \rho U/\sigma B^2$, which can be much smaller than the characteristic flow length scale when B is high. Reversion must occur if eddies of this size or smaller are dissipated by viscosity, i.e., if the Reynolds number based on U, namely,

$$(U/\nu)(U\rho/\sigma B^2) = (Re/Ha)^2 \qquad (5)$$

is sufficiently small. It is thus no surprise to find a critical value for the ratio Re/Ha: its square is the Reynolds number of the largest eddy that does not suffer ohmic dissipation.

This explanation implies that when the field is aligned with the flow, reversion occurs by dissipation -- ohmic for the large eddies, viscous for those not so large.

When the field is normal to the flow the same sources of dissipation are still effective, but now the body force $\bar{j} \times \bar{B}$ explicitly affects the mean momentum balance and acts to suppress the mean vorticity of the flow normal to $\bar{B}$. At large values of the Hartmann number, this leads to the formation of the well-known Hartmann layers at the surface. These layers have an exponential velocity profile, of the same kind as in non-MHD asymptotic suction flow. A detailed stability analysis of laminar MHD flows by Lock (1955) showed that their stability is affected chiefly because of the alteration of the mean velocity profile rather than by the direct action of magnetic field on the fluctuations. It is well known that the critical Reynolds number for the exponential profile is two orders of magnitude higher compared to the Blasius boundary layer. There is therefore an extraordinary stabilization provided by the magnetic field via its effect on the mean velocity profile.

Correspondingly, there is much indirect evidence that in the case of the normal field acting on turbulent duct flow, the effect on the mean velocity profile operates more quickly. We may note that while the critical values for the ratio Re/Ha given by different workers do not agree precisely, as we have already seen in Sec. I, they still remain an order of magnitude higher with normal fields than with aligned fields. Also, Branover and Gershon (1976) find that the critical values are higher when the side perpendicular to the field is shorter. Both of these features are consistent with the explanation that a major role in the laminarization of these flows is being played by the Hartmann effect.

Perhaps there is nothing new in this; according to Branover (1978), there is a "border" Reynolds number Re_b such that the "turbulence suppression effect" (presumably due to dissipation) predominates over the Hartmann effect at Re < Re_b, and vice versa at Re > Re_b. However, the value he estimates for Re_b (p. 141) is 1280, which appears too low: in non-MHD channel flow, the critical Reynolds number for transition from turbulent to laminar flow is itself higher (being about 1500, based on section-average velocity and half-height of channel, see I). Furthermore, the recent experiments of Hua and Lykoudis (1974) show a trend quite different from that observed in the earlier experiment of Branover and others. (We shall return to this point shortly.)

Now if the Hartmann effect plays a strong role in laminarization, then in light of the discussions in the previous section one should no longer look at dissipation as the chief driving force behind laminarization with normal fields. On the other hand, the mechanism must be closer to what happens in an accelerated boundary layer; the analog of the pressure gradient is the $\bar{j} \times \bar{B}$ force, whose domination over the Reynolds stress gradient becomes the key factor in laminarization. In analogy with the theory sketched in Sec. II and the parameter Λ of Eq. 3, we would then expect the correlating MHD groups to be,

at low fields: $\quad \dfrac{\sigma U B^2}{\tau/D} \simeq \dfrac{Ha^2}{c_f Re}$ \hfill (6)

at high fields: $\quad \dfrac{\sigma U B^2}{\tau_0/\delta} \simeq \dfrac{Ha^2}{c_f Re} \dfrac{\delta}{D} \simeq \dfrac{Ha}{c_f Re}$ \hfill (7)

where δ is the thickness of the Hartmann layer and c_f the skin friction coefficient.

The experiments of Hua and Lykoudis (1974) in a channel show that the decrease in turbulence intensity with increase in field correlates with Ha^2/Re at low field strengths; Gardner and Lykoudis (1971) similarly find the correlating parameter in pipes to be $Ha^2/Re^{0.75}$, which is the same as Eq. (6) if c_f is taken to obey the well-known Blasius law in turbulent flow, $c_f \simeq Re^{-1/4}$. The Blasius law should be a reasonable approximation at low fields and relatively low Re, but as the field and Re increase it appears from Gardner and Lykoudis' (1971) experiments that the c_f varies less with Re than the Blasius law suggests. The correlation of $(Ha/Re)_{cr}$ with Re, provided in Fig. 27 of their paper, shows two distinct regions, which we may describe approximately by

$(Ha/Re)_{cr} \simeq Re^{-0.05}, \quad 10^4 < Re < 8 \times 10^4$

$(Ha/Re)_{cr} \simeq Re^{-0.3}, \quad 10^5 < Re < 5 \times 10^5$

The high Re correlation is rather like Eq. (7).

In other words, the new correlating parameters found in the work of Lykoudis and his co-workers are of the kind that would be expected if the chief dynamic factor in laminarization is the domination of $\bar{j} \times \bar{B}$ forces over Reynolds stress

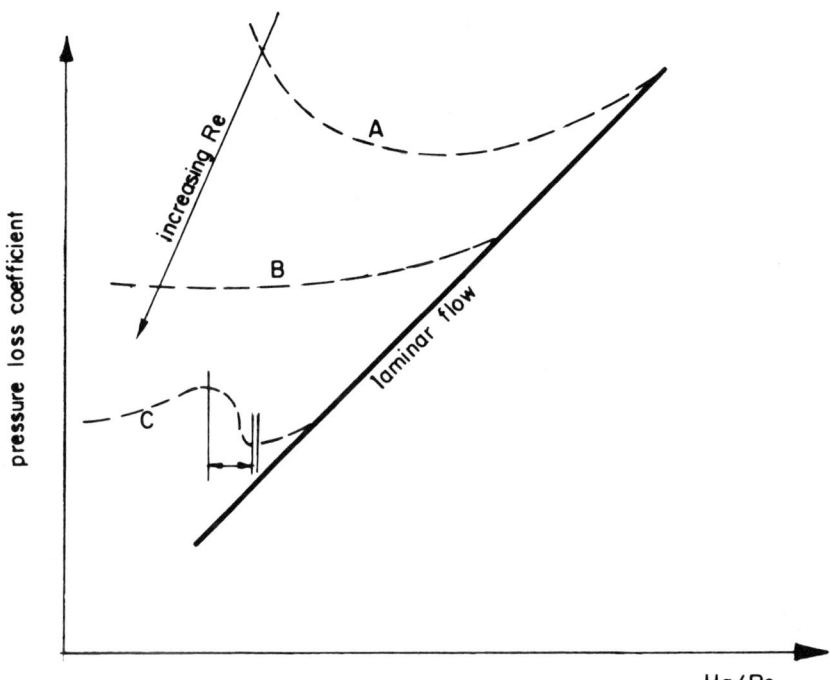

Fig. 4 Schematic diagram showing different possible kinds of variation of pressure loss coefficient in MHD duct flow. Dashed lines show experimental data; full line is laminar theory. The single vertical line marks the beginning of reverse transition, the double vertical line the end.

gradients. The residual turbulence found in the MHD experiments is not peculiar to MHD; as already pointed out in Sec. II, such turbulence is found in highly accelerated non-MHD laminarizing flows as well, although the precise character of such turbulence in the two flows may not be the same. An effectively laminar flow close to the wall is maintained because of the enhanced stability of the altered mean velocity profiles, due respectively to the Hartmann effect and the favorable pressure gradient. There thus appears to be a close parallel between MHD flows with high normal fields and non-MHD flows in strong favorable pressure gradients.

There is one final point that is relevant in this connection. The many different measurements that have been made of the pressure loss coefficient in duct flows exhibit three kinds of behavior with Hartmann number, as shown in Fig. 4. In all of them the coefficient goes up linearly with Ha at sufficiently high fields, as in laminar flow theory. However, at low fields, the coefficient first goes down in type A and varies little in type B. In type C, found particularly in the measurements of Hua and Lykoudis (1974), the coefficient first increases, then dips, and then increases once again. At first sight it may seem that these patterns of behavior are inconsistent with each other, but it may be worth pointing out that this need not necessarily be the case for the following reasons.

The values for the resistance coefficient during transition from laminar to turbulent flow can usually be obtained as a linear combination of appropriate values in laminar and turbulent flow using an intermittency factor γ, which is the fraction of time that flow is turbulent at any given station or flow condition. The behavior of mean flow during transition of the boundary layer on a flat plate is described satisfactorily using such a factor (Dhawan and Narasimha 1958). In pipe flow, the situation is a little more complicated (Rotta 1956, Pantulu 1962). At Reynolds numbers around the critical value, the intermittency is a function of streamwise distance along the pipe, i.e., the flow is not invariant in x. At somewhat higher Reynolds numbers the flow in the pipe takes on an oscillatory character, because the appearance of long turbulent slugs in the pipe slows the flow until the slug is washed out, following which the flow velocities increase, a slug forms again, and so on, repeating the cycle. Irrespective of Reynolds number, however, the length of the pipe becomes an additional parameter to be considered in the transition problem -- the diameter alone is not enough.

What forms these phenomena take in MHD duct flow is not known; in particular, the intermittency does not seem to have attracted much attention yet. Furthermore, how these factors affect the reverse transition is even less known.

In any case, the character of the variation of the actual skin friction coefficient during such transitions will depend (through the intermittency distribution) on the extent of the transition region. In Fig. 4 we have displayed the possible ways by which the observed friction coefficients would be obtained. For example, type C will be the logical result if the turbulent skin friction coefficient tends to increase with the application of the magnetic field, the critical value of the ratio Ha/Re is relatively low, and the extent of the transition zone is also relatively small.

It is clear that in this case a superposition of laminar and turbulent values through an intermittency factor spread over the transition zone would produce precisely the kind of behavior observed. (An increase in c_f with application of low magnetic fields is quite conceivable, and would occur for reasons similar to those that raise c_f in laminar flow. It is only with the onset of laminarization that c_f would come down. One may again compare this with the analogous effect in favorable pressure gradient non-MHD flow, illustrated in Fig. 1.)

On the other hand, if the increase of turbulent skin friction with magnetic field coincides with a relatively well spread out reverse transition zone, a curve of type B would result. Type A would be encountered if the application of even low magnetic field tends to bring down the friction coefficient, subsequent laminarization increasing it. We can expect type A only at relatively low Reynolds number, where viscous dissipation will reinforce the effect of the magnetic field to depress the friction coefficient rapidly with application of magnetic field. An examination of the data compiled by Branover (1978, p. 138) shows that the type A curve is indeed encountered only at channel Reynolds numbers of a few thousands -- indeed, often in the range where generally the flow is not fully turbulent at all, unless highly disturbed conditions prevail.

To make these explanations quantitative, however, would require more precise knowledge of the friction coefficient in turbulent flow and of the extent of the transition region at various values of the Re/Ha ratio. As Branover's review makes clear neither of these is known at present, but it should be worthwhile if experiments could be devised to shed light on these factors.

IV. Conclusion

The present comparison of laminarizing flows, in both MHD and non-MHD situations, shows that there are several parallels. The decaying turbulence in dissipative non-MHD reversion is strongly anisotropic. With an aligned field, the chief mechanism is a combination of ohmic and viscous dissipation. The persistence of a residual turbulence even when eddy transport is negligible is a feature of several flows, MHD as well as otherwise. With strong normal fields, the Hartmann effect plays a major role in laminarization; this situation, it is argued, is akin to the effect of strong favorable pressure gradients on a (non-MHD) turbulent boundary layer, which is a typical example of "soft" laminarization, where turbulence does not vanish but turbulent transport is negligible. The governing parameter, according to this view, should then be of the form Ha/Re^n, where n is appreciably lower than unity. These arguments are supported by recent measurements in channel flow. Future duct experiments should pay attention to intermittency in the flow, and investigate the effect of duct length on the observed pressure loss as well as the possibly oscillatory character of the flow under certain conditions.

Acknowledgment

I am indebted to Prof. K.R. Sreenivasan for many valuable discussions on the subject of this paper.

References

Back L.H., Massier P.F., and Gier H.L. (1964) "Convective heat transfer in a convergent-divergent nozzle," Int. J. Heat Mass Transfer, Vol. 7, pp. 549-568.

Badri Narayanan M.A. (1968) "An experimental study of reverse transition in two-dimensional channel flow," J. Fluid Mech., Vol. 31, pp. 609-623 (also PhD Thesis, Dept. of Aero. Eng., Indian Inst. of Science, Bangalore).

Batchelor G.K. and Proudman I. (1954) "The effect of rapid distortion of a fluid in turbulent motion," Q. J. Mech. Appl. Math., Vol. 7, pp. 83-103.

Blackwelder R.F. and Kovasznay L.S.G. (1972) "Large scale motion of a turbulent boundary layer during relaminarization," J. Fluid Mech., Vol. 53, pp. 61-83.

Bradshaw P. (1969) "A note on reverse transition," J. Fluid Mech., Vol. 35, pp. 387-390.

Branover H. (1978) Magnetohydrodynamic Flow in Ducts, John Wiley & Sons, New York.

Branover H. and Gershon P. (1976) "MHD turbulence study, Pt. 2," Dept. Mech. Eng., Ben-Gurion Univ., Beer-Sheva, Israel, Rept. ME6-76.

Businger J.A. and Arya S.P.S. (1974) "Height of the mixed layer in the stably stratified planetary boundary layer," Adv. Geophys., Vol. 18A, pp. 73-92.

Dhawan S. and Narasimha R. (1958) "Some properties of boundary layer flow during transition from laminar turbulent motion," J. Fluid Mech., Vol. 3, pp. 418-436.

Freim P.W. and Heiser W.H. (1968) "The effect of strong longitudinal magnetic field on the flow of mercury in a circular tube," J. Fluid Mech., Vol. 33, pp. 397-413.

Gardner R.A. and Lykoudis P.S. (1971) "Magneto-fluidmechanic pipe flow in a transverse magnetic field, Pt. 1: Isothermal flow," J. Fluid Mech., Vol. 47, pp. 737-764.

Globe S. (1961) "The effect of a longitudinal magnetic field on pipe flow of mercury," Trans. ASME, J. Heat Transfer, Vol. 83, p. 445.

Hartmann J. and Lazarus F. (1937) "Hg-dynamics II," K. Dan. Vidensk. Selsk., Mat.-Fys. Medd., Vol. 15, p. 7.

Hua H.M. and Lykoudis P.S. (1974) "Turbulence measurements in a magneto-fluid-mechanic channel," Nucl. Sci. and Eng., Vol. 54, pp. 445-449.

Hunt J.C.R. (1965) "Magnetohydrodynamic flow in rectangular duct." J. Fluid Mech., Vol. 21, p. 577.

Hunt J.C.R. (1973) "A theory of turbulent flow round two-dimensional bluff bodies," J. Fluid Mech., Vol. 61, pp. 625-706.

Laufer J. (1962) "Decay of non-isotropic turbulent field," Miszellaneen die Angewandte Mechanik, Festschrift Walter Tollmien, Akademie-Verlag, Berlin.

Launder B.E. (1963) "The turbulent boundary layer in a strongly negative pressure gradient," Gas Turbine Lab., Mass. Inst. of Tech., Cambridge, Rept. 71.

Lehnert B. (1955) "An instability of laminar flow of mercury caused by an external magnetic field," Proc. R. Soc. London, Ser. A, Vol. 233, pp. 299-302.

Lock R.C. (1955) "The stability of the flow of an electrically conducting fluid between parallel planes under a transverse magnetic field," Proc. R. Soc. London, Ser. A., Vol. 233, pp. 105-125.

Moffatt H.K. (1967) "On the suppression of turbulence by a uniform magnetic field," J. Fluid Mech., Vol. 28, pp. 571-592.

Moreau R. (1969) "On MHD turbulence," Turbulence of Fluids and Plasmas (ed. J. Fox), Brooklyn Polytechnic Press, New York, pp. 359-372.

Murgatroyd W. (1953) "Experiments on MHD channel flow," Philos. Mag., Vol. 44, pp. 1348-1354.

Narasimha R. (1977) "The three archetypes of relaminarization," Proc. 6th Canadian Congress of Applied Mechanics, Vancouver (also Aero. Eng., Indian Inst. of Science, Bangalore, Rept. 77FM7).

Narasimha R. and Sreenivasan K.R. (1973) "Relaminarization in highly accelerated turbulent boundary layers," J. Fluid Mech., Vol. 61, pp. 417-447.

Narasimha R. and Sreenivasan K.R. (1979) "Relaminarization of fluid flows," Adv. Appl. Mech., Vol. 19, pp. 221-309.

Nicholl C.I.H. (1970) "Some dynamical effects of heat on a boundary layer," J. Fluid Mech., Vol. 40, pp. 361-384.

Pantulu P.V. (1962) "Studies on the transition from laminar to turbulent flow in a pipe," MSc Thesis, Dept. Aero. Eng., Indian Inst. of Science, Bangalore.

Preston J.H. (1957) "The minimum Reynolds number for a turbulent boundary layer and the selection of a transition device," J. Fluid Mech., Vol. 3, p. 373.

Rajagopalan S. (1974) "Some experimental investigations of the fine scale structure of turbulence," PhD Thesis, Dept. Aero. Eng., Indian Inst. of Science, Bangalore.

Ramjee V. (1968) "Reverse transition in a two-dimensional boundary layer flow," PhD thesis, Dept. Aero. Eng., Indian Inst. of Science, Bangalore.

Richardson L.F. (1920) "The supply of energy from and to atmospheric eddies," Proc. R. Soc. London, Series A, Vol. 97, pp. 354-373.

Rotta J. von (1956) "Experimenteller Beitrag zur Enstehung turbulenter Strömung," Rohr. Eng. Arch., Vol. 24, p. 258.

Sajben M. and Fay J.A. (1967) "Measurement of the growth of a turbulent mercury jet in a coaxial magnetic field," J. Fluid Mech., Vol. 27, pp. 81-96.

Sibulkin M. (1962) "Transition from turbulent to laminar flow," Phys. Fluids, Vol. 5, pp. 280-284.

Sreenivasan K.R. (1974) "Mechanism of reversion in highly accelerated turbulent boundary layers," PhD Thesis, Dept. Aero. Eng., Indian Inst. of Science, Bangalore.

Sreenivasan K.R. and Narasimha R. (1974) "Rapid distortion of shear flows," Paper 2.3 presented at Aero. Soc. India Silver Jubilee Tech. Conf.

Sreenivasan K.R. and Narasimha R. (1978) "Rapid distortion of axisymmetric turbulence," J. Fluid Mech., Vol. 84, pp. 497-516.

Townsend A.A. (1958) "The effects of radiative transfer on turbulent flow of a stratified fluid," J. Fluid Mech., Vol. 4, p. 361.

Tsinober A.B. (1975) "MHD turbulence," Magn. Gidrodin., Vol. 11, pp. 7-22.

Viswanath P.R., Narasimha R. and Prabhu A. (1977) "Visualization of relaminarizing flows," J. Indian Inst. Sci., Vol. 60, pp. 159-165.

Experimental Studies on Liquid-Metal Two-Phase MHD Flow in Traveling and dc Magnetic Fields

Yoichi Fujii-e*
Nagoya University, Nagoya, Japan
and
Keiji Miyazaki,[†] Shoji Inoue,[‡] and Nobuo Yamaoka[‡]
Osaka University, Osaka, Japan

Abstract

A number of experimental studies on the magnetohydrodynamics of liquid-metal two-phase flows have been performed for the purpose of applying such flows to the advanced energy conversion system and/or the blanket cooling system of fusion power reactors. This paper reviews these research activities and the main results. The problems discussed are as follows: 1) effective electrical conductivity of two-phase fluid; 2) pressure drop and slip ratio of $NaK-N_2$ two-phase flow; 3) a trend that the gas bubble, or void, drifts inward and concentrates at the center of the channel. Void and velocity variations of liquid-metal two-phase flow along the induction MHD generator were analyzed. Experiments on the boiling of potassium in the presence of a transverse dc magnetic field were conducted to gain basic information on the application of boiling liquid-metal two-phase flow to fusion reactor blanket cooling.

Paper presented at Third Beer-Sheva International Seminar on Magnetohydrodynamic Flows and Turbulence, Ben-Gurion University of the Negev, Beer-Sheva, Israel, March 23-27, 1981. Copyright © American Institute of Aeronautics and Astronautics, Inc., 1982. All rights reserved.
  *Professor, Institute of Plasma Physics.
  †Associate Professor, Department of Nuclear Engineering, Faculty of Engineering
  ‡Department of Nuclear Engineering, Faculty of Engineering.

I. Introduction

The studies on the magnetohydrodynamics and the thermodynamics of liquid metals have been performed by the Nuclear Power Research Group of Osaka University in conjunction with the energy conversion and fusion reactor system and with nuclear safety.

The main items of the research are shown in Table 1 and the relative experimental facilities and their mutual relations in Fig. 1. Our main efforts have been concentrated on experiments with the NaK blowdown system and the forced-convection sodium boiling loop.

The purposes of the studies are to find the fundamental phenomena and to pursue and clarify the governing mechanisms. Therefore, small experimental devices are also provided to investigate the detailed fundamental phenomena found in the experiments with the large facilities. These investigations

Table 1 Main subjects of liquid-metal MHD experiments

Energy conversion	LMMHD	Single-phase dc mercury blowdown
		Single-phase ac NaK blowdown
		Two-phase ac NaK blowdown
	EGD	Mercury Rankine cycle
Heat pipe		Water
		Potassium and sodium
		In magnetic field
CTR blanket	Pressure drop	Single-phase mercury
		Single-phase NaK
		Two-phase NaK-N_2
	Boiling in magnetic field	Potassium pool boiling
Reactor safety	Sodium boiling	Direct heating
	FCI	Indirect heating
		Potassium pool boiling
		Sodium film boiling
	Pressure wave	Water steam loop
		Water rupture disk apparatus
Liquid-metal liner implosion		NaK liner implosion

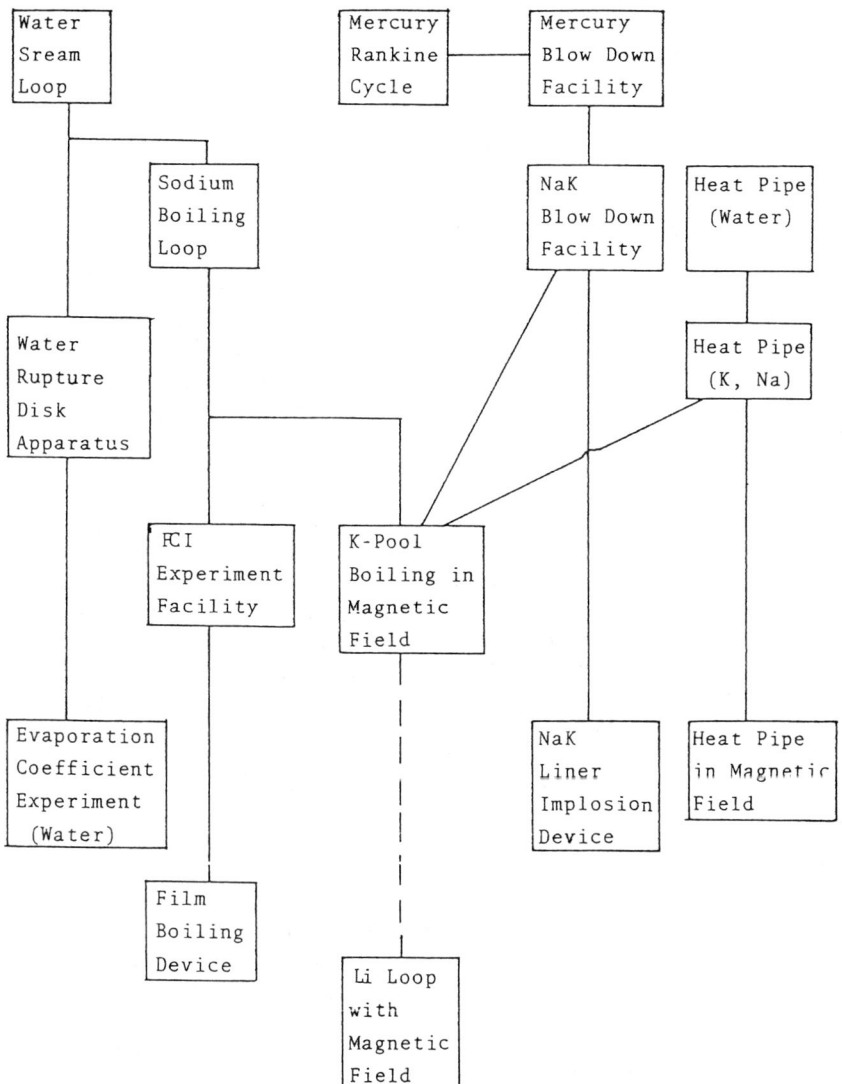

Fig. 1 Research activities of Nuclear Power Research Group, Osaka University.

contribute as well to the planning and checking of future experimental study. Some fundamental experiments were started later on the fuel/coolant interaction to Fast Breeder Reactor safety, and also on the potassium boiling and heat pipe experiments in the magnetic field for applying to the fusion reactor blanket cooling technology.

This paper presents a review and summary of the research activities on the liquid-metal magnetohydrodynamics (LMMHD) for the power generation and applications to the fusion reactor technology. Figure 1 shows the diagram of overall research activities of the Nuclear Power Research Group of Osaka University and Table 2 the breakdown of the MHD work.

II. Liquid-Metal Two-Phase MHD Power Generation

One of the most important problems in LMMHD power generation is the acceleration of liquid-metal flow. From thermodynamic consideration of the system, a higher thermal cycle efficiency would be expected for those cycles in which the two-phase flow is led into the generator without separation and condensation of the gaseous phase upstream of the generator than for those of single-phase flow. The gas or vapor plays an important role in the acceleration, and the performance characteristics of the generator have much dependence on a volumetric fraction of the vapor or gas, i.e., the void fraction. Therefore, these cycles may have some complicated problems such as 1) two-phase effective electrical conductivity, 2) two-phase frictional pressure drop, 3) variations in electrical conductivity and fluid velocity along the flow direction, 4) gas/liquid slip ratio, 5) distribution of the void fraction in the cross section perpendicular to the flow, 6) occurrence of oscillation or instability phenomena due to interaction between a compressible fluid and magnetic field, 7) boiling and/or condensation phenomena under the magnetic field in the case of one-component fluid, and so on.

The main experimental facility to study the items mentioned above came into operation in 1969. It consists of a NaK blowdown facility, dc and ac magnets, single- and

Table 2 Breakdown of NaK blowdown MHD experiments

Single-phase	Two-phase
Induction generator (40-250 Hz, 0.3 T)	Effective electrical conductivity

Pressure drop (dc, 2 T)
  Rec. cross section
  Pipe
	Flow Regime
	Liquid-gas slip ratio
	Pressure drop
	Distribution of void fraction

End effect	

two-phase MHD test channels, a various frequency type of motor-generator set for an ac magnet, liquid-metal boiling devices, two-phase effective electrical conductivity measurement devices, and instrumentation.

The problems already studied are discussed in the following sections.

Effective Electrical Conductivity of Two-Phase Fluid

It is difficult to avoid the introduction of the gas phase and the formation of a two-phase mixture with a low void fraction in the generator channel, even in single-phase cycles. Therefore, a problem of first importance is to evaluate the effective electrical conductivity of a two-phase fluid as a function of void fraction and its dependence on the flow pattern.

The effective electrical conductivity of two-phase mixtures was measured for the following combinations by the respective methods as schematically illustrated in Fig. 2:

1) H_2O-N_2 mixture in alternative electric fields.

2) $NaK-N_2$ mixture in static magnetic fields.

3) $NaK-N_2$ mixture in traveling magnetic fields.

4) $Hg-H_2O$ mixture in traveling magnetic fields.

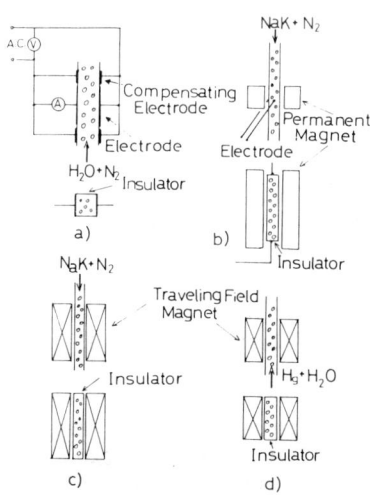

Fig. 2 Measurement of the effective electrical conductivity of two-phase flow.

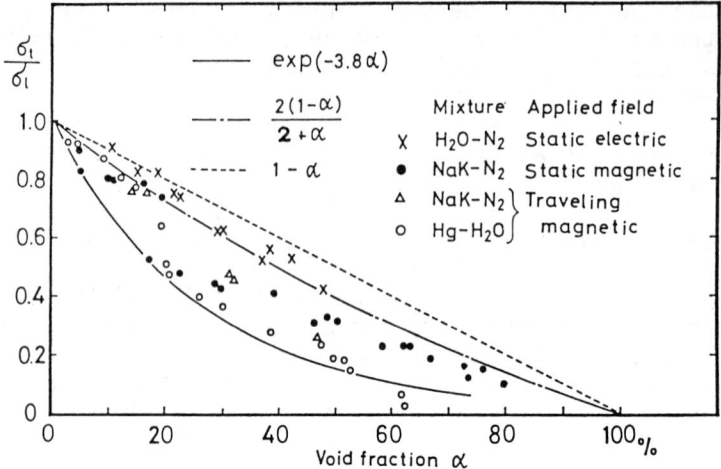

Fig. 3 Ratio of electrical conductivity of two-phase mixture to that of liquid alone.

The results are shown in Fig. 3. It is seen that the effective electrical conductivity of the two-phase mixtures agrees fairly well with Maxwell's theoretical value in the bubbly regime, but that a comparatively sharp reduction occurs around a void fraction of 0.2, which seems to result from the transition from the bubbly to sluggish flow regimes. And the transition of the flow pattern may depend on the strength of the MHD interaction.

Pressure Drop in Liquid-Metal Two-Phase Flow Under Transverse Magnetic Fields

The problem of liquid-metal two-phase flow becomes more complicated because of the superposition of a strong MHD interaction and compressibility of the two-phase fluid.

The experimental data on the MHD pressure drop of alkali metal two-phase flow are limited and an accumulation of data is necessary. This study is useful not only in designing a MHD generator channel but also in evaluating the technical feasibility of the CTR (controlled thermonuclear reactor) blanket cooling system employing single- or two-phase liquid-metal flow (Fujii-e and Suita 1974).

The results of $NaK-N_2$ two-phase flow experiments are shown in Fig. 4. The pressure drop ratio of $NaK-N_2$ two-phase to NaK single-phase flows for the same NaK mass flow

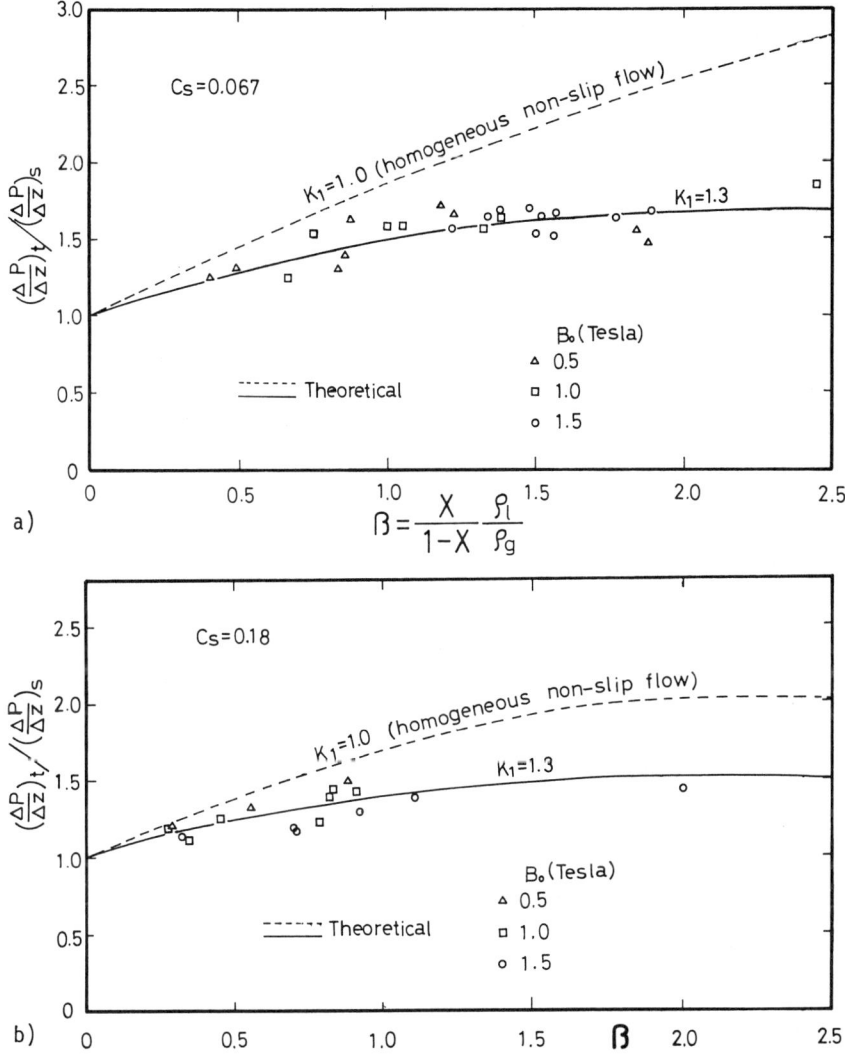

Fig. 4 Plots of two-phase to single-phase pressure drop ratio against gas/liquid volumetric flow rate ratio: a) wall-to-fluid conductance ratio $C_s = 0.068$, b) wall-to-fluid conductance ratio $C_s = 0.18$.

rate is plotted against the liquid-to-gas volumetric flow rate ratio β, with the applied magnetic field strength as a parameter. The parameter K_1 is defined as $K = \langle \alpha u_t \rangle / \langle \alpha \rangle \langle u_t \rangle$, where α and u_t are the void fraction and the two-phase flow velocity, respectively, and the symbol $\langle \ \rangle$ denotes the average value over the cross section. Therefore, $K_1 = 1$ corresponds to homogeneous nonslip flow. When compared in the same mass flow rate condition, the MHD pressure drop becomes larger for a two-phase flow than for a single-phase flow since the flow velocity should be higher for the former than for the latter.

Gas/Liquid Slip Ratio in Magnetic Field

The gas/liquid slip ratio in the magnetic field is treated in relation to the two-phase flow pattern, that is, the distribution of the gaseous phase and the fluid velocity which are strongly influenced by the applied magnetic field (Saito et al. 1978a). Discussion will be made from the viewpoint of the electromagnetic force balance in the cross section perpendicular to the flow direction.

The experiments on these subjects were performed under a dc magnetic field of 0.5 ∿ 1.5 T with a NaK-N_2 mixture as the working fluid. The results are described in the following paragraphs.

In Fig. 5 the ratio of the mean void fraction $\langle \alpha \rangle$ to the center void fraction α_c is shown as a function of the effective magnetic interaction number N_i, which is defined

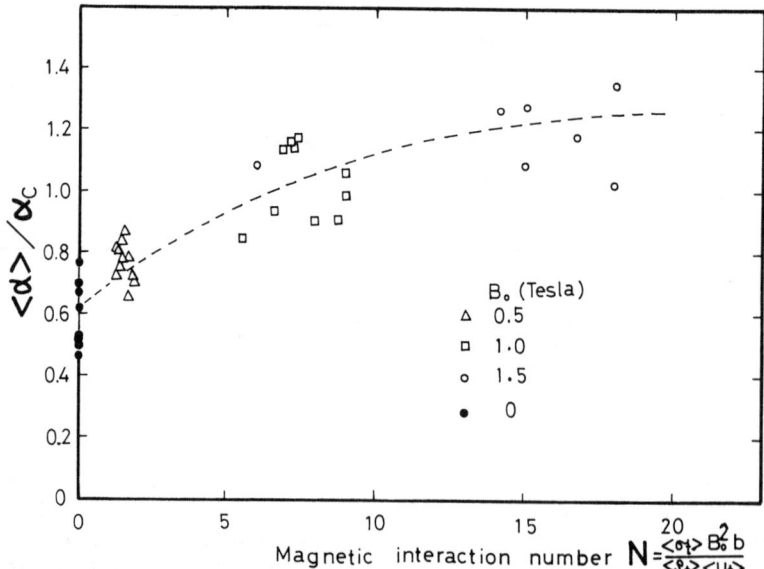

Fig. 5 Mean-to-center void ratio vs magnetic interaction number.

by

$$N_i = \frac{\text{electromagnetic force}}{\text{inertial force}} = \frac{\langle\sigma_t\rangle B_0^2 b \langle u_t\rangle}{\langle\rho_t\rangle\langle u_t\rangle^2}$$

It is clearly seen in Fig. 5 that the void distribution is much influenced by the applied magnetic field. The gas bubbles may be pushed away toward the channel walls on both sides, changing the mean-to-center void ratio $\langle\alpha\rangle/\alpha_c$ from below unity to over unity with the increasing applied magnetic field strength.

In Fig. 6 the gas/liquid slip ratios S_f are shown as a function of the liquid-gas volumetric flow rate ratio β for the case of C_s = 0.067 and 0.18 with the applied magnetic field strength as parameter, where $C_s = \sigma_w\delta_w/\sigma_f\delta_f$ or the wall-to-fluid electric conductance ratio in the case of single-phase fluid. From this figure it is seen that the effects of applied magnetic field strength and the channel geometry on the gas/liquid slip ratio are not significantly discerned and its dependence on the gas/liquid volumetric flow ratio is discernible.

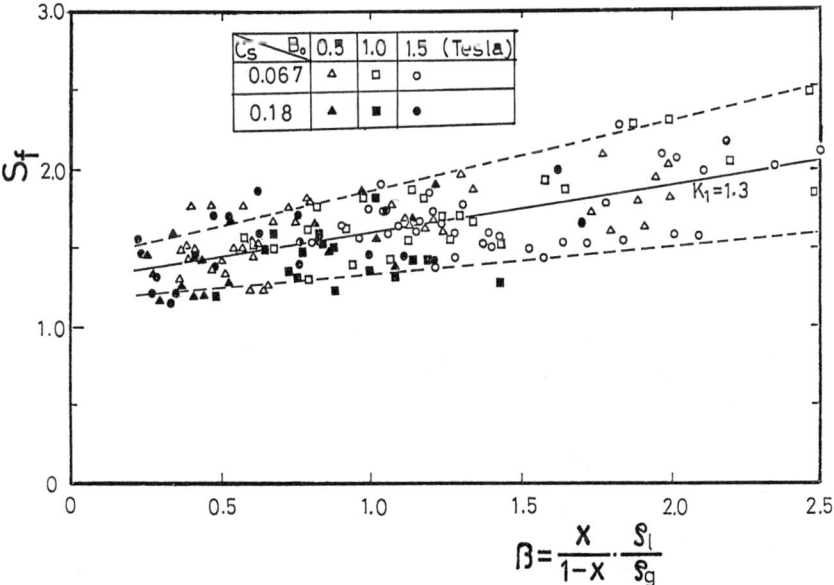

Fig. 6 Plots of slip ratio against gas/liquid volumetric flow rate ratio.

Change of Gaseous Phase Distribution of Liquid-Metal Two-Phase Flow due to Electromagnetic Pinch Effect

In order to demonstrate the electromagnetic pinch effects on the void distribution, the experiment was conducted in the rectangular test section, shown in Fig. 7, with the coils wound outside the channel wall. Thus, the parallel magnetic field B_{zs}, produced by the coils and superposed to the perpendicular field B_0, is expected to change the void distribution. The distribution of void fraction in the cross section across the flow channel was measured at the three locations along the flow with the x-ray attenuation technique.

When the parallel magnetic field is not superposed, the void distributions for the various perpendicular magnetic field are nearly symmetric, as shown in Fig. 8. With increasing perpendicular magnetic field strength, the void distribution makes the transition from near parabolic to M-shaped (double-peak) profiles, retaining the symmetry.

By superposing the parallel field, the symmetrical distribution is deformed as shown in Fig. 9, with the peak shifting toward the right or left, depending on the direction of the parallel magnetic field. This suggests the validity of the surmise that the change in void distribution results

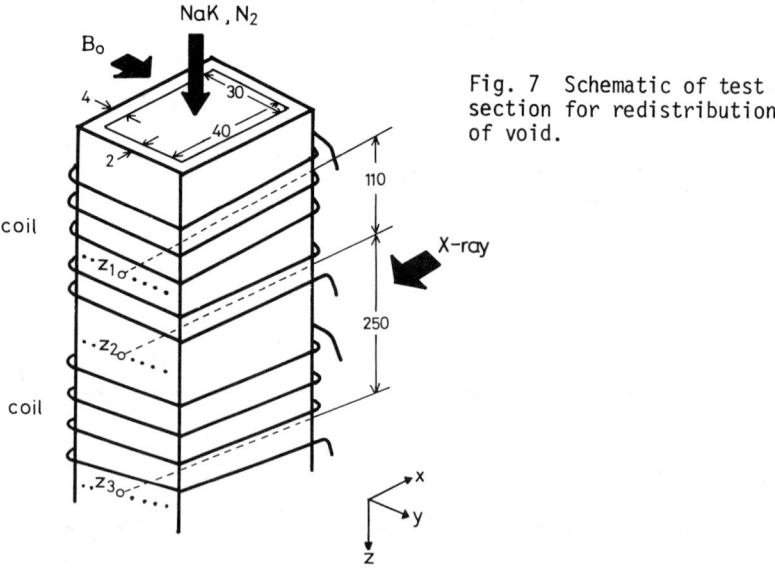

Fig. 7 Schematic of test section for redistribution of void.

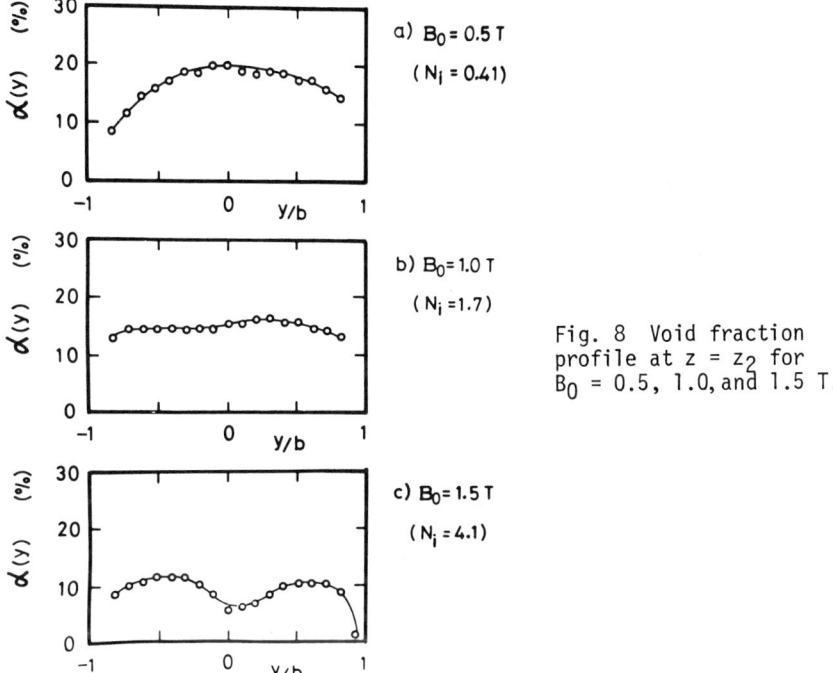

Fig. 8 Void fraction profile at $z = z_2$ for $B_0 = 0.5, 1.0,$ and 1.5 T.

from the electromagnetic interaction exerted on the liquid phase.

The analysis of the redistribution of void was made on the assumptions that 1) the shift of inverse peak, or trough, of the void distribution along the z direction corresponds to the drift of gas phase in the y direction; 2) the MHD force works only on the liquid phase; 3) the drag force, i.e., the gas/liquid interaction is proportional to the square of gas/liquid relative velocity

$$f_D = C_D \rho_\ell (u_{gy} - u_{\ell y})^2$$

and 4) the friction factor is a function of the magnetic interaction number N_i, i.e., the ratio between the electromagnetic and inertial forces. The coefficient C_D is expressed by

$$C_D = k N_i^n \quad \text{where} \quad N_i = \frac{J_x (B_{zi} + B_{zs}) b}{\rho_\ell u_{\ell y}^2}$$

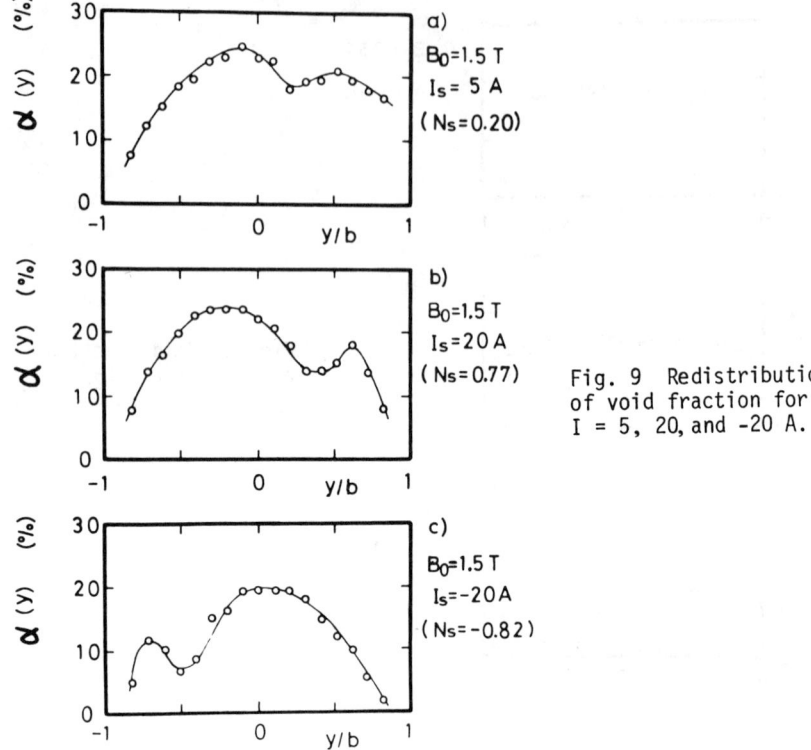

Fig. 9 Redistribution of void fraction for I = 5, 20, and -20 A.

and B_{zs} and B_{zi} denote the externally superposed and the induced magnetic fields, respectively, b the half-width of the flow channel, and $u_{\ell y}$ the drift velocity of liquid in the y direction.

The positions of inverse peak y_g can be obtained by solving the drift and continuity equations for the gas and liquid phases. By fitting the analytical result of y_g with the experimental we obtain the plots shown in Figs. 10 and 11. The index n and the proportional constant k for the drag coefficient C_D can be determined from the figures: k = 150 and n = 0.15 are estimated by the solid lines in the figures.

Figure 12 shows the relation between the shift of inverse peak and the superposed field strength, both being normalized. The dotted line indicates the estimation based on the drag force of the Stokes law. Thus, the present model of gas/liquid interaction with $C_D = kN_i^n$ gives a reasonable explanation for the redistribution of void fraction.

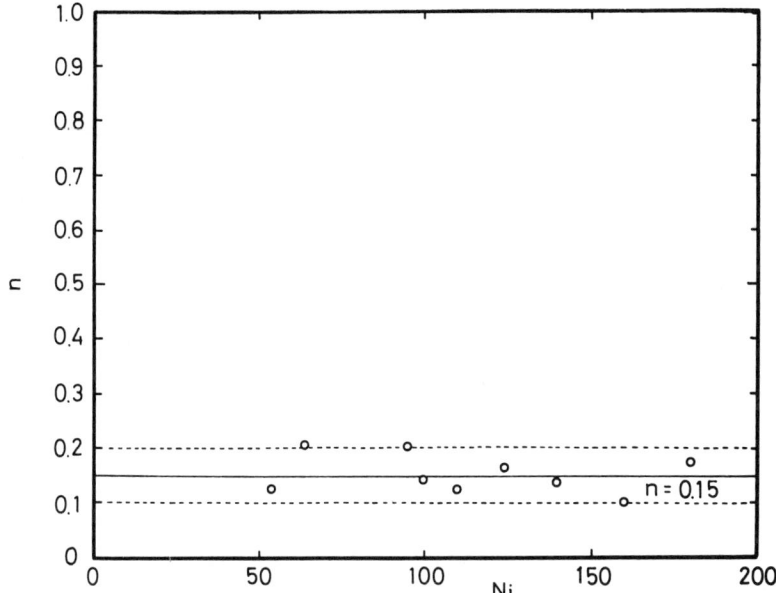

Fig. 10 Plots of power index n against magnetic interaction number N_i; drag coefficient $C_D = kN_i^n$.

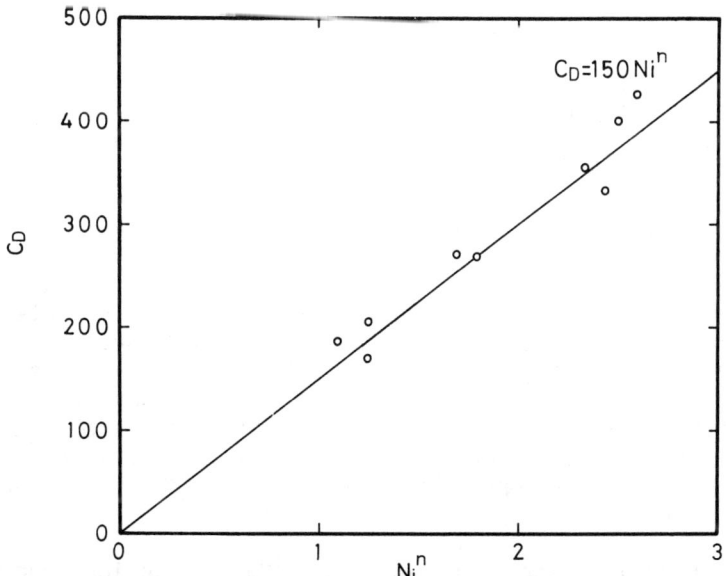

Fig. 11 Relation between drag coefficient C_D and magnetic interaction number N_i^n.

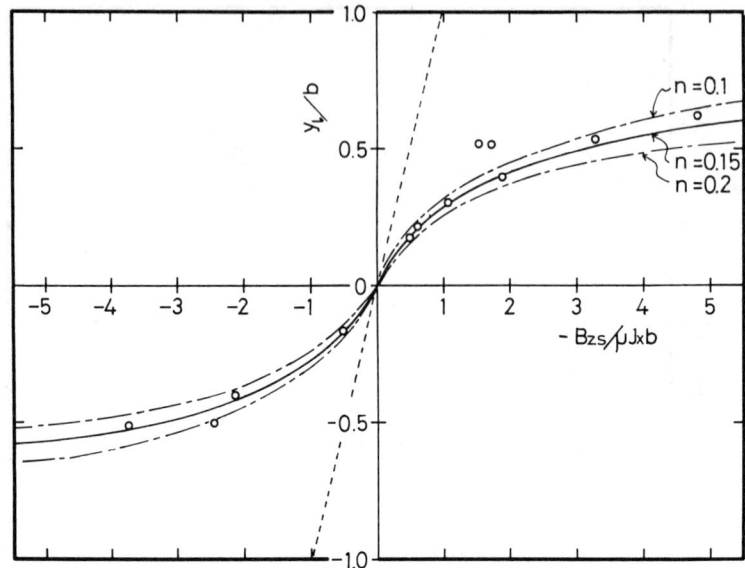

Fig. 12 Plots of void peak displacement against superposed-to-induced magnetic field ratio.

Influence of Void and Velocity Variation on Two-Phase LMMHD Induction Generator Characteristics

Nonuniformity of void and velocity along the flow will affect the performance characteristics of a two-phase MHD generator. When the variation in velocity is large, the portion of the fluid whose velocity exceeds that of the traveling magnetic field, or the synchronous velocity of the generator, does contribute to the power output, while the rest with a lower velocity is accelerated by the traveling field at the expense of part of the output. In addition, in the two-phase cycles, the thermal energy is converted first into the kinetic energy in the prestage nozzle or in the generator channel itself and subsequently into the electrical energy. This means that the gas phase expands in the generator channel and its expansion causes a change in the void fraction along the flow and consequently results in changes in the equivalent electrical conductivity and the fluid velocity.

These influences on two-phase LMMHD induction generator characteristics were treated analytically and experimentally in Fujii-e and Suita (1971), Tanatsugu et al.

(1974), and Fujii-e et al. (1975). The summary is presented here. In a generator, where the electrical conductivity and the fluid velocity vary only in the flow direction, the perturbed magnetic field $b_z(x,t)$ is obtained from the induction equation

$$b_z(x,t) = b_z(x)\exp(j\omega t)$$

$$b_z(x) = (A_1 + A_2 x)\exp(-jkx)$$

where

$$A_1 = \frac{-R_{mo}(\frac{u_b}{u_s} + J\frac{u_a}{u_s})B_m - [2j + R_{mo}(1 - S_o - \frac{u_a}{u_s})]\frac{A_2}{k}}{1 + R_{mo}\frac{u_b}{u_s} + jR_{mo}(S_o + \frac{u_a}{u_s})}$$

$$A_2 = \frac{jkR_{mo}(S_o R_{mo}\frac{u_a}{u_s} + \frac{u_b}{u_s})B_m}{1 + R_{mo}\frac{u_b}{u_s} + jR_{mo}(S_o + \frac{u_a}{u_s})}$$

$$u_a = \frac{\partial}{\partial x}\left(\frac{1}{\mu\sigma(x)}\right) \quad \text{and} \quad u_b = \frac{1}{k}\frac{\partial}{\partial x}u(x)$$

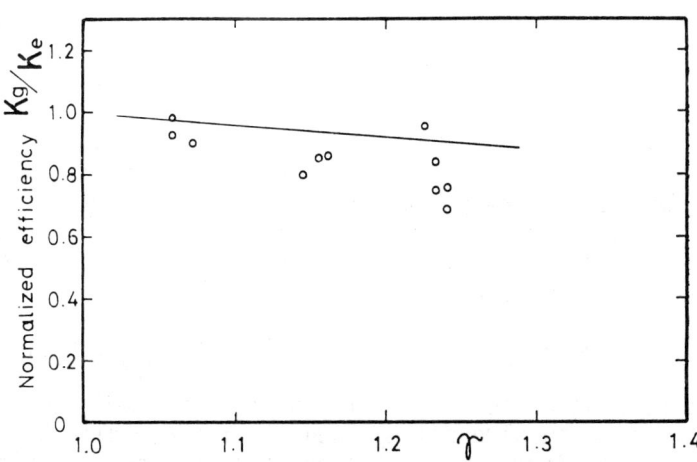

Fig. 13 Normalized generator efficiency vs parameter $1/\gamma$.

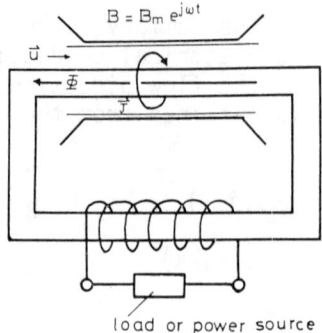

Fig. 14 Conceptual machine structure.

The condition of $b_z(x) = 0$, i.e., nonperturbation, leads to the following relation,

$$\gamma = \frac{-S_0}{1-S_0} \beta + \frac{1}{1-S_0}$$

where
$$\beta = \frac{\text{inlet electrical conductivity}}{\text{outlet electrical conductivity}}$$

$$\gamma = \frac{\text{outlet fluid velocity}}{\text{inlet fluid velocity}}$$

This relation prescribes the inlet slip ratio S_0 and consequently the mode of operation. The generator efficiency under this relation is dependent on γ and is expressed as

$$\bar{\eta}_g = \eta_{go}[1 - (\gamma-1)/2]$$

where $\bar{\eta}_g = P_e/P_g$, $P_e = \int P_g(x)dv$, and $P_g = \int P_e(x)\,dv$.

The experimental results for the channel with a constant cross section are shown in Fig. 13, where $K_g = P_e/P_{eo}$ and $K_g = P_g/P_{go}$ denote the perturbed-to-nonperturbed ratios of electromechanical and generator power outputs, respectively. With increasing the outlet-to-inlet velocity ratio γ, the normalized efficiency decreases.

Induction MHD Generator Using Alternating Magnetic Field

The machine proposed here is an induction MHD power generator using an alternating magnetic field, which has the principle of performance and the characteristics as explained in the following paragraphs and illustrated in Fig. 14.

The closed current loop is formed in the working fluid in the generator by means of the electromotive force $\vec{j} \times \vec{B}$ And then the alternating magnetic field induced by the current is guided out of the generator directly through the inner core, which has a high magnetic permeability, and coupled with the outer electrical circuit by the secondary coil for extracting the output power.

Strong magnetic coupling due to the high magnetic permeability of the inner core and high adaptability to a wide range of load, which results from the separation of the primary coil to excite the magnetic field for the electromotive force and the secondary coil to obtain the output power, bring about more flexible operating conditions than those for the conventional induction MHD generator using the traveling magnetic field.

For example, they make it possible to operate the MHD generator in an inductive mode with the use of the comparatively low electrical conducting fluid such as a two-phase liquid metal of higher void fraction (90%) or low-temperature plasma as well as liquid metal. Also, when the fluid with a high electrical conductivity and a high flow velocity such as fusion plasma is used as the working fluid, the machine is able to operate without the sacrifice of the load factor in a frequency region low enough to avoid the skin effect.

The primary theoretical analysis of the performance characteristics of this new type of machine is almost complete (Fujii-e et al. 1979).

NaK Blowdown Facility

The MHD experiments in the use of this facility are divided into two main groups. One is the experiment with a dc static magnetic field and the other is that with an ac traveling magnetic field. Moreover, they are classified into the experiments with single-phase flow and those with two-phase flow as the working fluid, respectively.

Figure 15 shows the flow diagram of the facility, in which the thick lines denote the piping for NaK and the thin lines those for nitrogen gas. The test section is interchangeable and we have various test channels, some for the dc experiment and others for the ac MHD experiment. The specifications of each channel are shown in Table 3 and

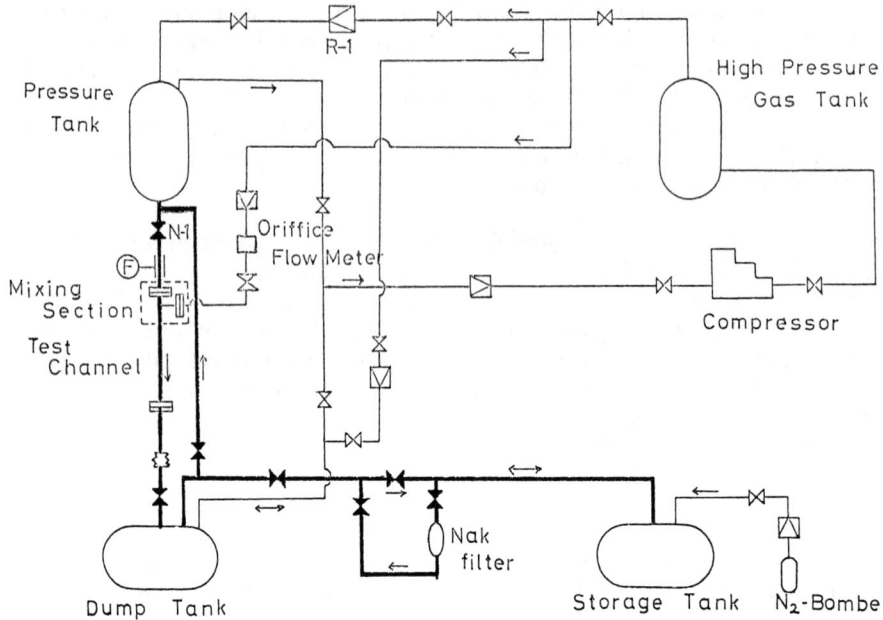

Fig. 15 Flow diagram of NaK blowdown facility.

Table 3 Test channels of NaK blowdown MHD experiments

Channel No.	Material	Cross section, mm^2	Length, m	Wall thickness, mm	Magnetic field, T	Frequency, Hz
DC 1	Stainless steel 304	30x40	1.7	4 and 2	2	dc
DC 2	Stainless steel 304	22.6 ID	1.7	2	2	dc
DC 3	Stainless steel 304	15.4 ID	1.7	2	2	dc
AC 1	Inconel alloy	30x3	1.7	4, copper side bar	0.3	40-250
AC 2	Epoxy resin	60x4	0.3	Insulated wall, silver side bar	0.1	40-250

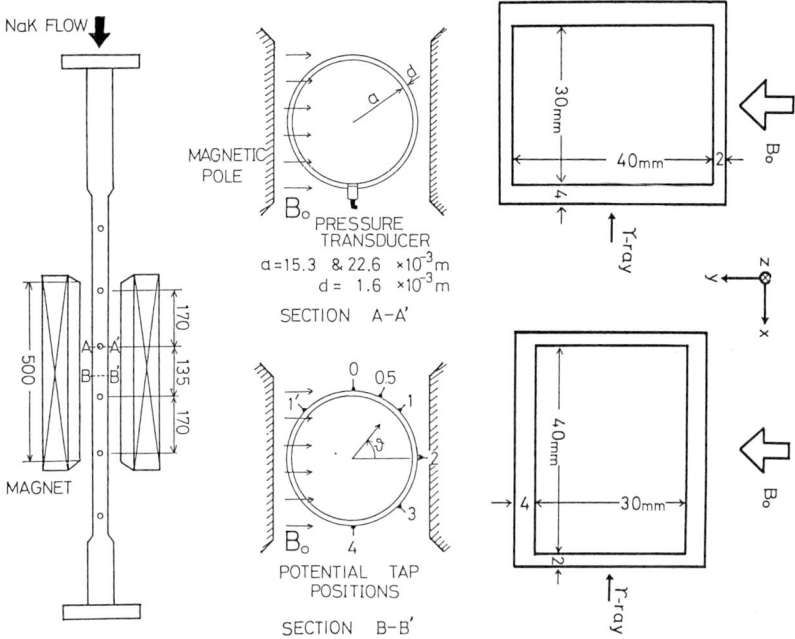

Fig. 16 Test sections for NaK-blowdown MHD (dc) experiment.

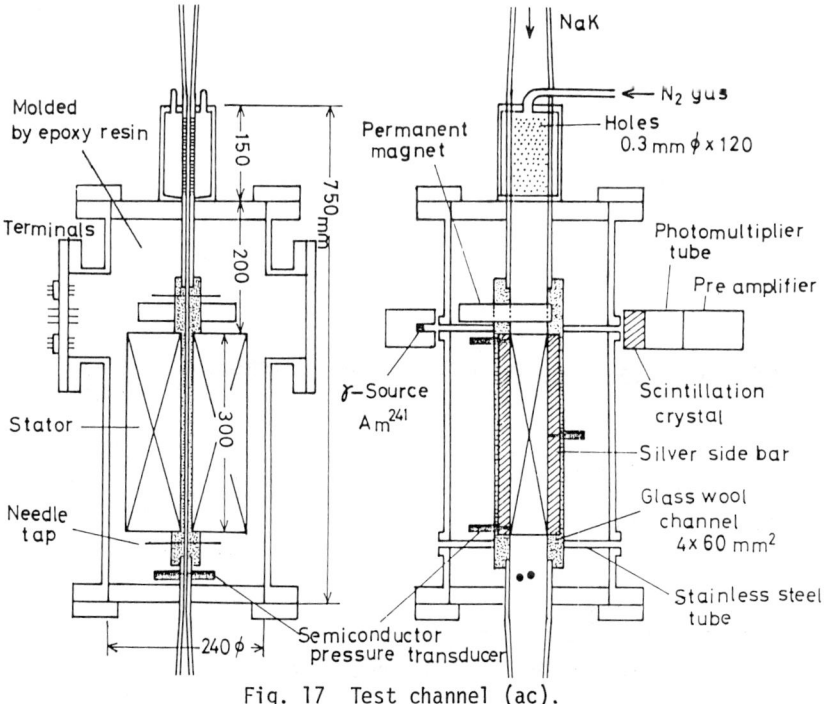

Fig. 17 Test channel (ac).

their schematic diagrams in Figs. 16-18, respectively. The system contains 230 ℓ of NaK-78 (22% Na/78% K) which is in liquid phase at room temperature. The maximum system pressure is limited to 10 Kg/cm^2, although the design limit is 50 Kg/cm^2, in order to meet Japan's national regulation on pressurized gas.

Blowdown Procedure. The NaK is transferred from the storage tank to the dump tank through the NaK strainer before each operation to remove any impurities in the NaK and then transfused into the pressure tank from the dump tank by compressed N$_2$ gas. First, the NaK in the pressure tank is pressurized at a prescribed value by the compressed N$_2$ gas provided from the high-pressure gas tank. In the next step, the valve N-1 is opened by the remote control and the high velocity flow of NaK passes through the test channel. During operation, the pressure of the pressure tank is automatically kept nearly constant by the pressure regulator R-1.

In the case of experiments with two-phase flow as the working fluid, nitrogen gas is introduced into the vertically flowing NaK stream through 120 perforations of 0.3 mm diam each, spaced uniformly in the channel wall in the mixing section which is installed in the upper section of the test channel.

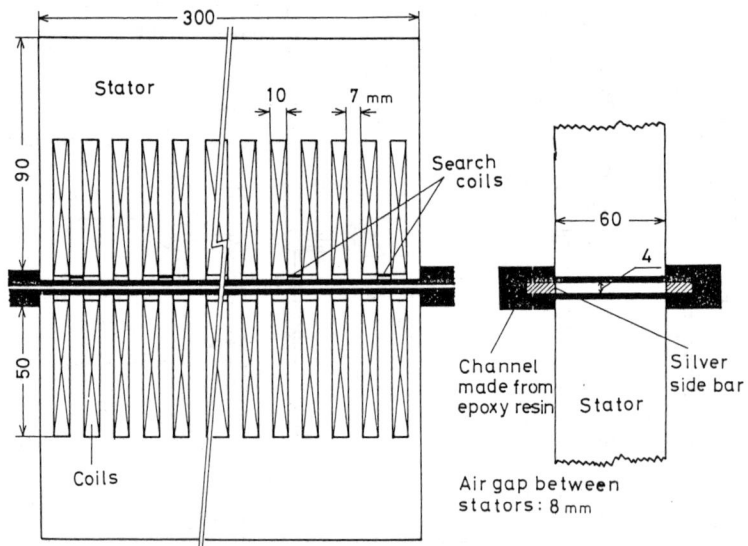

Fig. 18 AC induction generator channel.

Table 4   Instrumentation of blowdown procedure

Test	Instrument
Liquid flow rate	Electromagnetic flow meter
Gas flow rate	Orifice flow meter
Pressure	Semiconductor pressure transducer (at 5 9 points along test channel)
Void fraction	Attenuation technique of x-ray or γ-ray of Am^{241} (at inlet, middle, and outlet of test section)
Electrical output power	Power meter
Magnetic field	Gauss meter and/or six search coils which are corrected previously and set along the direction of flow

The blowdown procedures mentioned above are set into the prescribed sequence and the operation is performed with the remote control system in the control room.

The induction MHD generator is connected to a synchronous generator which is driven by an induction motor through a variable-speed coupling to provide a variable frequency of 40-250 Hz.

Instrumentation. The major experimental parameters to be measured are liquid flow rate, pressure, electrical power, electrical potential (in the case of the dc MHD experiment), and magnetic field; additional parameters in the case of the experiments with two-phase flow are gas flow rate and void fraction. The instrumentation is shown in Table 4.

III. Potassium Boiling under Magnetic Field

In the literature (Fujii-e and Suita 1974), we have proposed a fusion blanket cooling system with boiling two-phase flow of liquid metal as a working fluid. It was theoretically predicted that this cooling method reduced the MHD pressure drop which is considered to be a significant problem of blanket cooling with liquid metal. However, lack of experimental data on the boiling phenomena of alkali metals under the influence of magnetic fields prevents us from presenting a more detailed picture of the present concept of the cooling system. An experiment on liquid-metal boiling under the magnetic field has been performed using stagnant

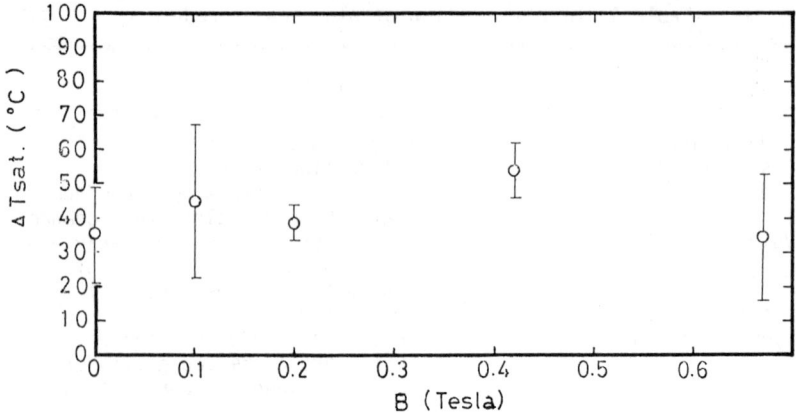

Fig. 19 Experimental plots of IB super heat vs B.

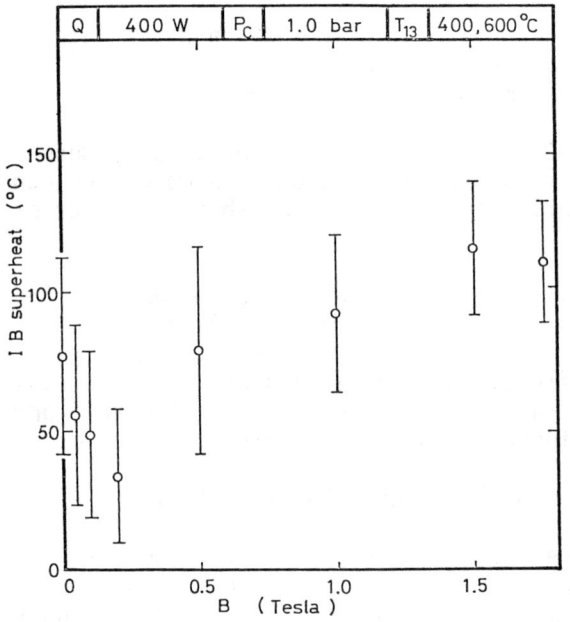

Fig. 20 IB super heat vs B.

potassium contained in a straight channel. The main objective was to assess the applicability of the boiling two-phase flow to the blanket cooling of a fusion reactor, as well as to develop a combined use for an advanced energy conversion system such as LMMHD of alkali metal vapor turbine. The experiments studied the influence of the transverse magnetic field on incipient boiling superheat, the motion of the liquid column above the vapor space, boiling pattern, etc.

Two types of devices were used: a rectangular channel (see Fig. 23) and a circular channel with an immersion-type heater pin (see Fig. 24). The details will be explained below.

Incipient Boiling Superheat

Incipient boiling (IB) superheat of stagnant potassium under the magnetic field was measured. Superheat is defined as the difference between the potassium temperature before the inception of boiling and the saturation temperature. Figure 19 shows the relation between the IB superheat and the strength of magnetic field, which was obtained in the rectangular channel. The plots of IB superheat scatter between 20 and 70°C. As seen from Fig. 19, IB superheat does not seem to be influenced by the magnetic field strength.

The MHD effect on the liquid motion in the nucleation cavity of the wall are considered to be negligible up to 13 T, as theoretically predicted, inferring from the experimental results that IB superheat is independent of the orientation of heating direction to the magnetic field.

The IB superheat under the strong magnetic field and the high heat flux is shown in Fig. 20, which shows the results obtained in the circular channel. This time the IB superheat seems to be dependent on the magnetic field strength. This might be an indirect effect of magnetic field. Accumulation of experimental data is necessary to draw a definite conclusion.

Motion of the Liquid Column above the Vapor Space

The typical signals at the initiation of boiling in the rectangular channel are shown in Fig. 21. As revealed by the signals from the potential taps and flow meter, the single-slug ejection takes place intermittently under the zero magnetic field strength (see Fig. 21a). However, the pattern of the signals at the boiling point under the magnetic field is quite different from that without the magnetic field because of the MHD effects on the motion of the liquid column above the vapor space. Furthermore, cases in which the vapor bubble did not condense perfectly in the conden-

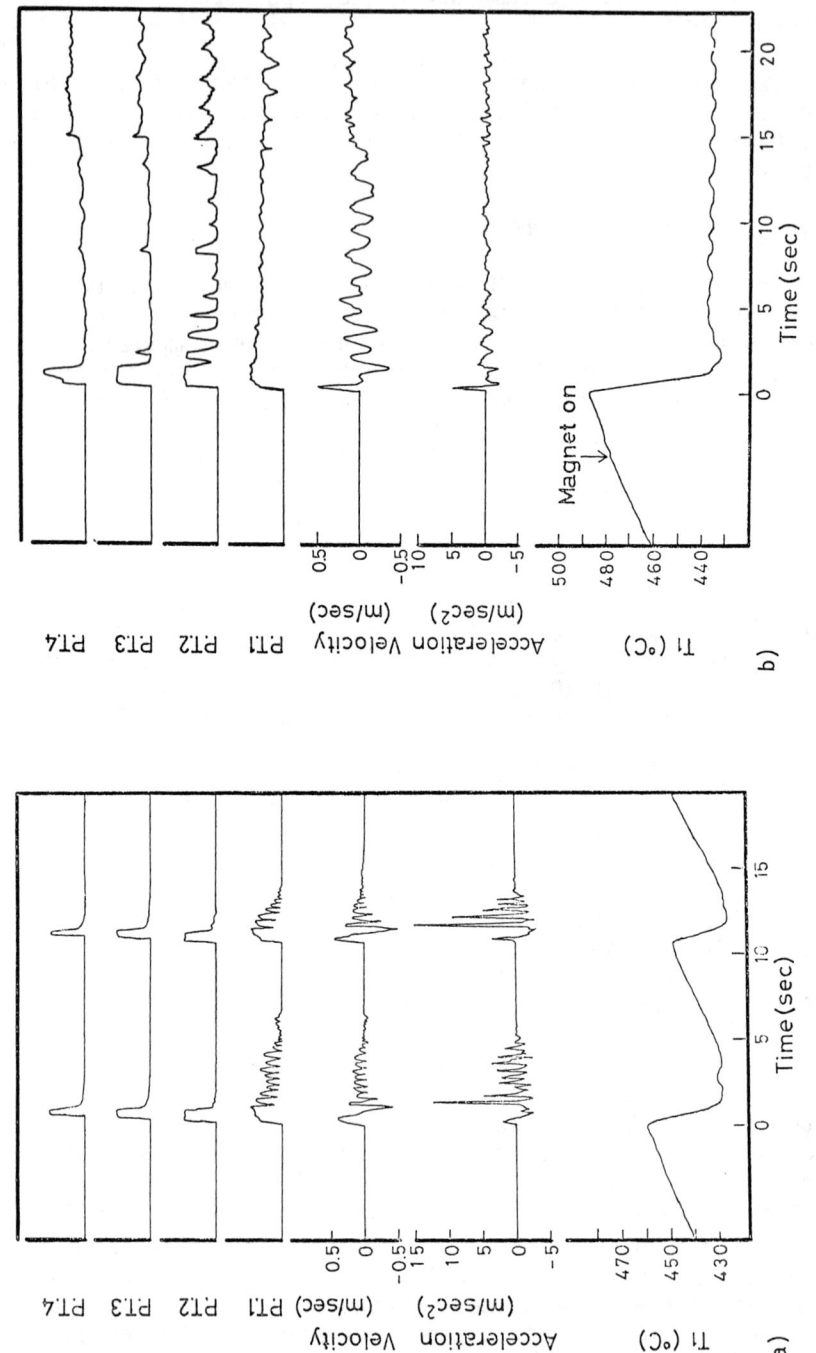

Fig. 21 Transient signals during boiling: a) 0 T, b) B = 0.6 T.

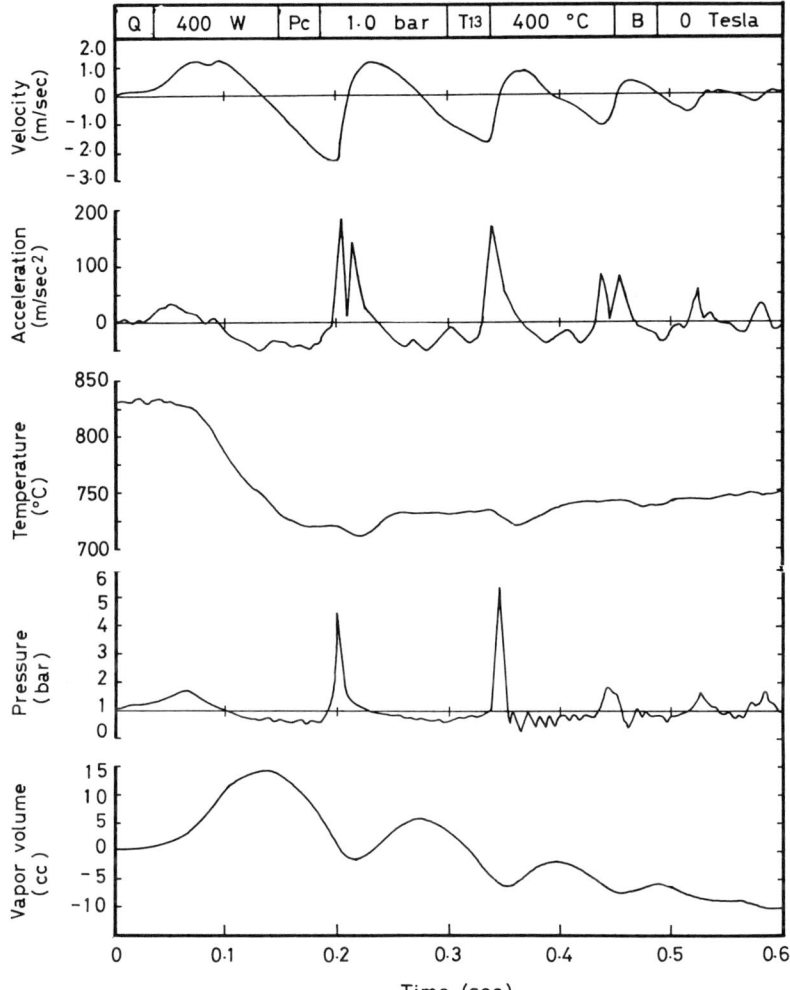

Fig. 22a Potassium boiling transient without magnetic field for circular test section.

sation process were often observed with an increase of the magnetic field. The typical examples are shown in Fig. 21b. In the case of imperfect condensation, the liquid column above the vapor space continued with a periodic motion and the temperature of the liquid below the vapor space changed periodically, corresponding to the motion of the liquid column.

The influence of the transverse magnetic field on the motion of the liquid column was observed in the reduction of

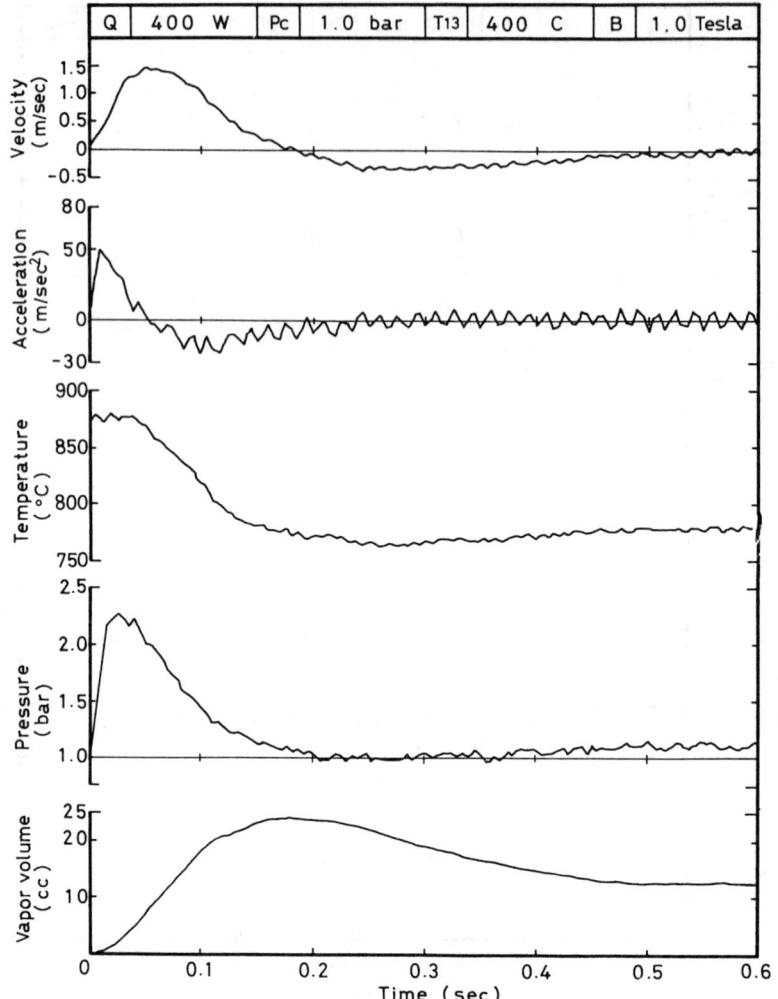

Fig. 22b Potassium boiling transient under magnetic field (B=1.0 T).

its velocity at condensation of vapor and consequently in the magnitude of condensation pressure pulse produced by the hammer action of the liquid column.

The typical signals at the initiation of boiling under the high heat flux (40 W/cm^2) in the circular channel are shown in Fig. 22. The results are similar to those shown in Fig. 21. In the absence of the magnetic field, the sharp condensation pressure pulse was produced, but it disappeared in the presence of the strong magnetic field (B = 1.0 T) and

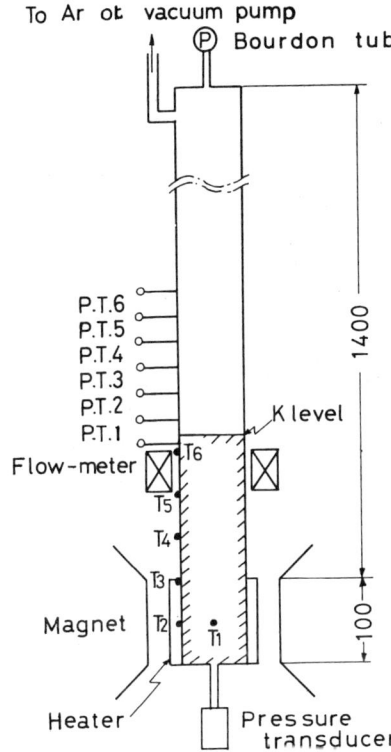

Fig. 23 Potassium boiling channel with a circular cross section.

the imperfect condensation that occurred. In addition, the dry-out was not observed during the imperfect condensation.

Experimental Apparatus for Potassium Boiling Under Magnetic Field

Rectangular Channel with Outside Wall Heater. A schematic diagram of the experimental setup is shown in Fig. 23. The test channel consists of a rectangular pipe made of 304 stainless steel with a 20 x 20 mm inner cross section and filled with potassium to a level of 270 mm.

A dc magnetic field with adjustable strength from 0 to 0.7 T is applied perpendicularly to the test channel to maximize the MHD effect. The pole piece of the magnet is 90 mm in diameter. Potassium is heated by a sheathed microelectric heater coil wound on the outside of the channel 10 mm from the bottom. However, the heat flux is small in comparison with that expected for an actual CTR blanket. The system pressure is controlled by the vacuum pump to a fixed value of about 10^{-3} mm Hg.

For measuring the bulk temperature of potassium, a thermocouple is immersed in the liquid inside the channel, while the others are mounted on the outer wall to measure the axial temperature distribution. For observing the motion of the liquid column and/or vapor bubble, potential taps are arranged on the channel wall at intervals of 1 cm in the axial direction and a constant current of 10 A is supplied to both ends of the boiling test section. Also, an electro-magnetic flowmeter consisting of a permanent magnet with a field strength of 0.1 T is provided to measure the velocity and acceleration of the potassium liquid column. A pressure transducer of the strain gage type is installed at the bottom of the channel for the detection of the condensation phenomenon.

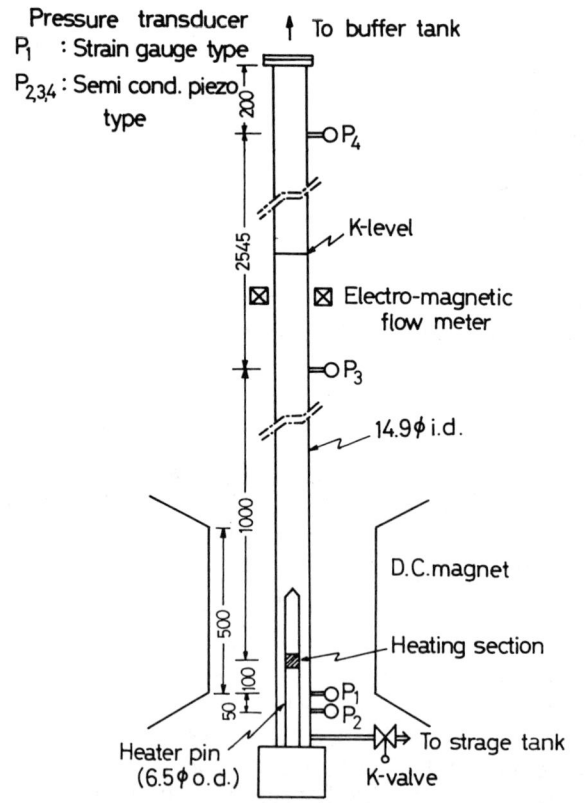

Fig. 24 Potassium boiling with a rectangular cross section.

Circular Channel with Immersion-Type Heater. A schematic diagram of the experimental setup for high heat flux and strong magnetic field is shown in Fig. 24. The test section is made of a 4 m long vertical tube of 316 stainless steel with 14.9 and 17.3 mm inner and outer diameters, respectively, and of a 55 cm long heater pin of 6.5 mm outer diameter inserted from the bottom of the tube and submersed in liquid potassium.

A dc magnetic field with an adjustable strength of 0-1.8 T is applied perpendicularly to the test channel and the pole pieces of the magnet are 15 x 50 cm^2. The measuring method for boiling signals are the same as in Fig. 23.

Signals from the thermocouples are fed to the pen recorders and those from the potential taps and flowmeter to a MTR of 14 channels.

Acknowledgments

Contributors to the work presented in this paer have been Tokuo Suita, Nobuhiro Tanatsugu, Tsutomu Matsuda, Tamotsu Sano, Yoshiharu Fukuzawa, Masaki Saito, Hiroshi Horiike, Hideyuki Takatsu, Takaki Emori, Hiroshi Nagae, Nobuo Yamanari, and Shoji Kotake.

References

Fujii-e Y. and Suita T. (1971) Proc. 5th Int. Conf. on MHD Electr. Power Generation, Vol. 3, pp. 41-48.

Fujii-e Y., Saito M., Inoue S., and Suita T. (1975a) J. Nucl. Sci. Technol., Vol. 12, No. 5.

Fujii-e Y., Saito M., Inoue S., and Suita T. (1975b) J. Nucl. Sci. Technol., Vol. 12, No. 5.

Fujii-e Y., Miyazaki M., Inoue S., Saito M., and Suita T. (1975c) Proc. 6th Int. Conf. on MHD Electr. Power Generation.

Fujii-e Y. and Suita T. (1974) Nucl. Fusion, Sp. Supp.

Fujii-e Y., Saito M., Nagae H., and Inoue S. (1979) Proc. 18th Int Conf. on MHD Electr. Power Generation.

Fukuzawa Y. and Fujii-e Y. (1978) J. Nucl. Sci. Technol., Vol. 15, No. 10.

Saito M., Inoue S., and Fujii-e Y. (1978a) J. Nucl. Sci. Technol., Vol. 15, No. 7.

Saito M., Nagae H., Inoue S., and Fujii-e Y. (1978b) J. Nucl. Sci. Technol., Vol. 15, No. 10.

Saito M., Inoue S., and Fujii-e Y. (1979) J. Nucl. Sci. Technol., Vol. 16, No. 3.

Tanatsugu N., Fujii-e Y., and Suita T. (1971a) J. Nucl. Sci. Technol., Vol. 8, No. 10.

Tanatsugu N., Fujii-e Y., and Suita T. (1971b) J. Nucl. Sci. Technol., Vol. 9, No. 4.

Tanatsugu N., Fujii-e Y., and Suita T. (1972) J. Nucl. Sci. Technol., Vol. 9, No. 12.

Tanatsugu N., Fujii-e Y., and Suita T. (1973) J. Nucl. Sci. Technol., Vol. 10, No. 4.

Tanatsugu N., Inoue S., Fujii-e Y., and Suita T. (1974) J. Nucl. Sci. Technol., Vol. 10, No. 4.

Some Peculiarities of Turbulence Suppression by Magnetic Fields and Their Use for Magnetic Flow Control in Liquid Metals

H. Branover*
Ben-Gurion University of the Negev, Beer-Sheva, Israel
and
S. Claesson†
University of Uppsala, Uppsala, Sweden

Abstract

A series of duct flow experiments have been performed with channels of the Hartmann type and channels with an azimuthal field. The unique electromagnet of the University of Uppsala was used. This magnet was built about 40 years ago, and its poles are made from an alloy containing an extremely high percentage of cobalt. In a gap of 7 cm between circular 12 cm poles, the field is about 2.7 T, while the non-homogeneity is much less than 1%. The first stage of the experiments was performed with mercury and Hartmann numbers up to 600 (based on hydraulic diameter) were reached. The results demonstrated a manifold difference in longitudinal pressure drop values between the Hartmann and the azimuthal cases. This offers a convenient possibility for noncontact flow rate control.

I. Introduction

Regarding the influence of a transverse magnetic field on initially turbulent flow of an electroconductive fluid in

---
Paper presented at Third Beer-Sheva International Seminar on Magnetohydrodynamic Flows and Turbulence, Ben-Gurion University of the Negev, Beer-Sheva, Israel, March 23-27, 1981. Copyright © American Institute of Aeronautics and Astronautics, Inc., 1982. All rights reserved.
*Professor, Department of Mechanical Engineering.
†Professor, Institute of Physical Chemistry.

a channel with a high-aspect-ratio rectangular cross section, two extreme cases have been pointed out (Branover 1978). When the magnetic field is perpendicular to the longer side of the cross section it corresponds to Hartmann flow, while when the magnetic field is perpendicular to the shorter side of the cross section, this is the flow in the quasiazimuthal field. In initially turbulent Hartmann flows, the increase of the intensity of the magnetic field results in an increase of the longitudinal pressure drop (excluding the case of very low Reynolds numbers, $Re < 5000$, where $Re = Vd/\nu$. V is the mean velocity of flow, D the hydraulic diameter, and ν the kinematic viscosity). It means that the Hartmann effect, leading to a more flat velocity profile, is stronger than the effect of direct suppression of turbulent flow disturbances. In the case of a flow in a quasiazimuthal field, there is no Hartmann effect and the pressure drop always monotonically decreases with the increase of magnetic field until the flow becomes completely laminarized.

II. Predictions

The critical Reynolds to Hartmann number ratio, corresponding to the completion of laminarization under influence of magnetic field, can be calculated from the empirical expression

$$(Re/Ha)_{cr} = [215 - 85 \exp(-0.35 \beta)] \qquad (1)$$

Here $Ha = BD\sqrt{\sigma/\rho\nu}$ is the Hartmann number, where B is magnetic flux density, and $\sigma\rho$ the electrical conductivity and density of the fluid. $\beta = b/a$ is the aspect ratio of the cross section of the Channel C and a the half-widths of the walls perpendicular and parallel to the magnetic field correspondingly. Obviously, for the extreme cases $\beta = \infty$ and $\beta = 0$ (Hartmann flow and flow in the azimuthal field), the critical conditions are

$$(Re/Ha)_{cr} = 215 \qquad \beta = \infty \qquad (2)$$

$$(Re/Ha)_{cr} = 130 \qquad \beta = 0 \qquad (3)$$

Beginning from the critical point and for all higher values of Ha at any fixed Re value, the longitudinal pressure drop dp/dx [or friction factor $\lambda = (dp/dx)/(\rho V^2/2D)$] corresponds to laminar Hartmann flow for $\beta = \infty$ and to Poiselle flow for

$\beta = 0$. As is well known, for the first case

$$\lambda_H = 8Ha/Re \qquad (4)$$

and for the second

$$\lambda_p = 96/Re \qquad (5)$$

All the above-mentioned lead to a conclusion that by applying a strong enough magnetic field to a flow of a liquid metal in a channel with high-aspect-ratio cross section perpendicular either to the wide wall or to the narrow wall it is possible to change the friction manifold.

Indeed, for a Hartmann number value corresponding to completion of laminarization in an azimuthal field according to Eq. (3), $Ha = Re/130$ and from Eq. (5) $\lambda_p = 96/Re$. Then also using Eq. (4), it is possible to find

$$\lambda_H/\lambda_p = Re/1560$$

This value is large even at moderate Reynolds numbers. For higher Ha value, the value of λ_H/λ_p still increases.

For a real flow, when instead of $\beta = \infty$ and $\beta = 0$ the real values $\beta \gg 1$ and $\beta \ll 1$ are considered, the λ_H/λ_p is smaller, since in the case $\beta \ll 1$ there are still some relics of the Hartmann effect. On the other hand, when the roughness of the wall is taken into consideration, the λ_H/λ_p ratio can be even higher (Branover 1978).

Thus the simple turning of a magnet around the axis of a channel can lead to an essential change of longitudinal pressure drop. This leads to the idea of a very simple magnetic valve for flow rate control in liquid-metal flows (Branover 1979).

For verification of the above predictions and to establish the practicality of the described magnetic valve, the present experimental work was undertaken.

III. Experimental Verification

A 0.4×4.0 cm^2 cross-sectional channel made of Perspex was used (Fig. 1). To make the study closer to practical cases, no stabilization length was provided (neither hydrodynamic stabilization nor magnetohydrodynamic stabilization). The total length of the prismatic part of the channel was 11.0 cm. The middle 5.0 cm of this length were used for

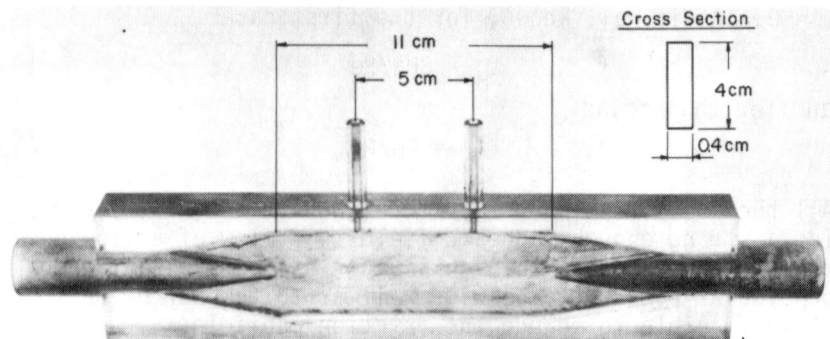

Fig. 1 Experimental channel.

Fig. 2 Magnet of Physical Chemical Institute, Uppsala. Pump and parts of mercury circulation loop in front of magnet.

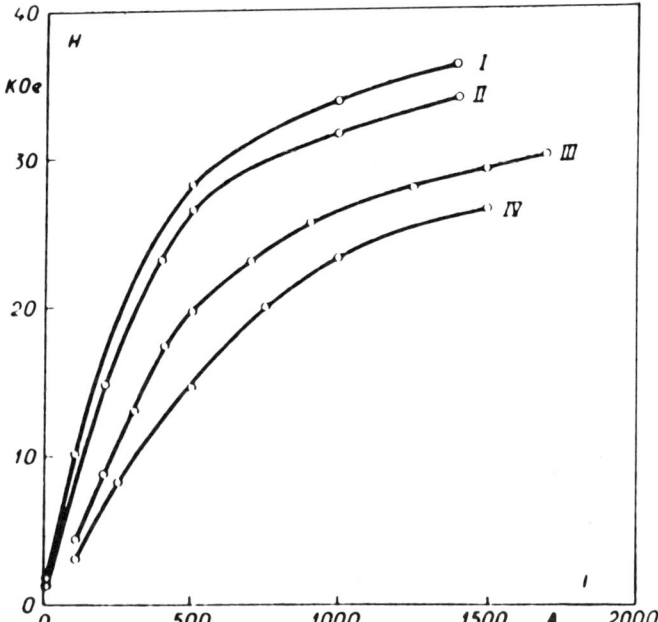

Fig. 3 Calibration curves of magnetic field (variation of field intensity Q_e vs electrical current A). (Pole shoe diameter 12 cm, air gap: I 3.8 cm, II 4.0 cm, III 8.8 cm, IV 12.1 cm.)

pressure drop measurements by means of two pressure taps. The pressure difference was measured by simple liquid manometers.

The channel could be placed in two positions ($\beta \gg 1$ and $\beta \ll 1$) in the gap of the electromagnet of the Physical Chemistry Institute of Uppsala University. This magnet was built more than 40 years ago for special experiments with magnetic double refraction (Snellman 1944) and still remains quite unique. In our experiments, conical pole shoes with a pole face diameter of 12 cm were used. Figure 2 shows the magnet prepared for the installation of the channel and Fig. 3 presents calibration curves of the magnetic field.

Results of two series of measurements are plotted in Figs. 4 and 5, for flows with Reynolds numbers of 25,000 and 105,000. It is easy to see that for both experimental Reynolds number values the λ_H/λ_p ratio at the laminarization point for the azimuthal field case is close to 10. It should be noted that in these experiments, the value of β was 10 and 10^{-1} for the Hartmann case and azimuthal field

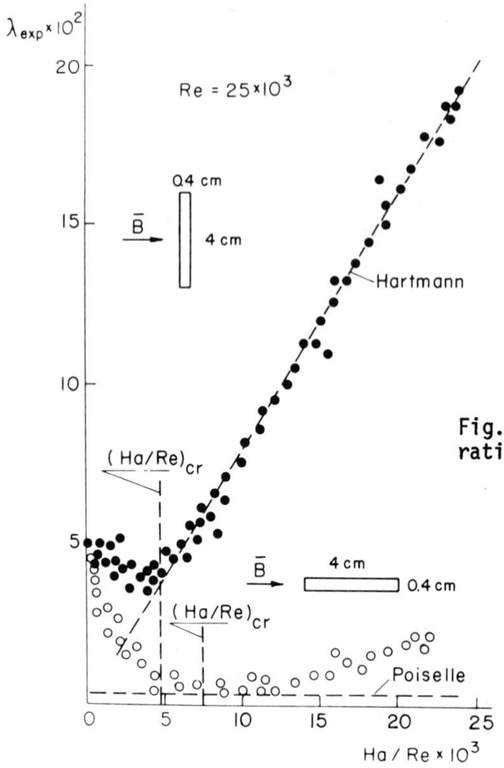

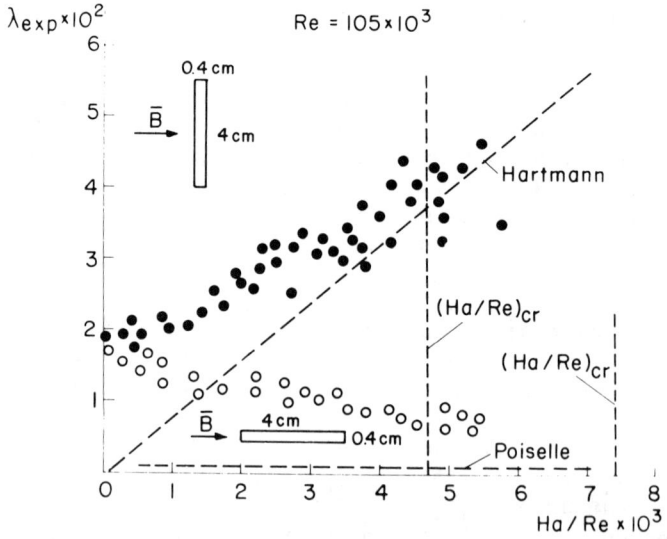

Fig. 4 Friction factor vs Ha/Re ratio for Re = $25 \cdot 10^3$.

Fig. 5 Friction factor vs Ha/Re ratio for Re = $105 \cdot 10^3$.

case, respectively. For Re = 25,000 the ratio of λ_H/λ_P remained approximately equal to 10 up to Ha/Re = w.5.10^{-2}.

The following note should also be made. The measured values of the friction factor without magnetic field λ_0 are essentially higher than according to the accepted empirical formulas. For Re = 25,000 λ_0 = 5·10^{-2}, while according to Blasius it should be 2.5·10^{-2}, and for Re = 105,000 λ_0 = 1.9·10^{-2}, while according to Blasius it should be 1.75·10^{-2}. This discrepancy can be easily explained by the fact that the experimental channel did not have a sufficient stabilization length. As expected, the discrepancy is higher at lower Reynolds numbers when the turbulence is less developed. Despite all this, in strong magnetic fields λ values follow theory exceptionally well. This is one more manifestation of the fact that in the transverse magnetic field stabilization lengths are extremely short. The practical conclusion from the obtained data is that a magnetic valve of the type considered is definitely feasible.

References

Branover H. (1978) Magnetohydrodynamic Flow in Ducts, John Wiley & Sons, New York, and Israel Universities Press, Jerusalem.

Branover H. (1979) "Methods and apparatus for controlling the flow of liquid metal," U.S. Patent 4,171,707.

Snellman O. (1944) "Some measurements of magnetic double refraction," Almquist and Wiksells Boktryceri, Uppsala, Sweden.

Geostrophic Turbulence, Stable Baroclinic Eddies, and Self-Exciting Dynamos in Rotating Fluid Systems: A Summary

R. Hide*
Geophysical Fluid Dynamics Laboratory, Bracknell, England

The motion of a fluid of low viscosity that departs but little from solid body rotation cannot in general be symmetric about the rotation axis, even when the boundary conditions are axisymmetric (theorem A). According to laboratory experiments there are two nonaxisymmetric regimes of thermal convection in a rotating fluid annulus subject to differential heating in the horizontal. One is highly regular (i.e., spatially and temporally periodic), while the other, which is reminiscent of large-scale flow in the Earth's atmosphere, is irregular ("geostrophic turbulence"). The flowfields are characterized by the presence of upper level jet streams, where intense concentrations of vorticity and strong concomitant horizontal temperature gradients are found.

The main features of the flow pattern can be interpreted by straightforward arguments based on general thermodynamic considerations and the requirement that the flow should be quasigeostrophic nearly everywhere, with the horizontal pressure gradient nearly in balance with Coriolis acceleration. Thus, when the distribution of applied heating and cooling is such that the corresponding gradient of the impressed radial temperature field has the same sign at all radii, the most conspicious feature of the upper level flow pattern in the regular nonaxisymmetric regime is a continu-

---

Paper presented at Third Beer-Sheva International Seminar on Magnetohydrodynamic Flows and Turbulence, Ben-Gurion University of the Negev, Beer-Sheva, Israel, March 23-27, 1981. Copyright © American Institute of Aeronautics and Astronautics, Inc., 1982. All rights reserved.
*Professor.

ous jet stream meandering in a wavy pattern between the bounding cylinders.

When, however, the impressed radial temperature gradient changes sign near midradius (as in the case when heat is introduced throughout the body of the fluid and is withdrawn at both side walls) the corresponding upper level flow consists of several closed eddies, each circulating "anticyclonically," with the horizontal motion largely confined to a narrow jet stream at the periphery of each eddy. In some respects these stable closed eddies are dynamically similar to the long-lived anticyclonic eddies, including the Great Red Spot seen in Jupiter's atmosphere in the southern hemisphere. Previous work on stable baroclinic eddies is now being extended in various directions and supporting numerical work is also being carried out. These findings have important implications for theories of atmospheric predictability (see Hide 1981a).

It is generally accepted that the magnetic fields of the Earth and sun, and of other magnetic planets (e.g., Jupiter and Saturn) and stars, are due to electric currents flowing within their interiors. It is also accepted that these currents are maintained against the effects of ohmic dissipation largely by electromotive forces due to motional induction, as Larmor pointed out. The fluid motions involved are produced in most (if not all) cases by the action of gravity on density variations. There have been other general proposals as to the nature of these emf's but, unlike Larmor's "self-exciting homogeneous dynamo" proposal, they invariably fail, usually by several powers of 10, to give the right magnitude. Thus, the dynamo mechanism is considered to hold out the greatest promise of final success in the explanation of cosmical magnetic fields, and it is for this reason that over the past two or three decades a great deal of mathematical work has been done on the difficult problem of investigating how such dynamos might work in detail.

The mathematical analysis of theoretical dynamo models is greatly complicated by the finding that suitable departures from axial symmetry seem to be required for dynamo action to occur. Indeed, Cowling was able to prove that no steady magnetic field with an axis of symmetry can be maintained by fluid motions, and it was later shown by Backus and Braginsky that no nonsteady axisymmetric magnetic field can be so maintained when the electrically conducting fluid is incompressible. But until recently the behavior of axisymmetric magnetic fields had not been established in the most general case, when the fluid can be compressible, the field nonsteady, and the scalar coefficients of electrical conductivity and magnetic permeability dependent of position and time. This led Todoeschuck and Rochester to speculate that it might be possible for a nonsteady axisymmetric magnetic field to be maintained by dynamo action provided that the fluid was sufficiently compressible. But it can be shown that under the most general conditions fluid motions cannot prevent the ohmic decay of an axisymmetric field (theorem B). Any search for axisymmetric dynamos is bound to be an unprofitable exercise. (For references see Moffatt 1978, Parker 1979, Hide 1981b.)

Another speculation which has motivated recent efforts to generalize Cowling's original "antidynamo" theorem is one by Hibberd. He proposed that nonsteady axisymmetric magnetic fields can be maintained by thermoelectric effects of the Nernst-Ettinghausen type, where the conductive flow of heat down a temperature gradient ∇T in the presence of a magnetic field $\vec{B}$ generates an emf proportional to $\nabla T \times \vec{B}$. Such effects can be included in the analysis of the behavior of axisymmetric fields and shown to have no significant influence. Thus, notwithstanding published speculations to the contrary, neither compressibility nor Nernst-Ettinghausen effects can prevent the ohmic decay of an axisymmetric field.

Theorems A and B indicate that rotation can promote the generation of magnetic fields by insuring that suitable departures from axial symmetry occur. It has been argued that the magnetic field might grow in strength until the flow becomes "magnetostrophic," with Coriolis and Lorentz forces comparable in magnitude. The study of magnetostrophic flows, particularly certain classes of wave motions, throws considerable light on the magnetohydrodynamics of the interiors of planets and stars, including pulsars, and the origin of planetary and stellar magnetism.

Details of this work and references to earlier studies can be found in the papers listed below.

References

Hide R. (1981a) "High vorticity regions in rotating thermally-driven flows," Meteorol. Mag., Vol. 110, pp. 335-344.

Hide R. (1981b) "The magnetic flux linkage of a moving medium: A theorem and geophysical applications," J. Geophys. Res., Vol. 86, pp. 11,681-11,687.

Moffatt H.K. (1978) Magnetic Field Generation by Fluid Motion, Cambridge University Press, Cambridge, England.

Parker E.N. (1979) Cosmical Magnetic Fields, Clarendon Press, Oxford, England.

Velocity and Particle Size Measurements by a Laser Technique

Y.M. Timnat*
Technion—Israel Institute of Technology, Haifa, Israel

In order to predict the performance of a system, for instance, an MHD power generator, a basic understanding of the flow phenomena and the processes taking place in it is required. This can be obtained either by use of experimental techniques or by computer modeling or by a combination of both. The last approach has proved often the most efficient and has been adopted in this work, which deals with a laser technique for measuring instantaneous velocity and particle size in multiphase flows, both reactive and nonreactive.[†] This nonintrusive approach may be utilized in flow studies in an MHD channel or in the combustion chambers of coal-fired furnaces and boilers, which are part of an MHD power-generating plant.

As a first step in the development of the method a forward scattering laser Doppler anemometer (LDA), using a He-Ne 15 mW source and standard DISA LDA components, was set up and used in an experimental facility, in which the influence of changes in inlet geometry on the performance of a dump combustor was studied (Greenberg and Timnat 1981, Shahaf et al. 1980). The experimental facility is shown schematically in Fig. 1. It permits investigation of a wide range of geometrical configurations, in this particular case quasi-two-dimensional channels using conventional and LDA measurements. The technique is adaptable for use with

---

Summary of paper presented at Third Beer-Sheva International Seminar on Magnetohydrodynamic Flows and Turbulence, Ben-Gurion University of the Negev, Beer-Sheva, Israel, March 23-27, 1981. Copyright © 1982 by Y.M. Timnat. Published by the American Institute of Aeronautics and Astronautics with permission.
*Professor, Department of Aeronautical Engineering.
†The main results reported here were published in Greenberg J.B. and Timnat Y.M. (1981).

solid, liquid, or gaseous fuel and can be operated over a wide range of mass flows. The facility is built around a channel with a square cross section (100 x 100 mm) attached to a metal table. Different configurations can be attached onto or inserted into the basic channel. An LDA system is also connected to the table on a frame that permits movement in three dimensions relative to the channel, with optical access to the test section under the channel being granted by two moveable windows on its sides. Relative motion between the optical system and the channel is made possible by an arrangement of linear bearings and hydraulic jacks operated by gears and power screws.

The air supply to the facility is provided by a high-pressure line (150 atm) which connects to the channel via a stagnation tank and is controlled by a suitable high-pressure control valve. For hot-flow experiments natural gas from a high-pressure bottle is injected into the airflow between the stagnation tank and the channel, which is operated as a combustion chamber. To prevent overheating of the structural elements of the latter during reacting flow experiments, it is equipped with double ceilings and floors to permit the injection of cooling air provided by a low-pressure line (15 atm).

A special requirement of the LDA system is seeding, which is provided here by a tank filled with small amounts of MgO powder excited by a tangential flow of secondary air. The air-suspended powder is sucked into the combustion chamber by the main flow via a plastic tube. Suction was employed in order to assure proportional seeding at higher flow rates. The results are recorded on an X-Y plotter after the signal processor was locked onto the signal, with the aid of a memory scope, on which the Doppler signal was displayed.

Experiments were performed with three different models, two with air injection at 45 deg at different heights above the centerline and the third at 90 deg with reacting and nonreacting flows. The results demonstrated the sensitivity of the Doppler velocity measurement, which allowed the detection of a small acceleration due to combustion for the 45 deg injection. Another feature revealed by the measurements was a 10 cm long recirculation zone, situated close to the upper wall, immediately downstream of the gas entrance. The experimental results were compared with calculations performed using a code developed by Greenberg and Presser (1981) and exhibited qualitative agreement.

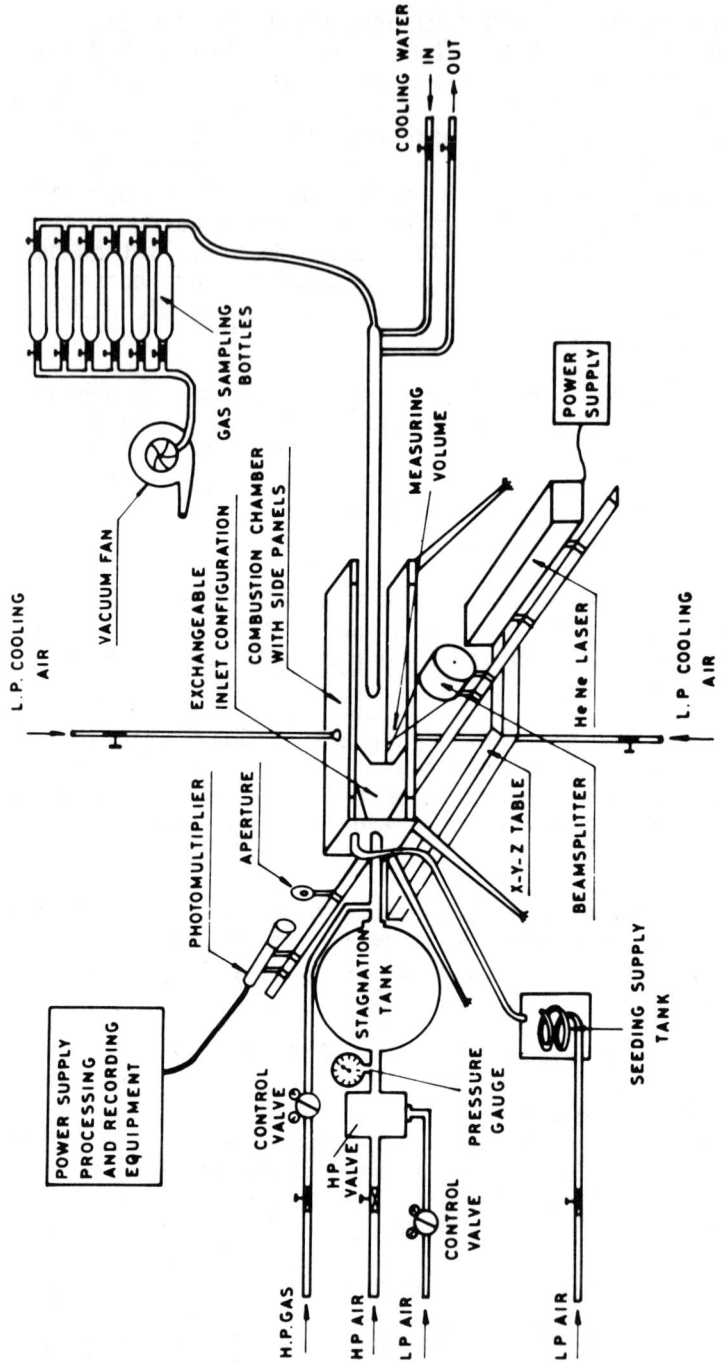

Fig. 1 Experimental facility.

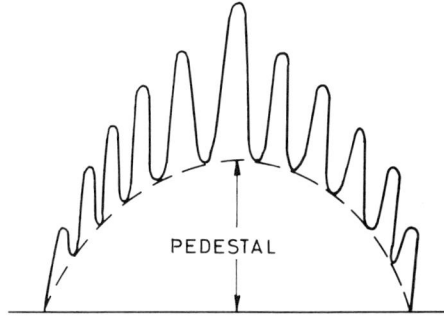

Fig. 2 Schematic of Doppler signal with pedestal.

A new data reduction system, connected directly to a PDP 11/34 minicomputer was installed, thus enhancing the capability of the system. A method for measuring the size of individual particles and to obtain their distribution is being developed. As a first target, kerosene droplets in the range between a few and a hundred microns diameter will be investigated. They will be injected by the system used before and it is expected to obtain a signal of the type shown in Fig. 2 (Levy and Lockwood 1980, Wang 1979). The Doppler frequency is again a measure of the flow velocity, while the height of the pedestal is proportional to the size of the individual droplets taking part in the interaction. It is, of course, necessary to calibrate this height. This will be done by employing an in situ technique of droplet size determination developed by Goldman et al. (1978).

References

Goldman Y., Lipschitz A., and Manheimer-Timnat Y. (1978) "Droplet vaporization in a confined burning jet," Paper A7, 6th Int. Conf. on Heat Transfer, Toronto, Canada.

Greenberg J.B. and Presser C. (1981) "A fully automatic method for predicting internal reacting flows," J. Comput. Phys., Vol. 40, pp. 361-375.

Greenberg J.B. and Timnat Y.M. (1981) "Sudden expansion injection for Ram-Rockets," Paper 29a, Proc. 5th Int. Symp. on Air-Breathing Engines, Bangalore, India.

Levi Y. and Lockwood F.C. (1980) "Two phase flow measurements in the free-board of a fluidized bed using L.D.A.," Isr. J. Technol., Vol. 18, pp. 146-151.

Shahaf M., Goldman Y., and Greenberg J.B. (1980) "An investigation of impinging jets in fuels with sudden expansion," Isr. J. Technol., Vol. 18, pp. 57-64.

Wang J.C.E. (1979) "Optical particulate size measurements using a small angle near forward scattering technique," Sandia Laboratories Rept. SAND 79-8246.

Interactions of Turbulent Eddies with a Free Surface

D. Naot*
Technion—Israel Institute of Technology, Haifa, Israel
and
W. Rodi†
University of Karlsruhe, Karlsruhe, Federal Republic of Germany

Abstract

The interactions between turbulent eddies and a free surface in turbulent open-channel flow are discussed. As the turbulent eddies are reflected by the water surface, processes similar to those which characterize wall proximity interactions are also induced near a free surface. The two main effects of a free surface on the turbulence are the reduction of the length scale of turbulence and the redistribution of fluctuating velocities. Models for simulating these effects are described, with an emphasis on the redistribution processes. Algebraic relations for the individual stresses are introduced which allow description of the main features of open-channel flow.

I. Introduction

There is clear evidence that the presence of a strong density interface (i.e., a water surface) strongly influences the turbulence structure beneath it. Recent experiments in open channels (Ueda et al. 1977 and Nezu 1977) have shown that the eddy viscosity and diffusivity are reduced consider-

---

Paper presented at Third Beer-Sheva International Seminar on Magnetohydrodynamic Flows and Turbulence, Ben-Gurion University of the Negev, Beer-Sheva, Israel, March 23-27, 1981. Copyright © American Institute of Aeronautics and Astronautics, Inc., 1982. All rights reserved.
*Senior Lecturer, Department of Nuclear Engineering.
†Professor, Department of Civil Engineering.

ably when the free surface is approached, yielding the well-known parabolic distribution with depth. This is in contrast to the distribution in closed channels, where at the corresponding plane of zero shear stress (the symmetry plane), the eddy viscosity has values which are only about 20% below the maximum value. Experiments have also shown that the turbulent kinetic energy k is lower at a free surface than at a symmetry plane in equivalent channel flows (Nakagawa et al. 1975) and that the macroscale of the turbulent eddies decreases considerably due to the presence of a free surface (Raichlen 1967 and McQuivey and Richardson 1969). According to the high Reynolds number formula for the dissipation

$$\varepsilon \propto k^{3/2}/\ell \qquad (1)$$

relating the dissipation rate ε to the turbulent kinetic energy k and to the dissipation length ℓ which is proportional to the macroscale, the dissipation rate ε increases toward a free surface. Since from dimensional analysis the eddy viscosity is proportional to $\sqrt{k}$ and ℓ, it is mainly the decrease in ℓ (equivalent to the increase in the dissipation rate) that is responsible for the reduction of the eddy viscosity near the free surface.

The measurements of Nakagawa et al. (1975) indicate further that the presence of a free surface has a significant influence on the distribution of fluctuating energy among the various components, namely that the vertical fluctuations are damped by the free surface, while the horizontal ones are enhanced. This mechanism is of great significance in relatively narrow open channels where a depression of the maximum streamwise velocity below the free surface has been observed. This depression is caused by transport of low-momentum fluid from the banks toward the channel center by secondary motions. These motions are now known to be caused by differences in the various fluctuating energy components (or turbulent normal stresses). Hence, the influence of a free surface on the difference between these components is the major cause for the observed depression of the maximum velocity below the surface.

There are therefore two effects of a free surface on the turbulence structure. First, the length scale of the turbulence is reduced by the free surface due to geometrical restrictions and consequently the dissipation rate is increased. Second, the free surface causes a redistribution of the velocity fluctuations, damping the vertical fluctuations and (owing to continuity) enhancing the horizontal ones. Both of these effects need to be accounted for in a model realistically simulating free-surface flows. The development of such a model for three-dimensional open-channel flow, where both free-surface effects are very important, is described in Naot and Rodi (1982). As the presence of a solid wall also reduces the length scale (to zero at the wall) and damps the fluctuations normal to the wall, a modeling approach similar to that used for simulating wall proximity effects was adopted. It should be mentioned here that the similarity is particularly close to shear-free turbulent near-wall layers as studied experimentally by Thomas and Hancock (1977). However, there is one basic difference in the conditions at a solid wall, namely that finite motions (primarily horizontal ones) are present at a free surface and that these have a finite length scale, whereas the length scale goes to zero at a solid wall. Naot and Rodi adopted the simple length scale proposal suggested by Hossain and Rodi (1980) who assumed that the length scale is proportional to the distance from some virtual origin above the surface

$$\ell = y + 0.07H \quad \text{or} \quad \ell_o = 0.07H \qquad (2)$$

where H is the channel depth. This empirical relation seems to work quite well for channel flow and will not be elaborated further in this paper. Rather, the remainder of this paper concentrates on a detailed discussion of the modeling of the other free-surface effect, namely the redistribution of the velocity fluctuations.

II. Free-Surface-Induced Velocity Redistribution

Turbulent eddies approaching a solid wall induce a kind of (fluctuating) stagnation pressure which slows down the motion toward the wall and, due to continuity, causes the fluid to move parallel to the wall. In a similar fashion, turbulent eddies approaching a free surface cause some upward bulging of the surface which in turn generates an excess hydrostatic pressure counteracting the vertical motion and thus reflecting the eddies. Again, due to continuity, the fluid approaching the surface has to move sideways parallel to the surface. This redistribution process by an interaction of fluctuating pressure and fluctuating velocities is described by the pressure-strain correlation appearing as source/sink term in the equations governing the individual fluctuating components and the cross correlations between them,

$$\pi_{ij} = \frac{1}{\rho}\overline{p\left(\frac{\partial u_i}{\partial x_j} + \frac{\partial u_j}{\partial x_i}\right)} = \frac{1}{4\pi}\iiint_V \overline{\left[\left(\frac{\partial u_i}{\partial x_j} + \frac{\partial u_j}{\partial x_i}\right)\left(\frac{\partial v_\ell}{\partial r_k}\frac{\partial v_k}{\partial r_\ell}\right)\right.}$$

$$\left.+ 2\overline{\left(\frac{\partial u_i}{\partial x_j} + \frac{\partial u_j}{\partial x_i}\right)\frac{\partial v_\ell}{\partial r_k}}\frac{\partial V_k}{\partial r_\ell}\right]G(\vec{r})dV + S_{ij} \qquad (3)$$

where u_i is the velocity fluctuation at $\vec{x}$, and v_i is the velocity fluctuation at $\vec{x} + \vec{r}$, and the integration is carried over $\vec{r}$ space. The surface integral S_{ij} is

$$S_{ij} = \frac{1}{4\pi}\iint_S \left\{\frac{\partial}{\partial r_n}\overline{\left(\frac{\partial u_i}{\partial x_j} + \frac{\partial u_j}{\partial x_i}\right)p(\vec{x}+\vec{r})}G(\vec{r})\right.$$

$$\left. - \overline{\left(\frac{\partial u_i}{\partial x_j} + \frac{\partial u_j}{\partial x_i}\right)p(\vec{x}+\vec{r})}\frac{\partial}{\partial r_n}G(\vec{r})\right\}dS \qquad (4)$$

Here G is the solution of the Laplace equation obeying $G \sim 1/r$ for vanishing radius and $G \to 0$ for large radii. Usually away from the surfaces the S_{ij} term can be omitted from Eq. (3) as the two-point correlations in Eq. (4) between the fluctuating strain rate and the fluctuating pressure vanish. Noting that the first volume integral in Eq. (3) involves only fluctuating quantities while the second one involves mean velocity gradients, it is clear that two distinct mechanisms contribute to the pressure-strain correlation. Launder et al. (1975) adopted Rotta's (1951) model for the first contribution

$$\pi_{ij,1} = -c_1 \frac{\varepsilon}{k}\left(\overline{u_i u_j} - \frac{2}{3}\delta_{ij}k\right) \qquad (5)$$

and proposed the following model for the second, mean strain contribution

$$\pi_{ij,2} = -\alpha\left(P_{ij} - \frac{2}{3}\delta_{ij}P\right) - \beta\left(D_{ij} - \frac{2}{3}\delta_{ij}P\right) - \gamma\left(\frac{\partial U_i}{\partial x_j} + \frac{\partial U_j}{\partial y_i}\right)k \qquad (6)$$

where

$$P_{ij} = -\left(\overline{u_i u_\ell}\frac{\partial U_j}{\partial x_\ell} + \overline{u_j u_\ell}\frac{\partial U_i}{\partial x_\ell}\right) \qquad (7)$$

$$D_{ij} = - (\overline{u_i u_\ell} \frac{\partial U_\ell}{\partial x_j} + \overline{u_j u_\ell} \frac{\partial U_\ell}{\partial x_i}) \tag{8}$$

and P is the production of turbulent kinetic energy k and hence the trace of P_{ij} and c_1, α, β, and γ are empirical constants in the pressure-strain model. This model is adopted in the present work.

Near surfaces the integral S_{ij} cannot be neglected, and it is precisely this term that describes the redistribution effects of surfaces. In models for the pressure-strain correlation this term is usually accounted for by an additive wall-proximity correction to the model for the pressure-strain correlation remote from surfaces. As Launder et al. (1975) have shown, such a correction should in general also have two contributions, one accounting for the interaction of turbulence quantities and the other for the presence of a mean rate of strain. These authors have proposed such a correction, which can be affected by making the empirical constants in the model Eqs. (5) and (6) functions of the dimensionless distance from the surface. This correction is not very suitable near a free surface where the mean velocity gradients are unimportant so that only the first contribution $\pi_{ij,1}$ is left. Simply changing the value of the coefficient c_1 as the surface is approached does not lead to an increase of the fluctuations parallel to the surface and to a damping of the fluctuations normal to it. Shir (1973) suggested the following additive surface-proximity correction

$$\pi_{ij,s} = c_3 \frac{\varepsilon}{k} (\overline{u_k u_\ell} n_k n_\ell \delta_{ij} - \frac{3}{2} \overline{u_k u_i} n_k n_j - \frac{3}{2} \overline{u_k u_j} n_k n_i) \tag{9}$$

which produces the effect just discussed. In this expression n_i is a unit vector normal to the surface and c_3 is an empirical coefficient that depends on the distance from the surface. Equation (9) accounts only for the turbulence-interaction part, and Gibson and Launder (1978) have extended Shir's suggestion by adding an equivalent mean-strain part. However, in the present work on free-surface effects this second part is not necessary and Eq. (9) will be used to model the redistribution effect of a free surface.

III. Model Equations for the Individual Reynolds Stresses

When the convective and diffusive transport terms in the equations governing the Reynolds stresses $\overline{u_i u_j}$ are neglected (assumption of local equilibrium) and the model suggestion of Launder et al. (1975) together with the additional free-surface correction of Shir (1973) are introduced, the following algebraic expression is obtained for determining $\overline{u_i u_j}$

$$(1-\alpha)P_{ij} - \beta D_{ij} - \gamma k\left(\frac{\partial U_i}{\partial x_j} + \frac{\partial U_j}{\partial x_i}\right) + \frac{2}{3}(\alpha + \beta)P\delta_{ij}$$

$$- \frac{\varepsilon}{k}\overline{u_i u_j} c_1 + (1-c_1)\frac{2}{3}k\delta_{ij} + c_3 \frac{\varepsilon}{k}(\overline{v^2}\delta_{ij} - \frac{3}{2}\overline{u_n u_i}\delta_{nj}$$

$$- \frac{3}{2}\overline{u_n u_j}\delta_{ni}) = 0 \qquad (10)$$

The free-surface proximity is expressed by the last term in Eq. (10) and is proportional to the parameter c_3 which depends on the distance from the surface. u_n is the fluctuating component normal to the surface. As is discussed in more detail in Naot and Rodi (1982), c_3 is a function of the distance from a virtual origin above the surface, made dimensionless with the turbulent length scale.

IV. Unidirectional Open-Channel Flow

In order to examine the influence of the free-surface proximity term on the various Reynolds stresses $\overline{u_i u_j}$, we consider the simplified case of unidirectional flow in a straight open channel. In that case the gradients of secondary velocities are neglected and only the gradients of the longitudinal velocity component in the z direction are retained in Eq. (10), which significantly simplifies the mathematics. Thus, Eq. (10) can be written for the individual components as

$$\frac{\varepsilon}{k}(c_1 + 2c_3)\overline{v^2} = \frac{2}{3}(\alpha - \frac{1}{2}\beta)P + \frac{2}{3}(c_1 - 1)\varepsilon + \beta[\overline{vw}\frac{\partial W}{\partial y} - \overline{uw}\frac{\partial W}{\partial x}]$$

$$(11)$$

$$\frac{\varepsilon}{k}c_1\overline{u^2} = c_3\frac{\varepsilon}{k}\overline{v^2} + \frac{2}{3}(\alpha - \frac{1}{2}\beta)P + \frac{2}{3}(c_1-1)\varepsilon - \beta[\overline{vw}\frac{\partial W}{\partial y} - \overline{uw}\frac{\partial W}{\partial x}]$$

$$(12)$$

$$\frac{\varepsilon}{k}(c_1 + \frac{3}{2}c_3)\overline{vw} = -[(1-\alpha)(\overline{v^2} + \overline{uv}\tan\psi) - \beta\overline{w^2} + \gamma k]\frac{\partial W}{\partial y}$$

$$(13)$$

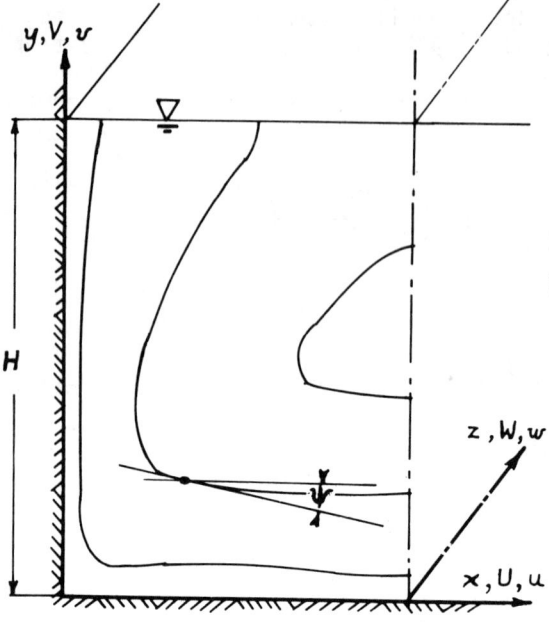

Fig. 1 Open-channel flow configuration with calculated isovels ($\bar{w}$ = 1.2, 1.1, and 0.9 m/s).

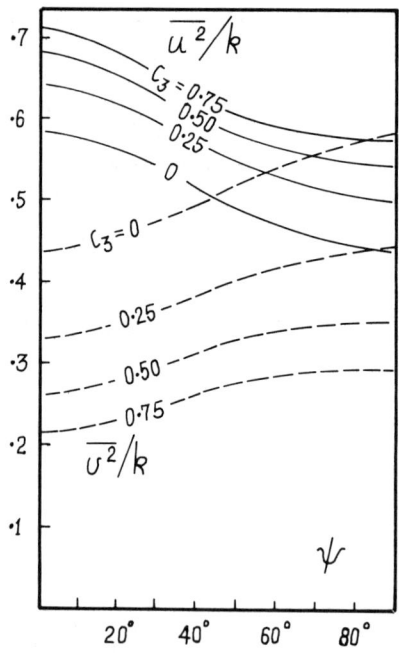

Fig. 2 Normal lateral stresses $\overline{v^2}$ and $\overline{u^2}$.

$$\frac{\varepsilon}{k} c_1 \overline{uw} = -[(1-\alpha)(\overline{u^2} + \overline{uv}\cot\psi) - \beta\overline{w^2} + \gamma k]\frac{\partial W}{\partial x} \quad (14)$$

$$\frac{\varepsilon}{k}(c_1 + \frac{3}{2}c_3)\overline{uv} = \beta(\overline{vw}\frac{\partial W}{\partial x} + \overline{uw}\frac{\partial W}{\partial y}) \quad (15)$$

with

$$\tan\psi = \frac{\partial W}{\partial x}/\frac{\partial W}{\partial y} \quad (16)$$

To understand the meaning of the angle ψ one should think of the isovels describing lines of constant streamwise velocity W given in the lateral x, y plane. In such a figure ψ denotes the angle formed between the tangent to the isovels and the x axis (see Fig. 1).

First applications of the model to two-dimensional open-channel flow have shown (Naot and Rodi 1982) that the coefficient c_3 in the surface-proximity term should take values in the range $0 < c_3 < 1$ over the top 30% of the channel depth, that is, for $0.7H < y < H$. In the following, a closer examination will be carried out as to how the free-surface correction, and in particular the value of the coefficient c_3, influences the various stresses.

To this end Eqs. (11-15) were solved iteratively for various values of c_3 in the range $0 < c_3 < 0.75$ and also for various values of the inclination angle ψ defined by Eq. (16) in the range $0 < \psi < 90$ deg. In accordance with the local equilibrium assumption already introduced in deriving Eq. (10), the dissipation rate ε has been put equal to the production of turbulence energy

$$P = -\overline{uw}\frac{\partial W}{\partial x} - \overline{vw}\frac{\partial W}{\partial y} \quad (17)$$

With this, Eqs. (11-15) can be brought into a form where only $\tan\psi$ appears, but not the individual longitudinal velocity gradients and also not P and ε. A numerical calculation of the individual stresses from the resulting equations was carried out by successive substitutions of values obtained from the former iterations to the right-hand side of the equations. In order to accelerate convergence, the new stresses obtained at each iteration were under-relaxed to 50%. [The parameters α, β, γ, and c_1 appearing in Eqs. (11-15) were taken from Launder et al. (1975).]

The results shown in Figs. 2-5 describe how the ratios of the individual turbulent stresses to the kinetic energy k depend on the free-surface parameter c_3 and the inclination angle ψ. The results for $c_3 = 0$ represent situations remote from the free surface and the results for $c_3 = 0.75$ situations close to the free surface. $\psi = 0$ represents results remote from the channel banks, that is, where the velocity gradient $\partial W/\partial y$ is dominant, while $\psi = 90$ deg represents results for the region adjacent to the channel vertical walls where the gradient $\partial W/\partial x$ is dominant. It should be mentioned, however, that no surface-proximity effects have been introduced for the channel walls, so that the stress ratios for $\psi = 90$ deg should not be compared with data near the vertical channel walls. The purpose of the present calculations was solely to examine the influence of the free-surface correction under various directions of the longitudinal velocity gradient.

Figures 2 and 3 describe the stresses which control the motion in the lateral plane, i.e., the secondary currents. Figure 2 shows that with decreasing distance from the free surface, here simulated by increasing values of c_3, the velocity fluctuations perpendicular to the free surface contributing to the turbulent stress $\overline{v^2}$ are reduced and the fluctuations parallel to the free surface contributing to $\overline{u^2}$ are increased. The difference between the normal turbulent stresses $\overline{u^2} - \overline{v^2}$, which was shown to be the main mechanism for inducing secondary currents in closed ducts (Perkins 1970), is thereby considerably increased. It may even become positive adjacent to the channel banks for $45 < \psi < 90$ deg, where it is negative for $c_3 = 0$.

The lateral shear stress $\overline{uv}$ shown in Fig. 3 is reduced by the free-surface proximity simulated by increasing c_3. The model implies that the decrease in the v fluctuations is accompanied by a smaller and statistically correlated in-

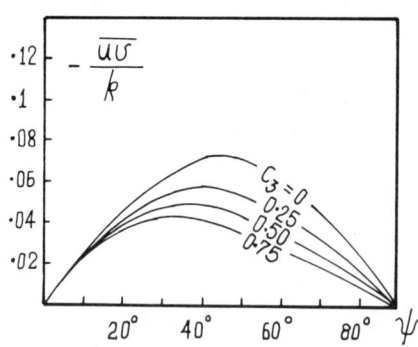

Fig. 3 Lateral shear stress $\overline{uv}$.

crease in the u fluctuations resulting in an overall reduction of $\overline{uv}$. The absolute value of this reduction is much smaller with respect to the effect of the free surface on the normal stresses. This observation implies that the main mechanism through which the free surface augments the secondary currents is the effect of the surface on the normal stresses.

Figures 4 and 5 show the results for the stresses $\overline{uw}$ and $\overline{vw}$ which control the streamwise mean motion W. The shear stress $\overline{vw}$ shown in Fig. 4, which is the dominant stress near the channel bed, is reduced adjacent to the free surface, here simulated by increasing c_3. However, since the $\overline{vw}$ stress itself goes to zero as the free surface is approached, the overall effect on the velocity profile is not very pronounced, as may be expected from Fig. 4. The model implies that the decrease in the v fluctuations is accompanied by a smaller and statistically correlated increase in the w fluctuations resulting in an overall reduction in $\overline{vw}$. The second main shear stress $\overline{uw}$ shown in Fig. 5 is the dominant stress near vertical channel banks where $\psi \approx 90$ deg. This also includes the important region where the free surface reaches the channel bank. Figure 5 suggests that $\overline{uw}$ is almost unaffected by the presence of the free surface simulated by increasing c_3. In this case the model suggests that the increase in both u and w fluctuations which follows the decrease in the v fluctuations is not statistically correlated, and therefore the increase in $\overline{uw}$ is relatively small.

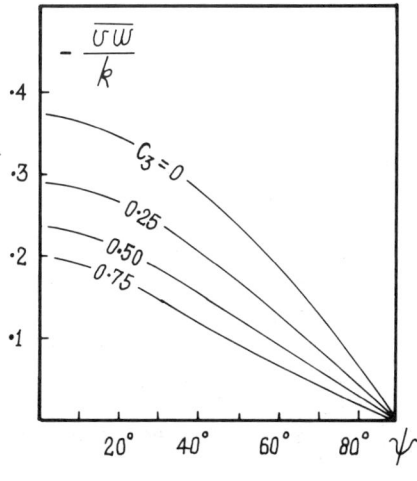

Fig. 4 Main shear stress $\overline{vw}$.

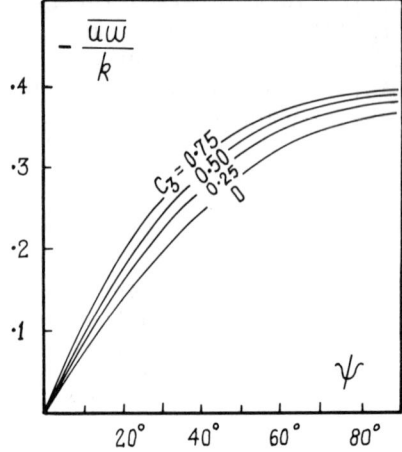

Fig. 5 Main shear stress $\overline{uw}$.

V. Approximate Formulas for the Main Shear Stresses

Previous calculations of closed-duct flows (e.g., Launder and Ying 1973) have shown that the refined treatment as given by Eqs. (13) and (14) is not needed for the main shear stresses $\overline{vw}$ and $\overline{uw}$. For this reason, Naot and Rodi (1982) developed the following approximate formulas which do retain the influence of the free surface on these stresses,

$$\overline{vw} = -c_\mu \frac{k^2}{\varepsilon} \left(\frac{c_1}{c_1 + \frac{3}{2}c_3}\right)\left(\frac{c_1}{c_1 + 2c_3}\right) \frac{\partial W}{\partial y} \qquad (18)$$

$$\overline{uw} = -c \frac{k^2}{\varepsilon} \left(\frac{c_1 + \frac{5}{2}c_3}{c_1 + 2c_3}\right) \frac{\partial W}{\partial x} \qquad (19)$$

With $c_\mu = 0.09$ suitable for calculations remote from surfaces, these expressions provide an approximation to the original Eqs. (13) and (14), accurate to within 10% for the c_3 and ψ range examined here. Remote from free surfaces, where $c_3 = 0$, these formulas simply yield the isotropic eddy viscosity relations employed in the widely used k-ε turbulence model. Introducing the stresses from Eqs. (18) and (19) into Eqs. (15), (11), and (12) for $\overline{uv}$, $\overline{v^2}$, and $\overline{u^2}$, respectively, results in calculations of the latter stresses showing errors of less than 5%. Equations (18) and (19) together with Eqs. (11), (12), and (15) can be solved directly for all turbulent stresses and are suggested here as an algebraic stress model for open-channel flow.

For a discussion on the limitations of the unidirectional solutions in the presence of secondary currents, the reader is referred to Naot and Rodi (1981).

VI. Numerical Simulation of Open-Channel Flow

Naot and Rodi (1982) have used the above algebraic relations for the individual turbulent stresses together with transport equations for the turbulent kinetic energy k and its dissipation rate ε to calculate two- and three-dimensional closed- and open-channel flows. The governing momentum, continuity, and k-ε equations were solved with the Patankar-Spalding (1972) finite-difference procedure for three-dimensional parabolic flows. The details of the complete mathematical model and of the results are given in Naot and Rodi (1982), and only the highlights of the results will be summarized here. It should be mentioned, however, that the redistributive effects of solid walls were accounted for by the wall proximity correction to the pressure strain model suggested by Launder et al. (1975), which is affected by making the coefficients c_1, α, β, and γ in the pressure-strain model functions of a dimensionless wall distance, as mentioned earlier. For simulating the wall effects, the model of Launder et al. (1975) was preferred to that of Shir (1973) because the former is unproblematic in the case of irregular wall boundaries (e.g., near corners), while the latter is difficult to apply in such cases because then a unit vector normal to the surface is not defined unequivocally.

The two main aspects of the results will now be discussed. First, the eddy viscosity distribution in both closed and open developed two-dimensional channel flow was calculated and compared with the measurements of Quarmby and Quirk (1972) for the closed channel and Ueda et al. (1977) for the open channel. The calculations follow closely the experimental observation that the eddy viscosity is reduced significantly near the free surface, leading to an almost parabolic distribution. In the calculations the reduction of eddy viscosity is brought about by two aspects of the model:

1) The increase of the dissipation rate near the free surface is achieved by fixing the length scale according to

Eq. (2) and hence also the surface value of the dissipation rate used as boundary condition in the ε equation.

2) The decrease of $\overline{v^2}$ due to the free-surface effect (see Fig. 2).

Preliminary calculations have shown that it is primarily the treatment of the dissipation rate near the surface that brings about the reduction in eddy viscosity. The good quantitative agreement with the measured eddy viscosity distribution supports the length-scale recommendation 2 of Hossain and Rodi (1980).

The second aspect of the channel flow calculations discussed here is the depression of the maximum of the longitudinal velocity below the free surface. This phenomenon, which was observed by Nikuradze (1926) and Raju (1932), is due to strong secondary currents which are driven mainly by the difference between the normal stresses, $\overline{u^2} - \overline{v^2}$. As this difference is influenced strongly by the free-surface proximity, the depression of the velocity maximum below the surface is, of course, also closely related to the presence of the surface. This effect is particularly strong because the surface-induced currents reach values of about 3% of the bulk channel velocity, while the secondary motions induced in closed ducts reach only about 1.5% of the bulk velocity. For this reason the effect of the secondary currents on the longitudinal velocity distribution is much more noticeable in open-channel flow, and hence the significant depression of the velocity maximum. Naot and Rodi (1982) have simulated the channel flows investigated experimentally by both Raju (1932) and Nikuradze (1926), which have different aspect ratios and hence significantly different velocity contours. In both cases, the depression of the velocity maximum below the surface was simulated very well by the model, supporting the assumptions that were made about the influence of a free surface on the turbulence.

VII. Conclusions

Experiments in open-channel flow indicate that the presence of a free surface has two main effects on the turbulence beneath it, namely to reduce the length scale of the turbulence and hence to increase the dissipation rate and to reflect the turbulent eddies, thus causing a redistribution of the velocity fluctuations. Models have been introduced to simulate these effects with an emphasis on the second, redistributive phenomenon, which is accounted for by a surface-proximity correction to the pressure-strain model. This correction was incorporated into local equilibrium forms of model equations for the individual turbulent stresses. The greatest influence was found for the vertical fluctuations $\overline{v^2}$, which were reduced by the presence of the surface, while the lateral fluctuations were increased, although to a smaller extent. However, the most important influence is the increase in the difference between the lateral normal stresses, $\overline{u^2} - \overline{v^2}$, as the surface is approached. The study has also shown that the shear stress $\overline{vw}$ is significantly affected by the free surface, while the shear stresses $\overline{uv}$ and $\overline{uw}$ are influenced relatively little. The algebraic modeling equations for the main shear stresses $\overline{vw}$ and $\overline{uw}$ were replaced by simpler approximations and the resulting algebraic stress model could be applied to the three-dimensional open-channel flow with reasonable computational effort. The results have shown that the model reproduces the main features of the flow, most prominent of which is the depression of the velocity maximum below the surface.

References

Gibson M.M. and Launder B.E. (1978) "Ground effects of pressure fluctuation in the atmospheric boundary layer," J. Fluid Mech., Vol. 86, Pt. 3, pp. 491-511.

Hossain M.S. and Rodi W. (1980) "Mathematical modelling of vertical mixing in stratified channel flow," Proc. 2nd Int. Symp. on Stratified Flows, Trondheim, Norway.

Launder B.E. and Ying W.M. (1973) "Prediction of flow and heat transfer in ducts of square cross-section," Proc. Inst. Mech. Eng. London, Vol. 187, pp. 455-461.

Launder B.E., Reece G.J., and Rodi W. (1975) "Progress in the development of a Reynolds stress turbulence closure," J. Fluid Mech., Vol. 68, Pt. 3, pp. 537-566.

McQuivey S. and Richardson E.Y. (1969) "Some turbulence measurements in open-channel flow," J. Hydraul. Div., HY 1, pp. 209-223.

Nakagawa H., Nezu I., and Veda H. (1975) "Turbulence of open channel flow over smooth and rough beds," Proc. Jap. Soc. Civ. Eng., Vol. 241, pp. 155-168.

Naot D. and Rodi W. (1981) "Applicability of algebraic models based on unidirectional flow to duct flow with lateral motion," Int. J. Numer. Methods in Fluids, Vol. 1, No. 3, pp. 225-235.

Naot D. and Rodi W. (1982) "Calculation of secondary currents in channel flow," J. Hydraul. Div., HY8, Vol. 108, pp. 948-968.

Nezu I. (1977) "Turbulence structure in open channel flows," Ph.D. Thesis, Civil Engineering Dept., Kyoto Univ., Japan.

Nikuradse J. (1926) "Turbulente Strömung im Innern des rechteckigen offenen Kanals," Forschungsarbeiten, Heft 281, pp. 36-44.

Patankar S.V. and Spalding D.B. (1972) "Calculation procedure for heat, mass and momentum transfer in three-dimensional parabolic flows," Int. J. Heat Mass Transfer, Vol. 15, pp. 1787-1806.

Perkins H.J. (1970) "The formulation of streamwise vorticity in turbulent flow," J. Fluid Mech., Vol. 44, Pt. 4, pp. 721-790.

Quarmby A. and Quirk R. (1972) "Measurements of the radial and tangential eddy diffusivities of heat and mass in turbulent flow in a plain tube," Int. J. Heat Mass Transfer, Vol. 15, pp. 2309-2327.

Raichlen F. (1967) "Some turbulence measurements in water," J. Eng. Mech. Div., EM 2, pp. 73-97.

Raju J. (1930) Cited in: "Reibungsverluste in Rohren und Kanälen, by Kirchner (1949), Die Wasserwirtschaft, Stuttgart, Vol. 39, No. 8, pp. 137-142 and 168-174.

Rotta J. (1951) "Statistische Theorie nichthomogener Turbulenz," Z. Phys., Bd. 129, pp. 547-572.

Shir C.C. (1973) "A preliminary numerical study of atmospheric turbulent flows in the idealized planetary boundary layer," J. Atmos. Sci., Vol. 30, pp. 1327-1339.

Thomas N.H. and Hancock P.E. (1977) "Grid turbulence near a moving wall," J. Fluid Mech., Vol. 82, pp. 481-490.

Ueda H., Möller R., Komori S., and Mizushina T. (1977) "Eddy diffusivity near the free surface of open channel flow," Int. J. Heat Mass Transfer, Vol. 20, pp. 1127-1136.

Diagnostics of Flowfields

S. Lederman,* P. Brown,† and T. Posillico†
Polytechnic Institute of New York, Farmingdale, N. Y.

Abstract

A novel arrangement which permits the utilization of a single Q-switched Ruby laser to obtain simultaneously CARS and spontaneous Raman signals is described. Two versions are indicated. In addition, with the incorporation of an LDV into the system a most versatile diagnostic apparatus for flowfields and combustion research can be made available. Experimental data obtained simultaneously using CARS and spontaneous Raman are given. Furthermore, it is shown that with proper gating, it is possible to obtain spontaneous Raman data even in flames containing carbon particles.

I. Introduction

The development of a universal probe capable of providing with reasonable accuracy and ease parameters of interest, nonintrusively, instantaneously, and remotely, in flowfields, combustion systems, and combustion-driven MHD flow has been the aim pursued at our laboratories for many years. The appearance of modern lasers redirected our efforts from visual methods such as interferometry and Schlieren, from radio frequency and microwave methods, and from electron beam and electromagnetic field methods to laser scattering methods. For a short time it appeared that a probe based on spontaneous Raman scattering principles

---

Paper presented at Third Beer-Sheva International Seminar on Magnetohydrodynamic Flows and Turbulence, Ben-Gurion University of the Negev, Beer-Sheva, Israel, March 23-27, 1981. Copyright © American Institute of Aeronautics and Astronautics, Inc., 1982. All rights reserved.
*Professor, Aerodynamics Laboratories.
+Research Assistant, Aerodynamics Laboratories.

could provide the desired results. As is well known (Lapp and Penny (1974), Widhopf and Lederman (1971), Lederman (1977)) a probe based on the spontaneous Raman principles can provide almost all of the important parameters in a flowfield or flame. Particularly if the short-duration, high-power laser pulse technique is utilized (Widhopf and Lederman (1971), Lederman (1977), Lederman (1980)), some of the derived parameters of major importance in turbulent combustion modeling can be obtained easily. However, difficulties appeared when diagnostic techniques based on the spontaneous Raman principles were applied to carbon-loaded flowfields such as may appear in jet engines and coal combustion-driven MHD channels. These difficulties were caused by the appearance of fluorescence, incandescence, and other interferring radiation. The basic difficulties could be traced to the unfortunately low Raman scattering cross section, which resulted in an inordinately low signal-to-noise ratio. While certain data acquisition techniques, such as gating, utilization of some field polarization properties, etc., may help in improving the inadequate signal-to-noise ratio, the basic difficulties remain. This prompted the development of a new technique based on the Raman effect, the coherent anti-Stokes Raman scattering (CARS). This technique can, under the same flow conditions, provide a signal several orders of magnitude higher than the spontaneous Raman technique, thus improving the signal-to-noise ratio immensely. CARS is not capable of replacing the spontaneous Raman scattering technique in terms of its generality. In addition, CARS generally requires two lasers to provide a single specie or temperature measurement.

In order to take advantage of the properties of both the spontaneous Raman and the CARS techniques, a novel arrangement has been constructed. This new diagnostic system utilizes a single high-peak-power Q-switched Ruby laser. The system can provide simultaneously a CARS signal of a particular specie and temperature of interest and spontaneous Raman signals of all the species in the flowfield. The latter provided the environment is not prohibitively hostile. In any case, upon an additional incorporation of a laser Doppler velocity (LDV) system as discussed in Lederman (1980), a most powerful diagnostic system emerged. After a short discussion of the theoretical background, a description of the apparatus including the necessary data acquisition and processing systems as well as the required computation facility will be indicated.

Some experimental procedures and results are described. Finally experimental results obtained in our laboratories on flames and flowfields are presented. A comparison is made between results obtained on carbon particle and soot-contaminated flames and clean flames.

II. Theoretical Background

The basic theoretical background of the formulation and operation of both the spontaneous Raman techniques and CARS has been discussed abundantly in the literature. It is therefore sufficient here just to cite some of the references: Eckbreth et al. (1978), Lapp and Penny (1974), Lapp and Penny (1977), Lederman (1976), Lederman (1977), Lederman (1980), Lederman et al. (1979), and Widhopf and Lederman (1971), and to point out a few of the major differences between the spontaneous Raman and coherent anti-Stokes Raman scattering (CARS) systems:

1) Spontaneous Raman is single ended; CARS is not.

2) Spontaneous Raman can resolve any number of Raman active species in a mixture simultaneously; CARS cannot.

3) Spontaneous Raman can provide the temperatures of any number of Raman active species in a mixture simultaneously and simply; CARS cannot.

4) Spontaneous Raman can provide a measure of the fluctuation of a number of species in the flowfield and thus a measure of turbulent intensity; CARS cannot.

5) Spontaneous Raman can provide a measure of the mixed parameters, autocorrelation or correlation of parameters of importance in a flowfield; CARS cannot.

6) Spontaneous Raman is linear; CARS is not.

These are some of the advantages of spontaneous Raman scattering over CARS.

However, one of the major drawbacks of the spontaneous Raman technique is its extremely low-differential scattering cross section. This feature is responsible for very low signal levels and therefore theoretically limits the application of spontaneous Raman scattering diagnostics to well-behaved, clean, low-noise systems, particularly those containing essentially no carbon particles or carbon soot. However, as will be shown later, this is not always the case. Proper timing and gating permits one to perform measurements in sooty flames. In those cases which are most important in a majority of combustion systems, CARS with its coherent signal several orders of magnitude higher than the spontaneous Raman signal is highly preferred in spite of its other limitations.

III. Experimental Apparatus

To take advantage of the positive properties of both systems, a unified diagnostic system has been designed, built, and tested. A schematic diagram of the system is shown in Fig. 1. This arrangement, in its simplest configuration, utilizes a Q-switched Ruby laser and a Raman cell filled with a gas of particular interest, whereby stimulation of the Stokes line of a particular gas of interest is caused and, when collinearly mixed with the primary beam and focused at a point, generates a CARS signal in the given flowfield under investigation. A part of the incident Ruby laser is simultaneously utilized to obtain spontaneous Raman signals. Thus, one is able to avail himself of the advantages of both the spontaneous and CARS techniques utilizing only one Ruby laser.

Figure 2 indicates a different version of the same apparatus. Here a doubler converts part of the Ruby pulse at 6943 Å to a 3471 Å pulse which is then separated by a dichroic mirror from the fundamental. Part of the 3471 Å laser energy is used to pump a broadband dye laser, which is then combined coaxially with the rest of the 3471 Å laser energy and focused at a point in the working fluid to

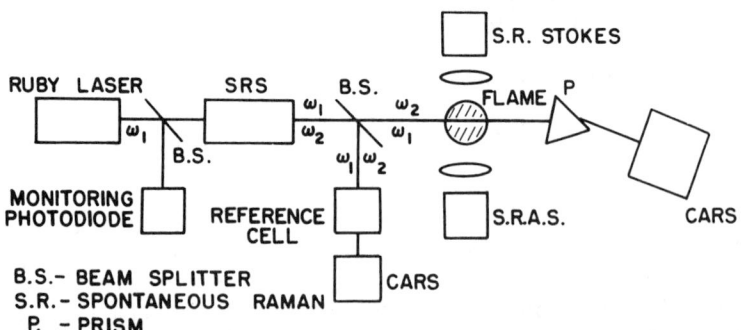

Fig. 1 Spontaneous Raman/CARS configuration utilizing a stimulated Raman scattering cell (SRS).

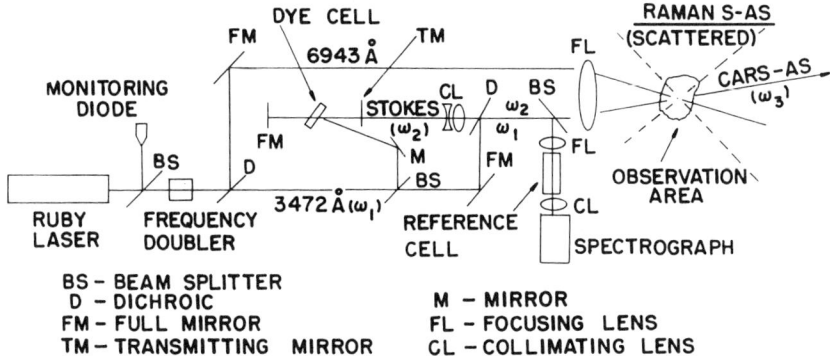

Fig. 2 Spontaneous Raman/CARS configuration utilizing doubler and dye cell.

obtain CARS. The total energy of the second harmonic (3741 Å) is approximately 500 mJ. Since the original Ruby laser pulse energy was over 3 J, well over 2 J of laser light is available after separation by the dichroic mirror to be focused on the working fluid to be used to excite spontaneous Raman scattering.

IV. Experimental Results and Conclusions

Preliminary experiments using these systems have been conducted on a methane diffusion flame. The purpose of these experiments was the demonstration of its feasibility, practicality, and advantages in terms of making this a more universal diagnostic system.

As is well known, the concentration of the unburned methane in an air-methane flame is generally very low and its measurement at a point of about 1 mm^3 is almost impossible using spontaneous Raman scattering. On the other hand, the simultaneous measurement of the concentration of the products of combustion (as well as of the heated nitrogen) and their temperature is impossible using CARS. This combined apparatus permits a complete diagnosis of the flame simultaneously and instantaneously.

Figure 3 indicates a preliminary survey of the concentration of the unburned methane using CARS and the concentration of nitrogen using spontaneous Raman. Figure 4 shows the temperature of the nitrogen in the flame using spontaneous Raman.

It is clear from the above that this system with the addition of the LDV capability which has been included previously in the diagnostic system using only spontaneous Raman (Lederman 1976) fulfills the requirements for a universal diagnostic system. It can provide data on most practical flowfields of interest, be they "clean" or "contaminated," be they environmentally "friendly" or "hostile." A system such as this that is portable and adjustable with respect to the flowfield of interest should be very useful.

As has been mentioned previously, the spontaneous Raman diagnostic process can, under certain conditions, enable one to obtain desired data signals from experimental configurations which are not necessarily well behaved, clean, or sootless. The ability to diagnose such types of flowfields is of particular importance for a number of reasons. The working fluids of coal combustion-driven MHD generators, as is well known, may contain, besides other contaminants, soot and other by-products of incomplete coal combustion.

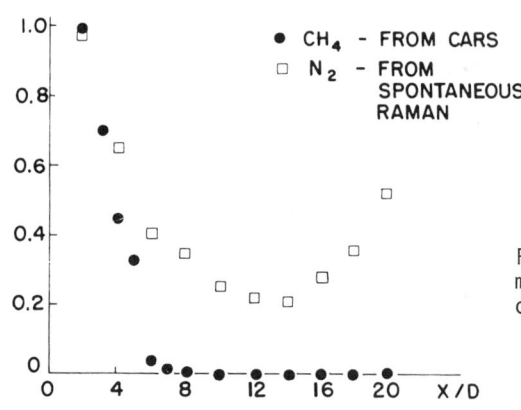

Fig. 3 Normalized axial methane and nitrogen concentration profiles.

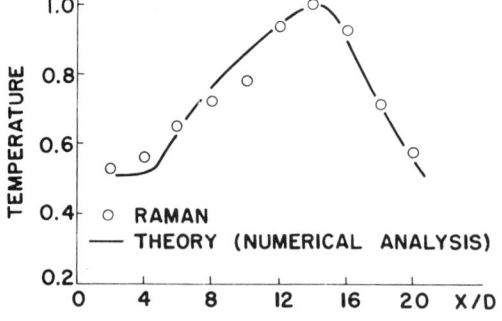

Fig. 4 Normalized spontaneous Raman axial temperature profile.

DIAGNOSTICS OF FLOWFIELDS

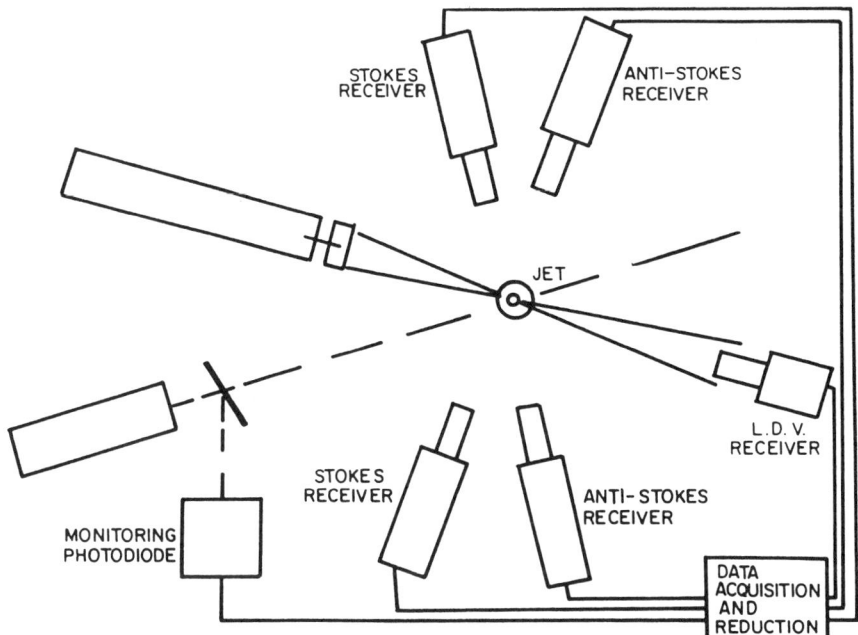

Fig. 5 Black diagram of experimental apparatus.

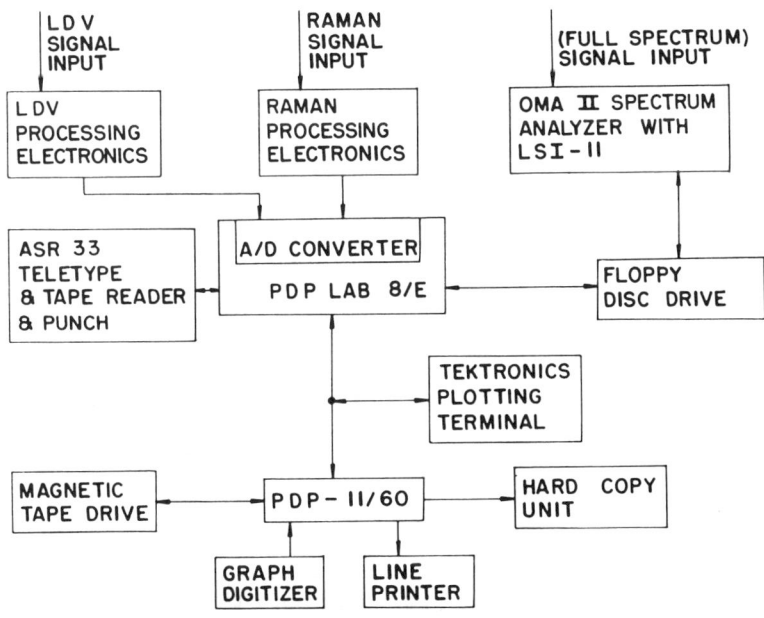

Fig. 6 Data processing system.

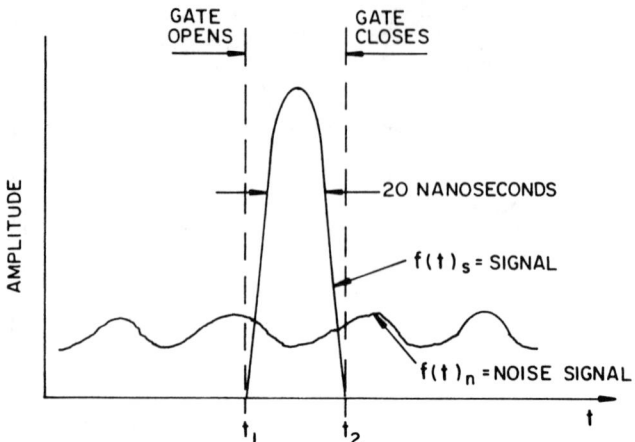

If the desired signal resulting from the laser pulse is represented by $f(t)_S$ and the undesired signal or noise by $f(t)_N$, then the signal-to-noise ratio for a gated operation will be larger than for an ungated operation. This can be represented as

$$\frac{\int_{t_1}^{t_2} f(t)_S \, dt + \int_{t_1}^{t_2} f(t)_N \, dt}{\int_{t_1}^{t_2} f(t)_N \, dt} > \frac{\int_{0}^{\infty} f(t)_S \, dt + \int_{0}^{\infty} f(t)_N \, dt}{\int_{0}^{\infty} f(t)_N \, dt}$$

$$\frac{\int_{t_1}^{t_2} f(t)_S \, dt}{\int_{t_1}^{t_2} f(t)_N \, dt} + x > \frac{\int_{t_1}^{t_2} f(t)_S \, dt}{\int_{0}^{\infty} f(t)_N \, dt} + x$$

in the limit as $t > \infty$ the ungated S/N $\to$ 0 and the gated S/N is

$$\frac{\int_{t_1}^{t_2} f(t)_S \, dt}{\int_{t_1}^{t_2} f(t)_N \, dt} = \frac{S}{N} > 0$$

Fig. 7 Schematic diagram of the data acquisition timing.

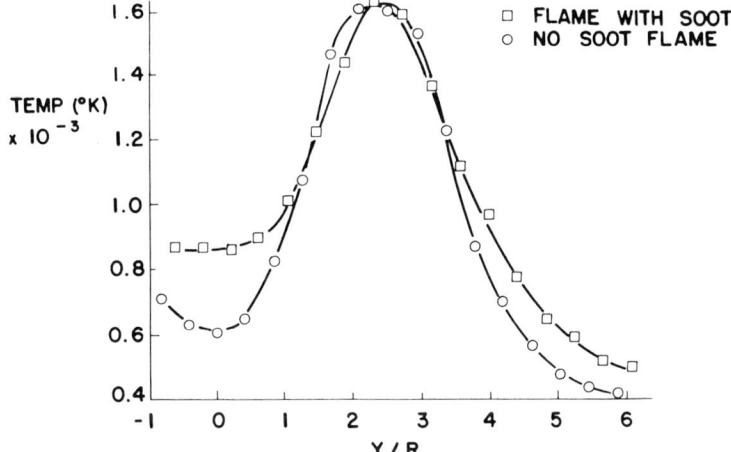

Fig. 8 Radial temperature profiles in the flame.

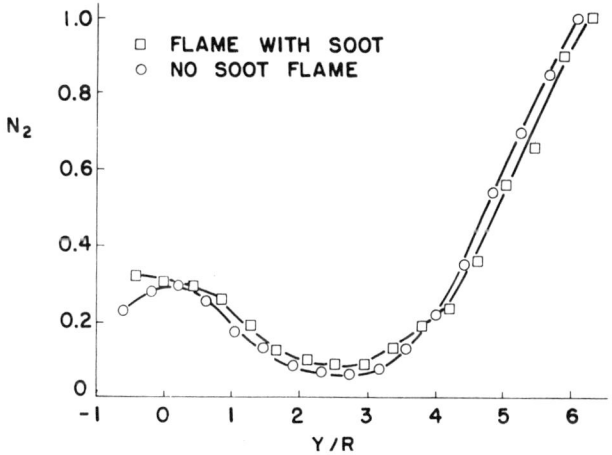

Fig. 9 Normalized radial nitrogen concentration profiles in the flame.

Flames in jet engines may contain soot. Practically all combustion systems utilizing hydrocarbon-based fuels may contain soot to some degree.

It was the purpose of the following experiments to determine whether combustion systems containing carbon impurities can be diagnosed utilizing spontaneous Raman scattering. To that end an air-methane flame, as described in Eckbreth et al. (1978) and Lapp and Penny (1977) has been uti-

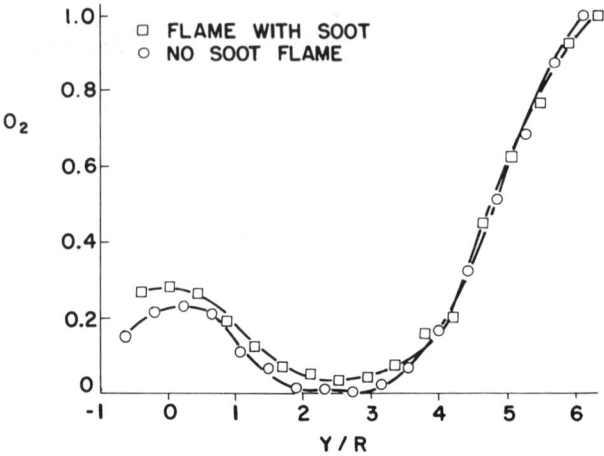

Fig. 10 Normalized radial oxygen concentration profiles in the flame.

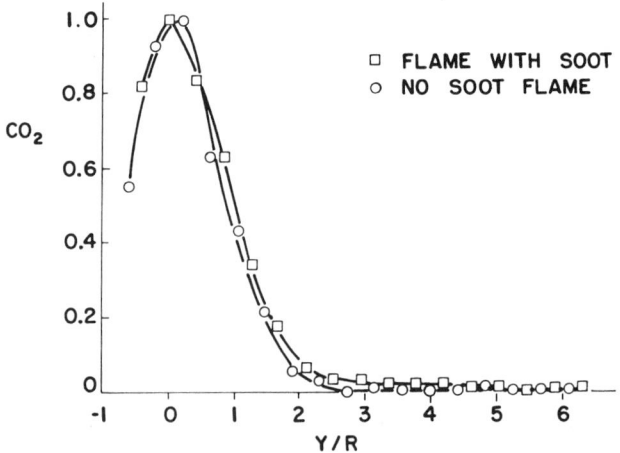

Fig. 11 Normalized radial carbon dioxide concentration profiles in the flame.

lized. The apparatus and the data acquisition and processing systems are shown in Figs. 5 and 6. While the schematic diagrams of Figs. 5 and 6 are self-explanatory, it should be pointed out that proper timing of the data acquisition modules (the so-called linear gates) to coincide with the arrival at the gate of the maximum Raman scattered signal (resulting from the Q-switched short-duration, high-power laser pulse) is the most critical point in this system. The proper timing of the opening and closing of the

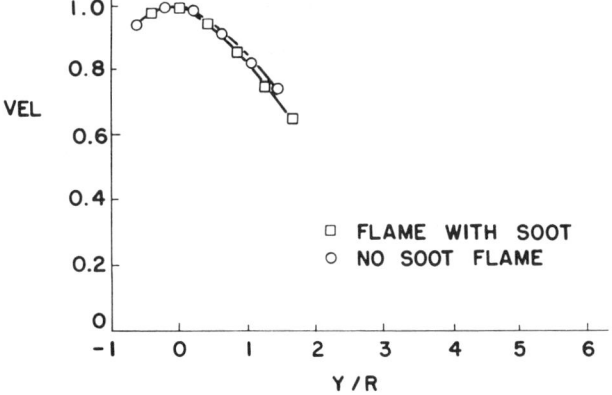

Fig. 12 Normalized radial velocity profiles in the flame.

gates assures a favorable signal-to-noise ratio of the individual pulses since the integration of the simultaneously acquired and superimposed desired and undesired signals would generally be favoring the desired signal. The desired signal duration corresponds to the duration of the Q-switched laser pulse, which is generally of the order of 10-20 ns, while the undesired signal is a continuous one. Without the proper gating, the integrated undesired signal would overwhelm the desired signal. This is visualized schematically in Fig. 7.

Tests have been performed on the pure uncontaminated flame and on the same flame with carbon particles. This was done in order to compare the obtainable results utilizing the same apparatus, the same laser, and the same data acquisition and processing schemes. The results are shown in Figs. 8-12 where the corresponding normalized profiles are shown. A first approximation indicates that this method of data acquisition utilizing the spontaneous Raman effect is quite applicable. Considering the fact that on the average only about 200 pulses/point have been utilized, the results are very good. A greater number of pulses per point would certainly reduce the standard deviations and improve the signal-to-noise ratio further.

Acknowledgment

This work has been supported by the U.S. Department of Energy under Contract AC01-78ET11056.

References

Eckbreth A.C., Bonchyk P.A., and Verdieck Y.F. (1978) "Laser Raman and fluorescence techniques for practical combustion diagnostics," Appl. Spectrosc. Rev., Vol. 13, No. 1, pp. 15-164.

Lapp M. and Penny C. (Eds.) (1974) Laser Raman Gas Diagnostics, Plenum Press, New York and London.

Lapp M. and Penny C. (1977) "Advances in infrared and Raman spectroscopy," Infr. and Raman Spectrosc., Vol. 3, p. 204.

Lederman S. (1976) "Some applications of laser diagnostics to fluid dynamics," AIAA paper 76-21.

Lederman S. (1977) "The use of laser Raman diagnostics in flow fields and combustion," Prog. Energy Combust. Sci., Vol. 3, pp. 1-34.

Lederman S. (1980) "Development in laser based diagnostic techniques," Proc. 12th Int. Symp. on Shock Tubes and Waves, A. Lifschitz and J. Rom (Eds.) The Magnes Press, Jerusalem, Israel, pp. 48-65.

Lederman S., Celentano A., and Glaser J. (1979) "Temperature, concentration and velocity in jets, flames and shock tubes," Phys. Fluids, Vol. 22, No. 6, p. 1965.

Widhopf G. and Lederman S. (1971) "Specie concentration measurements utilizing laser induced Raman scattering," AIAA J., Vol. 9, Feb., pp. 309-316.

Chapter II MHD Generation and Electromagnetic Flowmeters

Conceptual Design of a Coal-Fired Retrofit Liquid-Metal MHD Power System

E.S. Pierson,* H. Herman,† and M. Petrick‡
Argonne National Laboratory, Argonne, Ill.

Abstract

The application of the new, coal-fired open-cycle liquid-metal MHD (OC-LMMHD) energy conversion system to the retrofit of an existing oil- or gas-fired conventional steam powerplant is evaluated. The evaluation criteria are the net plant efficiency and the cost benefit relative to other options, i.e., continuing to burn oil, a conventional retrofit to burn coal (if possible), and an "over-the-fence" gasifier for boilers that cannot burn coal directly. The retrofit is described, including the significant fabrication and functional features. The substantial gains in net plant output power and efficiency for OC-LMMHD vs the reductions for a conventional coal conversion or a gasifier are discussed. A plant lifetime economic analysis for the OC-LMMHD retrofit relative to other options shows payback periods that are very good for utilities.

I. Introduction

The desire to develop a coal-fired energy conversion system with the high efficiencies and moderate temperatures

---

Paper presented at Third Beer-Sheva International Seminar on Magnetohydrodynamic Flows and Turbulence, Ben-Gurion University of the Negev, Beer-Sheva, Israel, March 23-27, 1981. Copyright © American Institute of Aeronautics and Astronautics, Inc., 1982. All rights reserved.
*Program Manager, Liquid-Metal MHD (presently Head, Department of Engineering, Purdue University Calumet, Hammond, Ind.).
†Deputy Manager, First Wall/Blanket/Shield Program.
‡Director, Fossil Energy Programs.

of two-phase generator liquid-metal magnetohydrodynamic (LMMHD) systems led to the open-cycle LMMHD (OC-LMMHD) concept. A liquid metal, most likely copper, which is compatible with combustion gases is used so that the combustion gas can be mixed with the liquid metal to form the two-phase mixture in the LMMHD generator, thereby eliminating the need for a primary heat exchanger, as shown in Fig. 1. Applications where OC-LMMHD appears to be particularly attractive include central powerplants larger than $\sim$10 MWe, retrofit of existing oil- or gas-fired central steam powerplants to burn coal, and cogeneration systems requiring high-temperature process heat. In particular, the latter two benefit from the clean combustion gas stream leaving the copper in the LMMHD system.

To explore the technical and economic feasibility of this new LMMHD concept, a conceptual design study of the retrofit of a coal-fired OC-LMMHD topping cycle to a conventional steam plant was selected, and extensive parametric studies carried out to establish the optimum parameter ranges for the retrofit cycle. A conceptual design was developed for the plant and the components with sufficient detail that a cost estimate for the retrofit could be readily made.

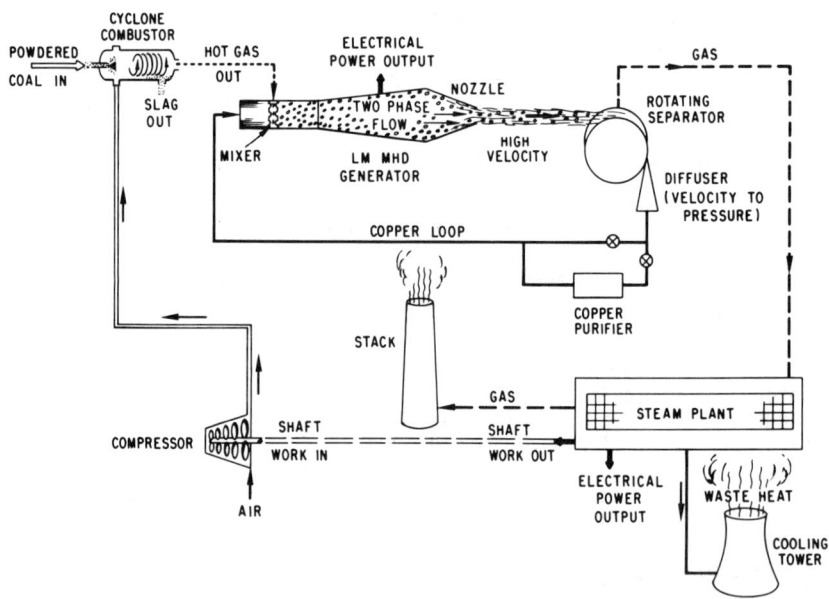

Fig. 1 Schematic of open-cycle coal-fired LMMHD cycle.

II. Retrofit Description

The OC-LMMHD plant flow diagram is shown in Fig. 2 and Table 1. The plant layouts, Fig. 3 (plan) and Fig. 4 (elevation), illustrate the manner in which the components fit together. The existing steam plant, not shown, is the Frank Bird plant of the Montana Power Company, constructed in 1952. It was chosen as representative of the many old steam plants of less than $\sim$100 MWe capability.

The LMMHD system is modular and consists of eight separate liquid-metal loops arranged in a ring around the combustor. The magnetic field for the eight generators is provided by a toroidal iron core with a rectangular cross section and eight air gaps, as shown in Figs. 3 and 4. The magnetic flux density of 1.0 T can be produced by either conventional water-cooled copper or superconducting coils. The superconducting version was chosen; it costs less (including the refrigeration equipment) and requires much less power. The eight LMMHD generators are connected electrically in series to yield higher output voltage. The dc-to-ac inversion is by means of a homopolar motor driving a synchronous generator. (Solid-state inverters cost much more and have only a slightly higher efficiency.)

The combustion gas leaving the separator goes to the knockout drum where it is cooled (by generating useful steam) so that copper vapor carried over (from the separator) is condensed and collected. Any liquid copper drops carried over are also collected. Copper vapor leaves the knockout drum with the gas at a rate of $\sim$12.5 lb/day ($\sim$5.7 kg/day). Combustion gas leaving the knockout drum powers gas turbines which drive the compressors, and then passes into the secondary combustor at the entrance to the radiant section of the existing boiler. (The primary combustor must operate at a stoichiometric ratio of less than unity to avoid oxidizing the copper.)

The significant fabrication and functional features of the plant are:

1) The coal is burned in an environmentally acceptable fashion because the liquid copper removes the particulates, sulfur, and nitrogen oxides while performing its normal function in the OC-LMMHD system. No stack gas cleanup or slag/ash removal from the boiler is needed.

Table 1 Heat and mass flow data for Fig. 2

No.	Stream	Temp, K	Flow, kg/s	Pressure, atm	Velocity, m/s	Line size, in.	Comments
1	Air	453	85.6	30	35	12	
2	Flue gas	1400	102.6	3.1	433	26	
3	Air	453	5.61	30.5	22.4	4.5	For coal feed
4	Air and coal	400	18.9	30	9.1	10	
5	Flue gas	1497	12.8	3.01	240	10	SR = 0.8
6	Flue gas	1497	102.6	3.01	600	23	SR = 0.8
7	Flue gas	1400	102.6	1.001	75.6	6 x 8 ft duct	SR = 0.8
8	Air	550	28.5	1.001	20	6 x 4 ft duct	For secondary combust.
9	Copper (liquid)	1497	25	3	5	1	To cleaning
10	Copper (liquid)	1497	200	1-3	5	3	To cleaning
11	Copper (liquid)	1497	200	60-65	5	3	From cleaning
12	Copper (liquid)	1497	4810	~63	6.1	-	Copper return line
13	Flue gas	2012	12.8	29.55	77	8	
14	Copper (liquid)	~1400	~0	---	---	3	Fill/drain only
16	Flue gas and air	589	5	2.0	72	12	34.7% SO$_2$
17	SO$_2$-rich gas	1497	2.5	1.1	175	8	
18	Air	1380	2.1	2.0	200	6	

COAL-FIRED LIQUID-METAL MHD POWER SYSTEM

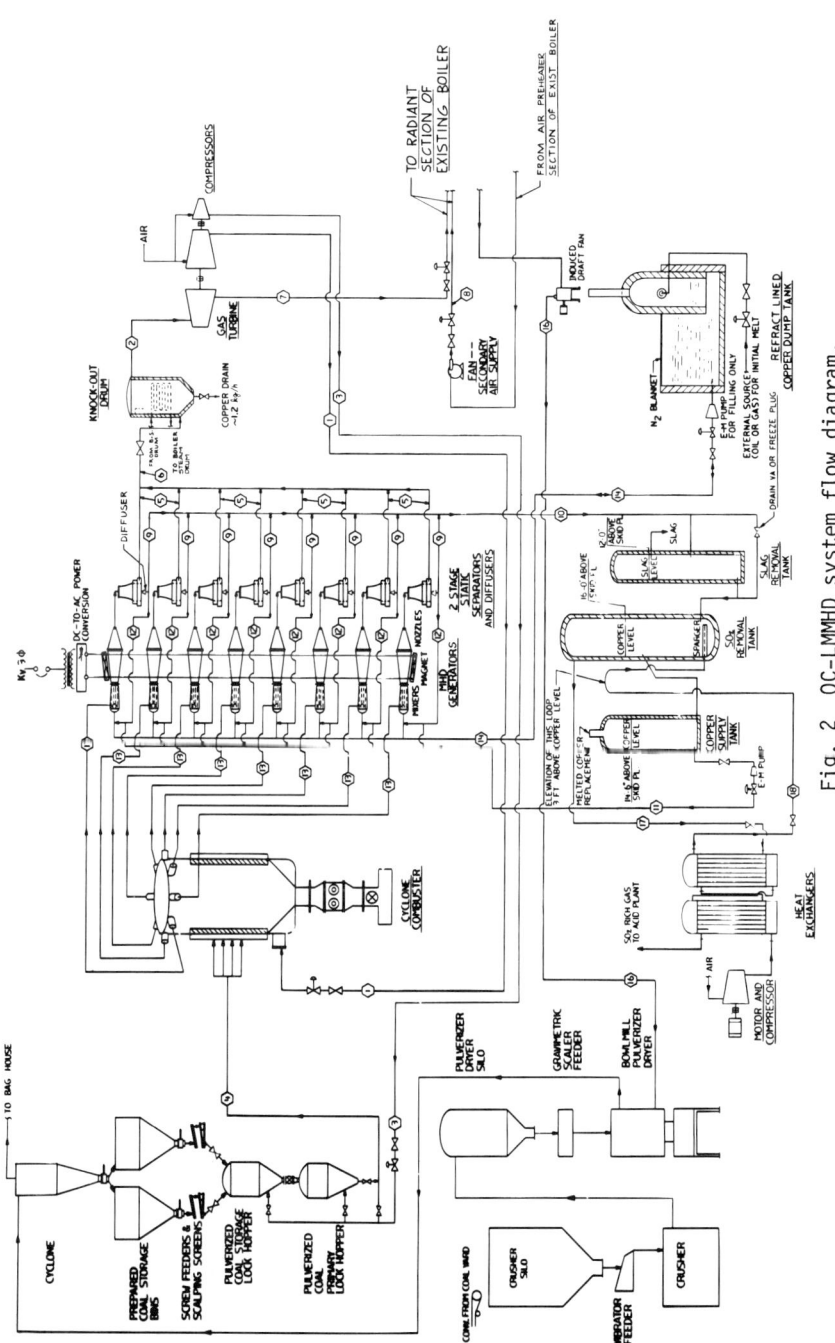

Fig. 2 OC-LMMHD system flow diagram.

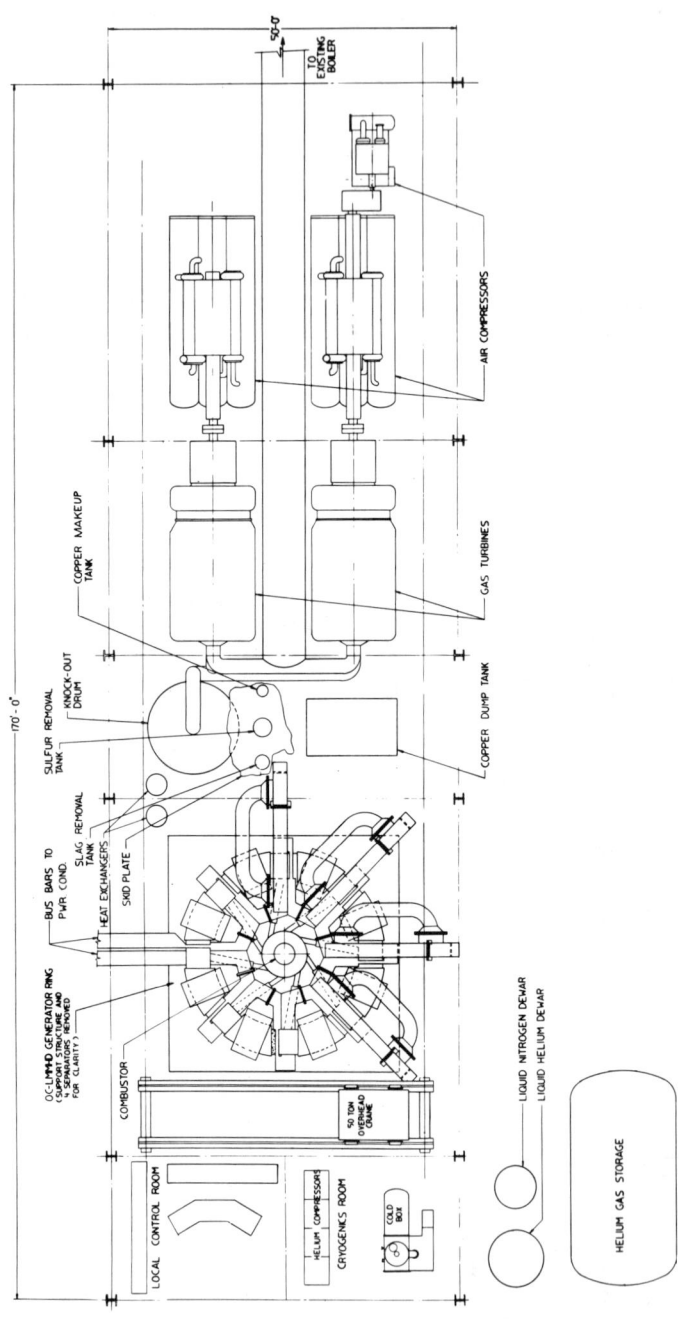

Fig. 3 OC-LMMHD plant layout, plan.

COAL-FIRED LIQUID-METAL MHD POWER SYSTEM

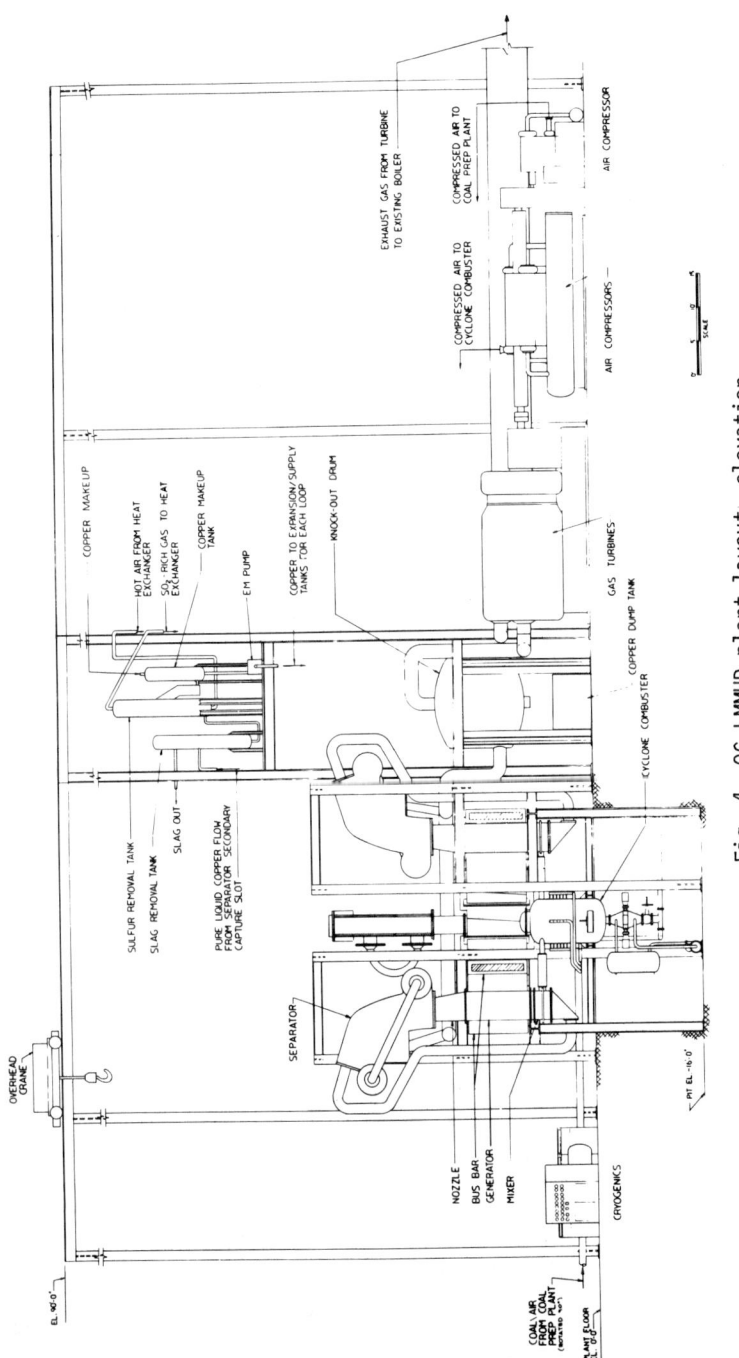

Fig. 4 OC-LMMHD plant layout, elevation.

2) The existing boiler can be used with, at most, minor modifications, because the combustion gas entering it is clean. The ability to burn oil and gas (if the OC-LMMHD system is not in operation) can be retained by relocating the existing burners.

3) The slag and sulfur are readily and inexpensively removed from the copper in forms that are easily handled by existing technology.

4) The combustor is a conventional cyclone design operating at 30 atm (3.04 MPa).

5) The LMMHD components and the magnet segments are small, can be fabricated and assembled in a factory, and require only minimal and easy field installation. Single components, or an entire liquid-metal loop, can be removed or replaced. Operation without one or more generators is possible by shunting the bus bar connections.

6) The OC-LMMHD components use combustion gas for preheating before plant startup; no trace heating is required for the components which contain liquid copper. Startup of the LMMHD system is accomplished by filling the loops with liquid copper, and then starting the flow of combustion gas so that the copper flow starts by natural circulation. On shutdown, the liquid copper is drained to the storage tank by gravity.

III. Efficiencies

The net plant output powers and efficiencies for the OC-LMMHD retrofit are listed in Table 2 for different operating conditions, i.e., liquid-copper temperatures. The lowest copper temperature was selected as the base case in order to be close to present copper industry practice. For comparison, the same data are shown for the existing boiler with oil firing (Gilbert/Commonwealth 1979), a conventional conversion to coal (Gilbert/Commonwealth 1979), and a coal conversion using an "over-the-fence" gasifier (for plants which cannot burn coal directly) based on a Combustion Engineering gasifier (Bechtel 1979). The OC-LMMHD retrofit cases illustrate the potential of the concept and the additional benefits if the copper temperature can be increased. At the lowest copper temperature, the output power is increased by 83% and the plant efficiency by 18% from the existing oil-fired case, compared with reductions of 6 and

Table 2 Plant efficiencies, existing Frank Bird and retrofits

Case	Existing, oil-fired[a]	Retrofit, coal-fired[a]	"Over-the-fence" gasifier[b]	Base	OC-LMMHD retrofits Var. 1	Var. 2	Var. 3
LMMHD conditions							
Stoich. ratio, main/secondary combustion				0.8/1.05	0.9/1.05	0.9/1.05	0.9/1.05
No. of compressor intercoolers				2	2	1	0
Copper temp. at mixer exit, °F				2242	2456	2566	2967
Power							
Gross LMMHD power, MWe		68.2	68.2	74.0	85.2	84.7	79.0
Gross steam power, MWe	68.2			57.3	49.6	55.2	65.1
Topping cycle losses, MWe	3.1	5.8	5.5	9.4	10.6	10.6	10.0
Bottoming cycle losses, MWe	65.1	61.4	62.7	2.5	2.2	2.4	2.9
Net plant output, MWe				119.4	121.9	126.9	131.3
Steam condition[c]							
Throttle steam flow, lb/h·10⁻³	597.9	597.9	597.9	484.	419.	466.	550.
Heat input							
Fuel type	Oil	Coal	Coal	Coal	Coal	Coal	Coal
Total heat input, MW	225.6	230.8	289.6	350.	350.	350.	350.
Performance							
Net plant efficiency, %	28.9	26.6	22.5	34.1	34.8	36.3	37.5
Net plant heat rate, Btu/kW·h	11,811	12,816	15,162	10,010	9,803	9,419	9,100

[a] Gilbert/Commonwealth (1979).
[b] Based on Combustion Engineering gasifier (Bechtel 1979).
[c] Condenser pressure = 2.5 in. Hg (absolute) (8.4 kPa), steam conditions at turbine inlet = 865 psia/900°F (6.17 MPa/756 K).

8%, correspondingly, for a conventional coal conversion and reductions of 4 and 22%, correspondingly, for a gasifier. At the highest copper temperature the increases are 102 and 30%.

IV. Economic Analysis

The design data were utilized by an architect/engineer for a plant lifetime economic analysis of the OC-LMMHD retrofit, in comparison with the other options listed in Table 2, to quantify the economic viability of the OC-LMMHD retrofit concept. The criteria used were:

1) The net present value of the extra OC-LMMHD modification capital, i.e., the present value of the cash flows over the life of the plant minus the extra modification capital of the OC-LMMHD retrofit ("extra" relative to the option being compared).

2) The yield on the extra OC-LMMHD retrofit modification capital, i.e., the percentage of return per year on the extra capital averaged over the lifetime of the plant.

Table 3 Financial benefit comparisons, OC-LMMHD vs other options

	OC-LMMHD Base		OC-LMMHD Variation 3	
Coal escalation rate	8%	11%	8%	11%
A) Net present value[a] ($ million) of OC-LMMHD modification relative to:				
Retrofit, coal-fired	36.6	80.1	62.2	128.7
"Over-the-fence" gasifier	126.6		165.5	
B) Yield[a] (%) on extra OC-LMMHD modification capital relative to:				
Retrofit, coal-fired	14.7	17.2	17.5	20.2
"Over-the-fence" gasifier	25.6		29.6	
C) Payback period[a] (years) for extra OC-LMMHD modification capital relative to:				
Retrofit, coal-fired	9.1	8.4	7.5	7.0
"Over-the-fence" gasifier	4.8		4.1	

Escalations:
  Operating and maintenance: 8%
  Fuel (coal): 8 and 11%
  Revenue: balance costs for comparison case, scale for OC-LMMHD
Discount rate: 11%
Plant life cycle: 31 years
Operation: base load, 65% utilization at full power

[a] Before taxes and depreciation.

3) The payback period for the extra OC-LMMHD retrofit modification capital, i.e., number of years required for recovery of that capital by the investor.
Positive values mean that OC-LMMHD is better than the alternative. For the first and second criteria, larger values are best; for the third criterion, smaller values are best.

All of the coal conversion alternatives showed very favorable results in comparison with continuing to burn oil, because of the high and escalating cost of oil. The differences between the coal options were small and OC-LMMHD compared favorably. However, the oil costs were so high that they dominated the study, making it difficult to compare the coal options. Thus, the OC-LMMHD cases were compared directly with the other coal options, the results of which are summarized in Table 3. In all cases, the base OC-LMMHD option appears advantageous relative to the other coal options, and variation 3 is better than the base case. Specifically, the payback period for the extra OC-LMMHD investment is 8-9 years compared with a conventional conversion to coal (where that is possible), and only 4-5 years compared with the more normal retrofit of an "over-the-fence" gasifier. These payback periods are very good vs the customary 15 year financing period for the electric utilities.

Acknowledgments

The work reported here was supported by the U.S. Department of Energy.

The authors gratefully acknowledge the contributions of R.W. Boom, L. Carlson, D. Cohen, G. Dubey, S.J. Grammel, F. Schreiner, B.K. Snyder, and T. Zinneman.

References

Bechtel National, Inc. (1979) "Economics of retrofitting power plants for coal-derived medium-btu gas," Electric Power Research Institute, Palo Alto, Calif., Rept. EPRI AF-1182.

Gilbert/Commonwealth (1979) "Feasibility study: MHD retrofit of steam power plants," ER-79-12, Reading, Pa.

Solar-Powered Liquid-Metal MHD Performance and Cost Studies

E.S. Pierson* and H. Herman†
Argonne National Laboratory, Argonne, Ill.

Abstract

The operation of the liquid-metal magnetohydrodynamic (LMMHD) gas or Brayton cycle and its application to solar power systems are described. Efficiencies are calculated for conditions appropriate to a solar central receiver, and the sensitivity to certain parameters explored. Performance limits for cogeneration systems are summarized. The economic advantages from coupling a 25 MW thermal high-temperature LMMHD system to a solar power tower are demonstrated by means of three representative comparison cases: 1) LMMHD vs photovoltaics for electric power generation alone, 2) LMMHD cogeneration vs photovoltaics combined with a solar power tower for equivalent cogeneration, and 3) LMMHD cogeneration vs photovoltaics combined with fossil fuel heat for equivalent cogeneration. For case 1, the LMMHD capital costs are clearly much lower, $22.1 million vs $28.3 million for photovoltaics for existing technology, and $10.6 million vs $18.4 million for mature technology. The case 2 costs are $19.0 million vs $20.9 million, respectively, for existing technology, and $8.4 million vs $11.9 million, respectively, for mature technology. For case 3, the total cost savings

---

Paper presented at Third Beer-Sheva International Seminar on Magnetohydrodynamic Flows and Turbulence, Ben-Gurion University of the Negev, Beer-Sheva, Israel, March 23-27, 1981. Copyright © American Institute of Aeronautics and Astronautics, Inc., 1982. All rights reserved.
 *Program Manager, Liquid-Metal MHD, Engineering Division (presently Head, Department of Engineering, Purdue University Calumet, Hammond, Ind.).
 †Deputy Manager, First Wall/Blanket/Shield Program, Fusion Power Program.

for LMMHD over the plant life are $12-29 million for existing technology and $21-39 million for mature technology.

I. Introduction

The two-phase generator, liquid-metal magnetohydrodynamic (LMMHD) energy conversion concept was initially developed at Argonne National Laboratory to meet the anticipated need for an energy conversion system compatible with liquid-metal-cooled heat sources, such as the liquid-metal fast breeder reactor (LMFBR) and the controlled thermonuclear reactor (CTR). In particular, the use of two working fluids, a thermodynamic fluid (gas or vapor) and an electrodynamic fluid (liquid metal) to provide the electrical conductivity, gives LMMHD great versatility in coupling to differing heat-source temperatures and meeting material constraints, e.g.,

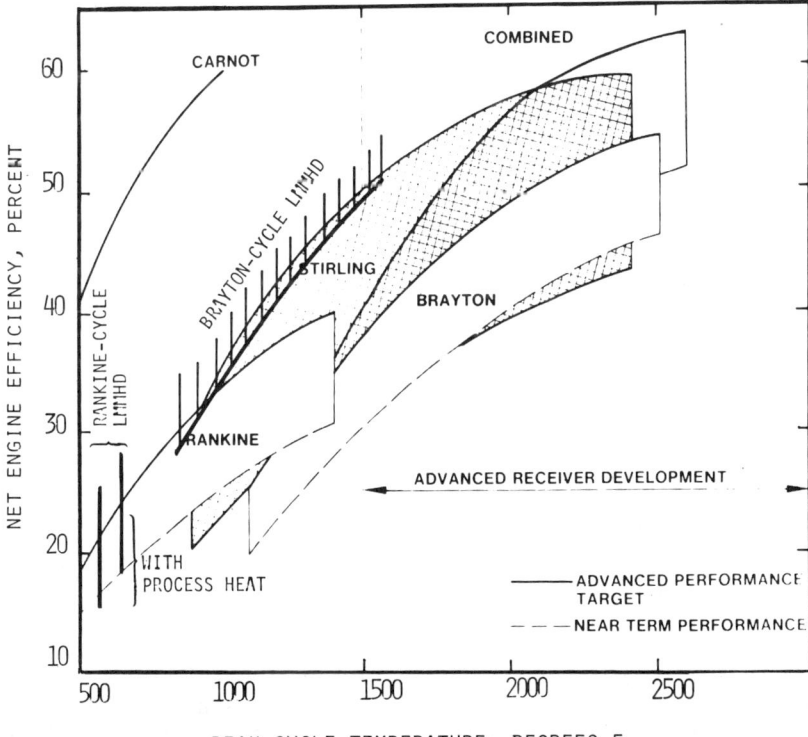

Fig. 1 Performance of LMMHD and other engines modified figure from U.S. DOE (1979a).

heat addition can be to the liquid metal, the gas or vapor, or both. Three LMMHD versions are under investigation:

1) A Rankine cycle version best suited to low heat-source temperatures, 370-850 K (212-1070°F), because there is no compression work. It yields calculated efficiencies several points higher than those of conventional steam cycles for the same top steam temperature.

2) A Brayton cycle version that gives good performance at source temperatures above ∼900 K (1160°F) with coal and at lower temperatures with other energy sources.

3) A new, open-cycle (OC-LMMHD) version suitable for coal or other fossil fuels (Pierson et al. 1982).

The factors that make LMMHD appear particularly attractive for solar applications are: 1) the almost constant temperature expansion in the LMMHD system can potentially provide either a higher efficiency than alternative conversion systems for the same source and sink temperatures, or both electricity and high-temperature process heat (cogeneration); 2) the MHD interaction is a volume effect, so that efficiency and cost are almost independent of size; and 3) the use of liquid metals in the collector and/or a direct-contact boiler or gas heater means that temperatures are higher in the conversion system (hence, conversion efficiency is higher) and/or the solar collector is more efficient.

Solar-powered LMMHD systems have been analyzed previously (Pierson et al. 1979, Pierson 1980, Pierson et al. 1980a), and the resulting efficiencies are very good relative to other conversion systems operating at the same temperatures, as summarized in Fig. 1. The results are given in Sec. III for one LMMHD concept - a high-temperature Brayton cycle system coupled to a liquid-metal-cooled solar power tower, as such towers are currently part of the U.S. solar program (Hildebrandt and Dasgupta 1980, Pomeroy et al. 1980). However, the application of solar LMMHD systems will be decided on life-cycle cost, not efficiency. Thus, the economic analysis of Sec. IV was completed.

Solar LMMHD cogeneration (Pierson 1980, Pierson et al. 1980a) has potential applications in the metals processing and electrochemical industries, such as aluminum production, electroplating, etc. The petroleum refining, chemical, and pulp and paper industries use large quantities of steam at

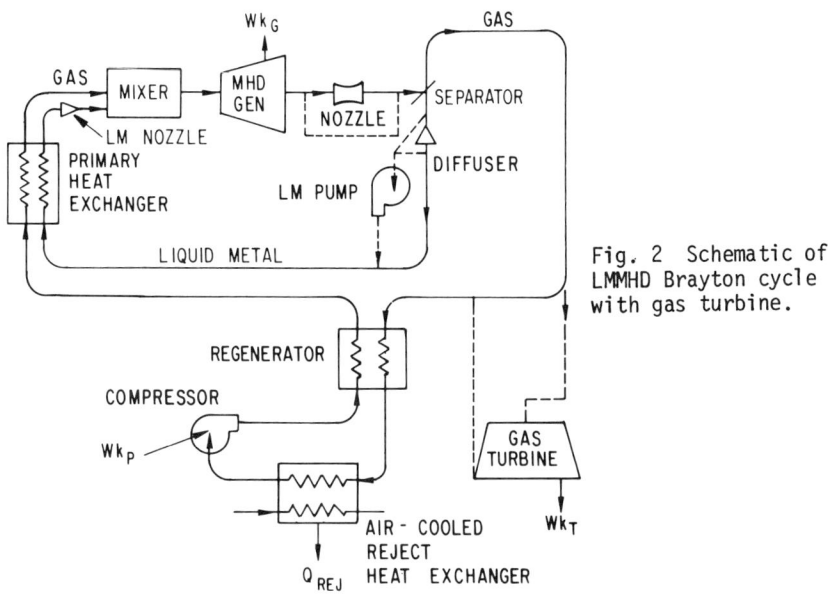

Fig. 2 Schematic of LMMHD Brayton cycle with gas turbine.

temperatures above ∿500 K (∿450 F), according to a 1977 study by Thermo Electron Corporation (Fam and Nydick 1977); these would fit well with the LMMHD system. According to Fam and Nydick (1977), "the industrial mix of energies is generally in the range of 0.8 to 0.08 units of electrical or mechanical energy per unit of heat (i.e., approximately 230-23 kWh/10^6 Btu, respectively)." LMMHD can cover much of that range, but because of the large number of potential applications no specific application was adopted for this study.

II. Description of LMMHD Cycle

A schematic diagram of the Brayton cycle LMMHD concept with a gas turbine is shown in Fig. 2. An inert gas (e.g., helium) is the thermodynamic working fluid and a liquid metal (e.g., sodium or lithium) is the electrodynamic fluid in the MHD generator. In operation, the gas and liquid are combined in the mixer and the resulting two-phase mixture enters the MHD generator. The MHD generator acts as a turbine and electric generator in one unit: the gas expands, drives the liquid across the magnetic field, and thus generates electrical power. Because the liquid has a high heat content, the expansion occurs at almost constant temperature and considerable available energy remains in the gas exiting the MHD generator. (The liquid acts as an "infinite-reheat" source for the gas, thermal energy is continuously transferred from the liquid to the gas, and most of the enthalpy change in the generator comes from the liquid.) It is this almost constant temperature expansion that accounts for the potentially higher thermodynamic efficiency of the two-phase-generator LMMHD concepts. From the MHD generator, the two-phase mixture enters a nozzle, where additional gas/liquid energy is used (as in the generator) to accelerate the liquid; the resulting high-speed flow is separated in a separator (possibly rotating to minimize losses), and the liquid pressure required to return the liquid through the primary heat exchanger to the mixer is obtained in the diffuser. The nozzle/diffuser system can be replaced by a liquid-metal pump.

Because the gas leaving the separator still has considerable thermal energy, it must be used effectively in order to obtain the highest efficiency for the system. It can be transferred from the hot gas to the colder gas in a regenerator, extracted with a gas turbine, extracted with a steam boiler (which would replace both the gas turbine and regenerator of Fig. 2), or used for process heat (cogeneration).

The Brayton LMMHD cycle as modified for a solar power tower is shown in Fig. 3 (without a gas turbine). Here the heat addition is solely to the liquid metal in the receiver (collector), and the gas (helium) is heated by the liquid metal in the mixer. This eliminates the need for a separate gas heater, at the insignificant penalty of having to heat the liquid 2-8 K above the desired mixer exit temperature. More significantly, a liquid-metal-cooled receiver may be smaller, less expensive, and more efficient than one that is gas cooled. The box labeled "Heat Exchanger" in Fig. 3 contains a regenerator and a reject heat exchanger for electrical generation only, or a process heat application for cogeneration.

III. Performance

Central Power Generation

Efficiencies of the pure LMMHD (no gas or steam turbine) and the LMMHD/gas turbine Brayton cycles are shown in Figs. 4 and 5, in that order, for mixer exit temperatures of 700-1255 K (800-1800°F). The efficiencies are calculated by means of a computer code that includes all major losses and the best available data for component performances; the va-

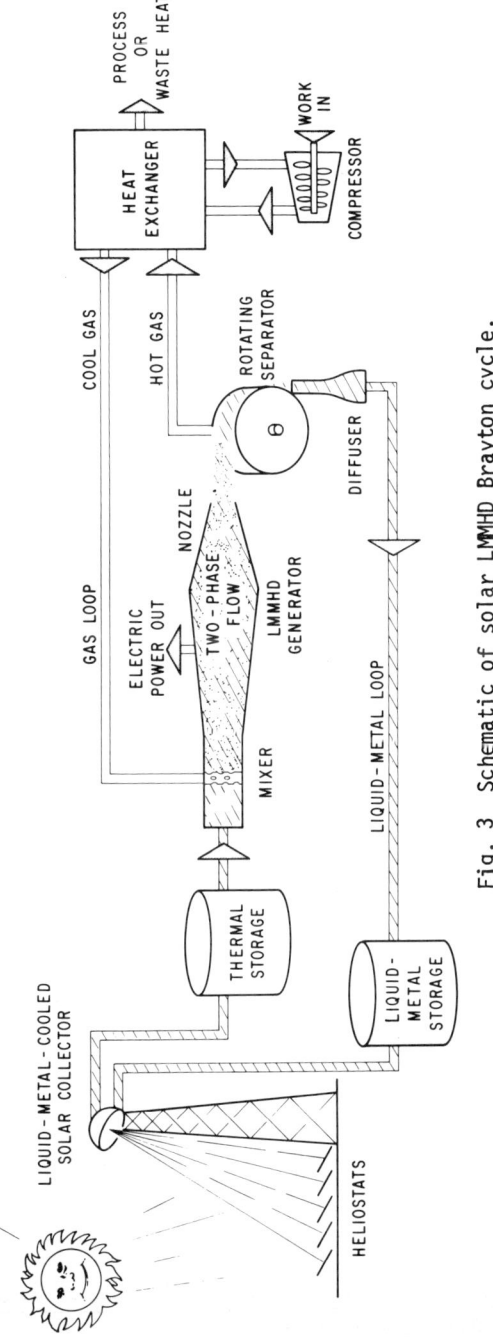

Fig. 3 Schematic of solar LMMHD Brayton cycle.

lues used (see Table 1) are the same as for phase 1 of the Energy Conversion Alternatives Study (Brunsvold and Pierson 1976), except for increasing the regenerator effectiveness from 0.85 to 0.9 and decreasing the number of compression stage intercoolers from four to two. The code and models have been documented by numerous studies. The efficiencies do not include solar receiver efficiency, magnet power (negligible for a superconducting and zero for a permanent magnet), and auxiliary power. The mixer exit pressure is 5.07 MPa (735 psia); changing this pressure by ±2.53 MPa changes the efficiency by less than ±0.01. The cycle used is as shown in Fig. 3, with the addition in some cases of a gas turbine in the hot gas stream after the separator. The efficiencies of both LMMHD systems are very attractive [(i.e., approaching 0.5 at ∼1089 K (1500°F)].

With proper working fluids, conversion efficiency will increase as temperature is increased. The efficiency increase with temperature that is shown in Figs. 4 and 5 becomes slight, and it may even decrease (e.g., see the curve for 1311 K in Fig. 4) because of the carryover of liquid-met-

Table 1 Component performance parameters, LMMHD Brayton cycle

Pressure drop[a]

Δp_{mixer} = 34.5 kPa (5 psi)
Δp_{primhx} = 68.9 kPa (10 psi)
Δp_{sep} = 0
$\Delta p_{reg\ hot}$ = 0.015 p_{turb}
Δp_{rejhx} = 0.025 $p_{reg\ hot}$
$\Delta p_{reg\ cold}$ = 0.015 p_{rejhx}

Efficiency

Compressor	0.88	Diffuser	0.9
Turbine	0.9		
Nozzle	0.9	LMMHD generator	0.8

Heat rejection conditions

Ambient temperature   297.2 K (75°F)
Pinch point   11.1 K (20°F)

Other

Compressor intercoolers   2
Regenerator effectiveness 0.9
Separator loss, 10% of inlet kinetic energy

[a] Indicated pressures are at exit of referenced component.

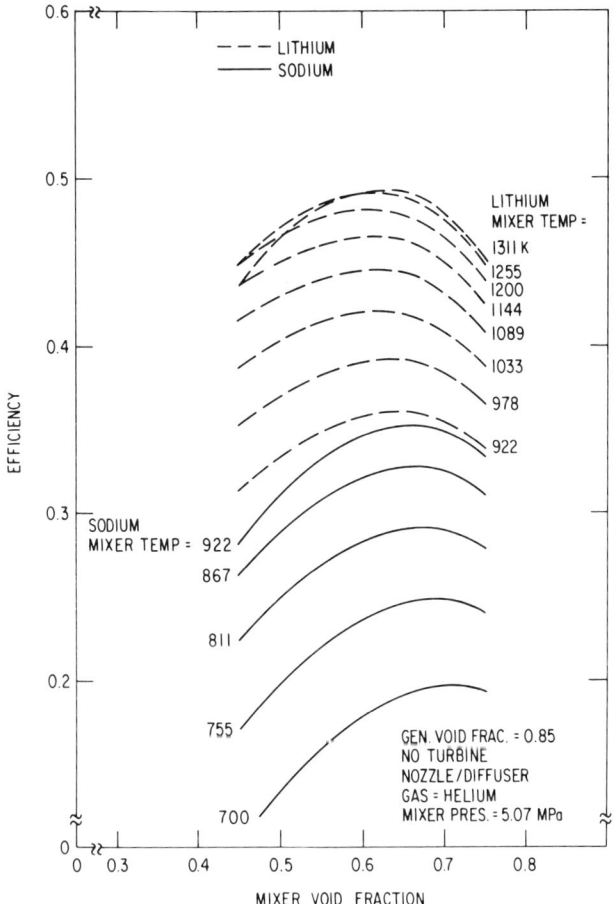

Fig. 4 Efficiency vs mixer void fraction for pure LMMHD gas system.

al vapor with the gas from the separator. Vapor carryover is the reason for changing from sodium to lithium above ~867 K. Operation at temperatures higher than 1200 K, resulting in higher efficiencies, is feasible with other liquid metals, for example, aluminum or copper.

The efficiencies shown in Figs. 4 and 5 can be increased by improving the performance of individual components and reconfiguring the components, as summarized in Table 2 for 1089 K (1500°F), an attractive temperature for point-focus solar receivers that is too hot for steam plants and too cold for conventional Brayton cycles. The base case is the

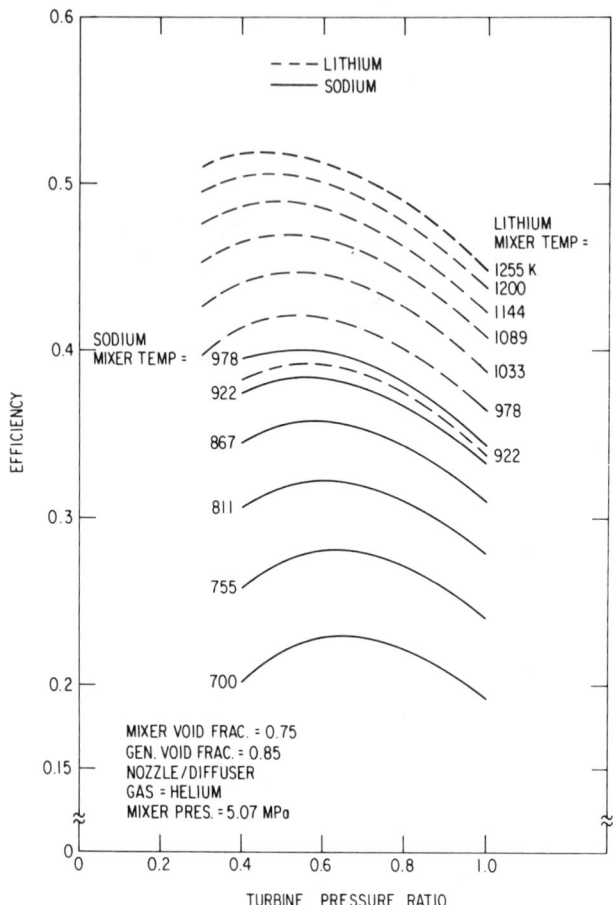

Fig. 5 Efficiency vs turbine pressure ratio for LMMHD gas turbine system.

pure LMMHD gas system with the component performance parameters of Table 1 and an efficiency of 0.45 (Fig. 4). The efficiency can be increased to 0.47 either by adding a gas turbine (Fig. 5) or by increasing the regenerator effectiveness to 0.95. The latter is more appealing, because a large rotating component is eliminated and the resulting powerplant is simpler. An efficiency of 0.47 can also be obtained with four compressor intercoolers. The effects of increasing LMMHD generator efficiency, void fraction at the generator exit, and separator performance are also shown. The changes in efficiency are additive (within ±0.01), so that efficiencies well in excess of 0.5 are attainable in

Table 2 Effect of component parameters on system efficiency

Base Case: Component parameters in Table 1
Void fraction at generator exit = 0.85
No turbine
Nozzler/diffuser
Efficiency = 0.45

Parameter change	Efficiency change
Add turbine	+ 0.02
Regenerator effectiveness = 0.95	+ 0.02
Regenerator effectiveness = 0.975	+ 0.04
Generator efficiency = 0.85	+ 0.03
Void fraction at generator exit = 0.9, 0.95, 0.98	+ 0.01
Number of compressor intercoolers = 4	+ 0.02
Kinetic energy loss in separator = 0.05	+ 0.01

principle, e.g., an efficiency of 0.52 for a regenerator effectiveness of 0.95, a generator efficiency of 0.85, and four compressor intercoolers. Increasing the void fraction at the generator exit will yield a reduced liquid-metal flow rate and inventory and thus a less expensive system, but has relatively little impact on the efficiency for these conditions. (If the nozzle/diffuser system is replaced by a liquid-metal pump the void fraction at the exit has more impact.) Table 2 does not include the options of using multiple LMMHD stages, which has been shown to improve cycle performance (Brunsvold and Pierson 1976), or a steam bottoming plant.

Cogeneration

For cogeneration applications, the LMMHD conversion system can be optimized so as to maximize the temperature available for process applications, i.e., there could be no regenerator and no gas turbine (Fig. 3). This results in a simpler system, with higher total energy utilization but lower electrical efficiency (ratio of useful electrical output power to thermal input power to the LMMHD system), because the thermal energy in the gas exiting the separator is neither converted to electrical energy in a turbine nor returned to the cold gas in a regenerator. The liquid metal is heated by either the separate receiver fluid in a heat exchanger or, better, directly in the solar central receiver; the gas is heated by the liquid metal in the LMMHD mixer and generator.

Electrical efficiencies and temperature drops (maximum liquid-metal temperature, i.e., at the exit from the solar receiver or heat exchanger, minus temperature of the gas leaving the separator) are listed in Table 3 for several conditions. Electrical efficiencies of 0.22-0.25 are attainable with no compression stage intercoolers, and all of the heat rejected is available for process applications. The temperature drops ($\sim$10 K) are probably less than can be attained with a gas in a pure solar process heat system because of the superior heat-transfer properties of liquid metals; in addition the LMMHD version provides valuable electrical power.

Higher electrical efficiencies are attainable by either rejecting some of the heat at temperatures too low for process use (e.g., from compressor interstage coolers) or decreasing the temperature of the gas made available to the process application. (The highest electrical efficiency occurs when none of the thermal energy is used for process heat.) An example of the former is the use of two compressor intercoolers; this results in electrical efficiencies of 0.30-0.33, but 40% of the heat rejected from the cycle is removed in the interstage coolers. Examples of the latter are the addition of a gas turbine or a regenerator of low effectiveness.

IV. Economic Analysis

Scope

This study is aimed at applications which require $\sim$10 MW of electric power, or a lesser amount of electric power but large quantities of high-temperature [1100-1250 K (1500-1800°F)] process heat, using solar energy. This could be a remote application which operates only when there is insolation, or an application which uses solar energy when possible and other energy sources the rest of the time. In neither case is thermal or electric energy storage provided. The LMMHD system costs are compared with a conventional solar option for the former case, and with solar for the electric power and fossil fuel for the thermal power in the latter case, i.e., solar energy displaces or saves fossil fuels. For the latter case, a hybrid solar/fossil fuel LMMHD system would be a natural combination because of the high energy utilization efficiency and the high plant capital utilization.

The LMMHD system is sized in all cases to fit a solar power tower providing 25 MW thermal input to the liquid-metal coolant. The electric and thermal powers produced, and the costs, are calculated for both existing technology (i.e., an early plant to be built within the next few years) and mature technology which incorporates the cost savings resulting from experience and quantity production and allows operation at higher temperatures. No provision is made for technology improvements except for an increase in homopolar motor efficiency from 0.9 to 0.95. Design and engineering costs, except for installation of the LMMHD system and its solar equipment, are not included.

The comparison case selected is a photovoltaic system to provide the electric power, with costs taken from a 1979 U.S. Department of Energy program plan document (U.S. DOE 1979b). The thermal power is provided by either a solar power tower similar to that used for the LMMHD system or burning a fossil fuel. The photovoltaic price goals for an intermediate load center (intermediate-size commercial, institutional, community, and industrial on-site systems) are used (1986 application, cost in 1980 dollars). The existing technology LMMHD is compared with the moderate-volume photovoltaic price goal, and the mature technology LMMHD is compared with the high-volume photovoltaic price goal.

Electric and Thermal Powers

The electric and thermal powers are calculated as described in the previous section (except for increasing the regenerator effectiveness to 0.95 and varying the number of compression stage intercoolers). The cycle used is as shown in Fig. 3, with the addition for only electric power of a gas turbine. The working fluids are helium and lithium.

Table 3 Performance of 1089 K cogeneration system[a]

No. comp. intercoolers	Gen. eff.	Elec. eff.	Temp drop (max liq. metal minus max heat rej.), K	Fraction heat rej. to process
0	0.80	0.22	10.0	1.0
0	0.85	0.25	9.6	1.0
2	0.80	0.30	8.5	0.6
2	0.85	0.33	8.8	0.6

[a]Mixer pressure = 5.07 MPa, nozzle/diffuser system, helium and lithium, no regenerator, no turbine, generator exit void fraction = 0.85. Component performance parameters not stated are in Table 1.

Table 4 LMMHD solar plant performance summary, existing technology, 1089 K (1500°F)

Case	No co-generation	Partial co-generation	Full co-generation
LMMHD conditions			
No. compressor intercoolers	4	1	0
Turbine pressure ratio	0.55	-	-
Powers			
Heat input, MW	25.0	25.0	25.0
Gross LMMHD power, MW	12.7	15.7	11.5
Turbine power, MW	10.8	-	-
Compressor power, MW	10.7	8.8	6.5
Performance			
Net electrical output power, MW	10.9	4.8	3.5
Net useful heat output power, MW	0	13.6	20.4
Net plant electrical efficiency	0.44	0.19	0.14
Net plant total efficiency	0.44	0.73	0.96

Calculated powers are shown in Table 4 for existing technology, 1089 K (1500°F), and in Table 5 for mature technology, 1255 K (1800°F). In both tables, the first case is pure electric power (no cogeneration), i.e., a regenerator and four compressor intercoolers to maximize the electrical efficiency. The third case in Tables 4 and 5 is full cogeneration, i.e., all of the rejected heat is available at a temperature ~17 K (30°F) lower than the mixer temperature. The second case in each table shows the results of making only part of the rejected heat available at the maximum temperature. For the second case one compressor intercooler is used to reduce the compressor work, and the net electrical output powers and efficiencies are thus higher than the full cogeneration cases. Also, the net plant total efficiencies (energy utilization) are lower because of the heat rejected by the intercooler that is not available to the process. Other partial cogeneration cases are also possible, e.g., driving the compressor with a turbine which results in a lower process heat availability temperature. Also, the electrical efficiency can be increased with partial regeneration, e.g., if the heat rejected by the process can be returned to the helium after the compressor.

The compressor is driven by a dedicated gas turbine for the no-cogeneration cases and by a homopolar motor for the

Table 5 LMMHD plant performance summary, mature technology, 1255 K (1800°F)

Case	No co-generation	Partial co-generation	Full co-generation
LMMHD conditions			
No. compressor intercoolers	4	1	0
Turbine pressure ratio	0.71	-	-
Powers			
Heat input, MW	25.0	25.0	25.0
Gross LMMHD power, MW	13.4	13.1	13.0
Turbine power, MW	8.7	-	-
Compressor power, MW	8.6	6.0	6.8
Performance			
Net electrical output power, MW	11.5	5.9	5.0
Net useful heat output power, MW	0	14.3	18.3
Net plant electrical efficiency	0.46	0.24	0.20
Net plant total efficiency	0.46	0.81	0.93

other cases. The difference between the gross LMMHD power and the net electric output power is the sum of the compressor power for the cogeneration cases (including motor losses), the losses in converting the dc output from the LMMHD generators to ac (if required for the desired application), and the power required for the magnet for the LMMHD generators. The conversion from dc to ac is accomplished by means of a homopolar motor driving a synchronous generator; a solid-state inverter could also be used, it would have slightly higher efficiency but a significantly higher cost (Pierson et al. 1980b). Efficiencies are 0.90 for the homopolar motor (Greene 1980) (0.95 assumed for the mature technology) and 0.98 for the synchronous generator (Greene 1980). The magnet is a conventional design with an iron core and water-cooled copper windings; the magnet power is estimated to be 2.6% of the gross LMMHD power based on the previous OC-LMMHD retrofit study (Pierson et al. 1980b). (Superconducting magnets appear better suited to systems of this size operating continuously or larger systems operating continuously or intermittently; permanent magnets, desirable for smaller systems, are probably too expensive for this size.)

The net electrical and total efficiencies, the latter defined as the net electrical plus heat output powers divided by the heat input, are both very attractive for these systems, with electrical efficiencies (including all losses and auxiliary powers) to 0.46 and total efficiencies of 0.44-0.96. The cases were selected to minimize cost rather than maximize electrical or total efficiency, i.e., minimum compressor power is very important because this is an expensive item. Also, optimization was not possible within the constraints of this study. Nevertheless, the efficiencies are high and the costs (see below) are low.

LMMHD Capital Cost

The costs of the major components are based on the recently completed evaluation of a coal-fired open-cycle LMMHD topping cycle for the retrofit of an existing steam power plant (Pierson et al. 1980b, Pierson et al. 1982). That study was done in sufficient detail so that a cost estimate could be made from a conceptual design and plant layout, including design concepts for all major components.

The solar systems employ four liquid-metal loops in a single iron-core magnet with water-cooled copper coils. (The retrofit employed eight loops in a magnet with superconducting coils; however, costs for a conventional magnet with water-cooled copper coils were also estimated.) Gross LMMHD powers of 11-16 MWe peak (vs 74 MWe for the retrofit) are produced. The solar LMMHD costs are scaled from those of the retrofit, based on the lower power level, lower temperature, and smaller physical size. No "first-of-a-kind" engineering or design costs are included. The mature technology at higher temperatures includes the cost benefits of experience and quantity production.

The costs of the major plant components are listed in Tables 6-9 in 1980 dollars for four cases - existing technology electric power generation (no cogeneration), existing technology with full cogeneration, mature technology electric power generation (no cogeneration), and mature technology with full cogeneration - all based on peak power. The installed cost of each item uses a multiplier (the installation factor in the tables) which varies with the complexity of the installation and experience. Instrumentation and control costs are estimated at 10% of the LMMHD plant subtotal, as for the retrofit. In addition, 10% of the plant subtotal is added to the overall cost to account for mis-

Table 6 LMMHD plant capital costs with existing technology and no cogeneration, $ thousand

Equipment	Unit cost	Units	Sub-total	Installation factor	Total
Mixer	16	4	64	1.3	83
LMMHD generator	30	4	120	1.3	156
Magnet	225	1	225	1.3	292
Nozzle	11	4	44	1.3	57
Separator/diffuser	37	4	148	1.3	192
Power conversion equipment	1090	1 set	1090	1.3	1417
Bus bars	190	1 set	190	1.3	247
Compressor with drive	1650	1 set	1650	1.3	2145
Piping	6	1 set	6	1.3	8
Reject heat exchanger	195	1	195	1.3	254
Regenerator	1020	1	102	1.3	1326
Subtotal					6177
Instrumentation and controls (10% of subtotal)					618
Miscellaneous items (10% of subtotal)					618
LMMHD total					7413
Collectors	270/m^2	41,667 m^2	11,250	1.0	11,250
Tower	810	1	810	1.3	1053
Receiver	1800	1	1800	1.3	7340
Power tower total					14,643
Total cost					22,056

cellaneous equipment not costed in detail (5% was used in the retrofit study).

The costs for the solar power collection systems are also listed in Tables 6-9. The heliostat existing technology cost used is $270/m^2, based on the Barstow Pilot Plant near-term projection (Hildebrand and Dasgupta 1980). A heliostat cost of $72/m^2 is used for mature technology, i.e., mass production (Hildebrand and Dasgupta 1980). The power tower costs were estimated from a study of 17 steel and 16 concrete towers varying in height, receiver weight, and local atmospheric and ground conditions (Sandia/Stearns Roger 1979). The solar receiver/lithium heater cost was based on a sodium-cooled absorber concept (Pomeroy et al. 1980). The cost estimate assumed similar construction and was scaled to 25 MW thermal. An optical efficiency of 60% was assumed.

Table 7 LMMHD plant capital costs with existing technology and full cogeneration, $ thousand

Equipment	Unit cost	Units	Sub-total	Installation factor	Total
Mixer	16	4	64	1.3	83
LMMHD generator	30	4	120	1.3	156
Magnet	225	1	225	1.3	292
Nozzle	11	4	44	1.3	57
Separator/diffuser	37	4	148	1.3	192
Power conversion equipment	366	1 set	366	1.3	476
Bus bars	70	1 set	70	1.3	91
Compressor with drive	910	1 set	910	1.3	1183
Piping	5	1 set	5	1.3	7
Process heat exchanger	810	1	810	1.3	1053
Subtotal					3590
Instrumentation and controls (10% of subtotal)					359
Miscellaneous items (10% of subtotal)					359
LMMHD total					4308
Collectors	270/m^2	41,667 m^2	11,250	1.0	11,250
Tower	810	1	810	1.3	1053
Receiver	1800	1	1800	1.3	2340
Power tower total					14,643
Total cost					18,951

Comparison Capital and Fuel Costs

LMMHD systems generating electrical power only, i.e., no cogeneration, are compared with standard photovoltaic systems supplying the same net electrical power. According to a 1979 DOE program plan (U.S. DOE 1979b), for intermediate load centers (intermediate-size commercial, institutional, community, and industrial on-site systems), the 1986 commercial readiness system price goals (in 1980 dollars) are $2.60 per peak watt at moderate volumes and $1.60 per peak watt at high volumes. For this comparison, the former is used as the existing technology cost and the latter as the mature technology cost.

Two comparison cases with cogeneration are used. For the first, photovoltaics supply the same net electrical powers, and solar power towers supply the same net useful heat outputs. The solar power tower costs are taken from the appropriate values in Tables 7 and 9 for existing and mature technologies and scaled proportional to the heat rating.

Table 8 LMMHD plant capital costs with mature technology and no cogeneration, $ thousand

Equipment	Unit cost	Units	Sub-total	Installation factor	Total
Mixer	14	4	56	1.1	62
LMMHD generator	22	4	88	1.1	97
Magnet	170	1	170	1.1	187
Nozzle	9	4	36	1.1	40
Separator/diffuser	30	4	120	1.1	132
Power conversion equipment	760	1 set	760	1.1	836
Bus bars	130	1 set	130	1.1	143
Compressor with drive	1100	1 set	1100	1.1	1210
Piping	5	1 set	5	1.1	6
Reject heat exchanger	145	1	145	1.1	169
Regenerator	1400	1	1400	1.1	1540
Subtotal					4422
Instrumentation and controls (10% of subtotal)					442
Miscellaneous items (10% of subtotal)					442
LMMHD total					5306
Collectors	72/m^2	41,667 m^2	3000	1.0	3000
Tower	700	1	700	1.1	770
Receiver	1350	1	1350	1.1	1485
Power tower total					5255
Total cost					10,561

(The scaling is only approximate. For the non-LMMHD cases, a more expensive, air-cooled receiver would probably be required but this difference is neglected.) For the second comparison, photovoltaics again supply the electrical power but the heat is supplied by a fossil fuel. No solar-related capital cost is assigned to the fossil fuel combustion system because it must be present for nonsolar operation. The annual fuel cost, used below in a life-cycle cost analysis, assumes oil, which is suitable for systems of this thermal capability, at a current cost of $0.02/kW thermal.

Results

Three cost comparisons are made:

1) LMMHD vs photovoltaics for electric power only (no cogeneration).

2) LMMHD cogeneration vs a combined system using photovoltaics for the electric power and a solar power tower for the heat.

Table 9 LMMHD plant capital costs with mature technology and full cogeneration, $ thousand

Equipment	Unit cost	Units	Sub-total	Installation factor	Total
Mixer	14	4	56	1.1	62
LMMHD generator	22	4	88	1.1	97
Magnet	170	1	170	1.1	187
Nozzle	9	4	36	1.1	40
Separator/diffuser	30	4	120	1.1	132
Power conversion equipment	478	1 set	478	1.1	526
Bus bars	130	1 set	130	1.1	143
Compressor with drive	680	1 set	680	1.1	748
Piping	5	1 set	5	1.1	6
Process heat exchanger	610	1	610	1.1	671
Subtotal					2612
Instrumentation and controls (10% of subtotal)					261
Miscellaneous items (10% of subtotal)					261
LMMHD total					3134
Collectors	$72/m^2$	$41,667\ m^2$	3000	1.0	3000
Tower	700	1	700	1.1	770
Receiver	1350	1	1350	1.1	1485
Power tower total					5255
Total cost					8389

3) LMMHD cogeneration vs a combined system using photovoltaics for the electric power and a fossil fuel (oil) for the heat.

Operating and maintenance costs are assumed to be the same for all systems. There are no fuel costs for the first and second comparisons, so only capital costs are used. The third comparison included fuel costs and is made on a life-cycle basis.

Capital costs for the first and second comparisons are listed in Table 10 and plotted in Fig. 6. Clearly, LMMHD has substantial economic benefits for both the existing and the anticipated mature technologies. The cost savings are greatest without cogeneration and decrease with the degree of cogeneration because, in the limit of no electric power, the power tower cost is the same for both systems.

For the third comparison, the capital costs for the LMMHD cogeneration system are shown in Table 9. The capital

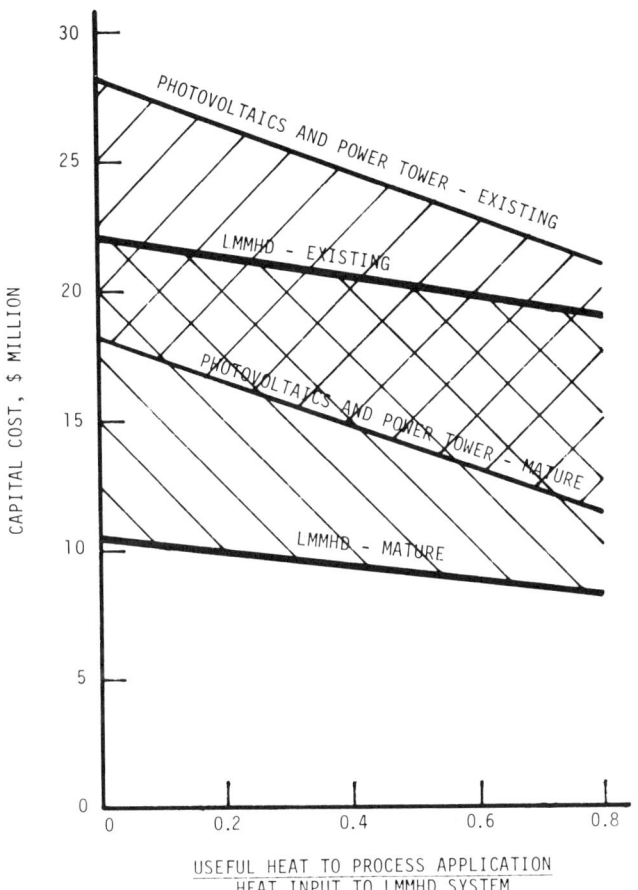

Fig. 6 Capital cost vs heat fraction.

costs of the photovoltaics alone are $9.0 million for existing technology and $8.1 million for mature technology. No capital cost is included for the fossil fuel system because it must exist even without a solar system. The annual fuel cost saved by the LMMHD system process heat is the product of the peak net heat output power from Tables 4 or 5, the number of hours per year (8760), the estimated capacity factor of ∼0.2 for the southwestern U.S., and the 1980 U.S. oil price of ∼$0.02/kW·h. The resulting costs are $0.72 and $0.64 million in 1980 dollars, respectively, for existing and mature technologies.

The plant life-cycle costs are the best basis for comparing the value of the fuel saved by the LMMHD system

Table 10 Capital cost summary, no fossile fuel, $ thousand

	No cogeneration		Full cogeneration	
	LMMHD	Photovoltaic	LMMHD	Photovoltaic plus tower
Existing technology	22.1	28.3	19.0	20.9
Mature technology	10.6	18.4	8.4	11.9

against the extra capital cost of the LMMHD system. The cost calculation uses the above capital and fuel costs, a 30 year assumed plant lifetime, a discount rate of 11%, and oil cost inflation rates of 11 and 15%. [The life-cycle cost analysis for the recently completed open-cycle LMMHD retrofit study (Pierson et al. 1980, Pierson et al. 1982), performed by an architect/engineer, used a 30 year plant life, a discount rate of 11%, and oil cost inflation rates of 15 and 18%.] The total cost savings with the LMMHD option over the 30 year lifetime for existing technology are $11.7 and $29 million, respectively, for the 11 and 15% oil cost inflation rates. For mature technology, the savings are $21 and $39 million, respectively. Clearly, there are substantial economic benefits associated with the use of solar LMMHD systems.

V. Conclusions

The high-temperature LMMHD Brayton cycle efficiencies are potentially higher than other energy conversion systems for the same solar collector and heat sink temperatures. In addition, LMMHD can effectively and safely use liquid-metal-cooled receivers and is uniquely suited to solar cogeneration applications which require high-temperature process heat. The LMMHD systems demonstrate major economic benefits compared to photovoltaic systems for electric power generation alone, and photovoltaics combined with a solar power tower or fossil fuel to supply the heat for cogeneration.

Acknowledgment

The authors thank Roger Cole for his help with the solar aspects of this study.

The work reported here was supported by the U.S. Department of Energy.

References

Brunsvold A.R. and Pierson E.S. (1976) "Liquid-metal MHD coupled to coal-fired fluidized-bed combustors," Proc. 15th Symp. on Eng. Aspects of MHD, Philadelphia, pp. III.8.1-III.8.8.

Greene D. (1980) Telephone discussion with E.S. Pierson, Oct. 3.

Fam S.S. and Nydick S.E. (1977) "Assessment of technologies for research, development, and demonstration of industrial cogeneration and waste heat recovery in the near term," Vol. 1, Thermo Electron Corp., Waltham, Mass., Rept. TE4214-75-77.

Hildebrandt A.F. and Dasgupta S. (1980) "Survey of power tower technology," J. Sol. Energy, Vol. 102, pp. 91-104.

Pierson E.S. (1980) "New liquid-metal MHD concepts for solar and coal," Proc. Am. Power Conf., Vol. 42, pp. 379-385.

Pierson E.S., Branover H., Fabris G., and Reed C.B. (1979) "Solar powered liquid-metal MHD power systems," ASME Paper 79-WA/Sol-22 (also Mech. Eng., Vol. 102, Oct., pp. 32-37).

Pierson E.S., Cohen D., and Grammel S.J. (1980a) "Liquid-metal MHD for solar and coal," Proc. 7th Int. Conf. on MHD Electr. Power Generation, Boston, Mass., pp. 150-157.

Pierson E.S. et al. (1980b) "Coal-fired open-cycle liquid-metal magnetohydrodynamic topping cycle for retrofit of steam power plants," Argonne National Laboratory, Argonne, Ill., Rept. ANL/MHD-80-18.

Pierson E.S., Herman H., and Petrick M. (1982) "Conceptual design of a coal-fired retrofit liquid-metal MHD power system," Liquid-Metal Flows and Magnetohydrodynamics, Progress in Astronautics and Aeronautics, Vol. 84, AIAA, New York, pp.127-137.

Pomeroy B.D., Roberts J.M., and Narayan T.V. (1980) "A high flux sodium-cooled absorber concept for central receiver power plants," Proc. 1980 Annual Meeting of Am. Sec. Int. Solar Energy Soc., Phoenix, Ariz., Vol. 3.1.

Sandia/Stearns Roger Eng. Co. (1979) "Final report for tower cost data for solar central receiver studies," SAND 78-8185.

U.S. DOE (1979a) "Third semi-annual advanced technology meeting - A review of advanced solar thermal power systems," DOE 5102-127.

U.S. DOE (1979b) "Photovoltaics program multi-year plan," June 6, draft.

An Operating Model of Two-Phase Heavy-Metal MHD Power Systems

H. Branover,* A. El-Boher,† and A. Yakhot‡
Ben-Gurion University of the Negev, Beer-Sheva, Israel
and
S. Claesson§
University of Uppsala, Uppsala, Sweden

Abstract

Complete small-scale power conversion two-phase liquid-metal MHD systems have been built and tested in Beer-Sheva (Israel) and Uppsala (Sweden). Two such systems have been assembled and tested by the MHD Laboratory of the Ben-Gurion University of the Negev and one was developed jointly with the Institute of Physical Chemistry of Uppsala University and tested in Uppsala. It should be emphasized that although much work was done in studying the LMMHD system analytically and in experimental studies of components, a complete system performing the full cycle has never been built before. At the present stage both systems use mercury, while in the next stage of experiments, the Uppsala model will operate with gallium and gallium-indium-tin alloy. The volatile liquids are Refrigerant-113 in Beer-Sheva and n-pentane in Uppsala. Both systems were at mixer temperatures between 65 and 80°C. The condensation temperatures in Beer-Sheva are about 35°C, while the Uppsala system performs on open cycle. The flow characteristics and the electrical

---
Paper presented at Third Beer-Sheva International Seminar on Magnetohydrodynamic Flows and Turbulence, Ben-Gurion University of the Negev, Beer-Sheva, Israel, March 23-27, 1981. Copyright © American Institute of Aeronautics and Astronautics, Inc., 1982. All rights reserved.
 *Professor, Department of Mechanical Engineering.
 †Engineer, Department of Mechanical Engineering.
 ‡Senior Lecturer, Department of Mechanical Engineering.
 §Professor, Institute of Physical Chemistry.

current output have been measured and the efficiencies calculated.

I. Introduction

The concept of a liquid-metal MHD generator with two-phase flow is already well known (Amend and Petrick 1971). In an MHD generator of this type, a highly electroconductive two-phase mixture composed of a liquid metal and a gas (or vapor) moves across a magnetic field, thus generating electrical power. The two-phase flow is propelled by the expanding gas bubbles and the gas goes through the thermodynamic cycle.

A large number of studies of the concept described have been carried out during the last decade at Argonne National Laboratory in the United States (Fabris and Hantman 1981). These studies concentrated mainly on the performance of the MHD generator, the characteristics of the two-phase flow in the channel of the generator, etc. However, a complete system, i.e., performing the full cycle, was not built.

Since 1976, the use of this concept for direct conversion into electricity of heat from very low-temperature sources has been studied in the MHD Laboratory of Ben-Gurion University (Branover 1977, Branover et al. 1980a, Branover et al. 1980b). Applications to the utilization of solar energy, waste industrial heat, and geothermal energy at temperatures of 80-300°C have been considered.

The purpose of the present work was to design, construct, and test a small two-phase liquid-metal MHD power system operating at a low temperature. As mentioned above, a complete energy conversion system with a liquid-metal MHD generator had not heretofore been built. Therefore, it was believed that the construction and testing of the system described in the present paper would contribute to solving the problem of the feasibility of liquid-metal MHD power generation in general.

II. Description of the Systems

A general schematic diagram of the investigated systems is presented in Fig. 1. The system operates as follows. The liquid is heated in the heat source to a temperature a few degrees higher than T_H. For waste heat and geothermal energy a heat exchanger is used, while for solar

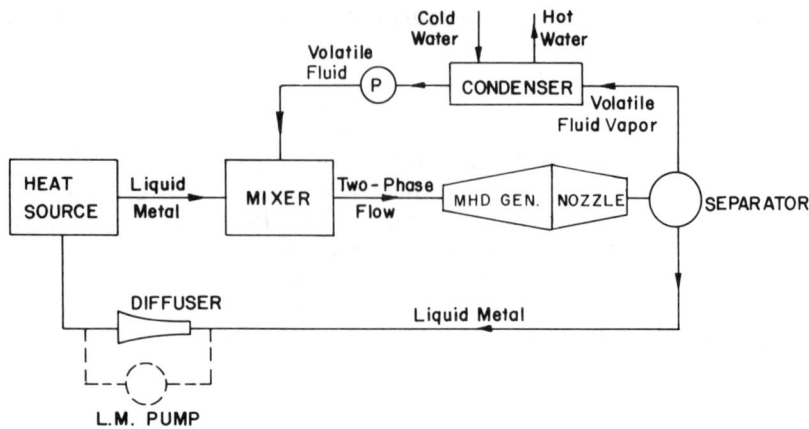

Fig. 1 General schematic diagram of the system.

heat the liquid metal is circulated directly through the solar collector. Special studies performed at Ben-Gurion University indicated that the use of liquid metals in solar collectors improves their performance considerably (Branover et al. 1980a). From the heat source, the hot metal proceeds to the mixer, where a volatile fluid is injected. The droplets of the volatile fluid are heated and boiled, because of direct-contact heat exchange with the hot metal, and the expanding bubbles propel the two-phase mixture into the MHD channel located in the gap of a permanent magnet. After the MHD channel the liquid metal and vapor are separated. The vapor goes to a desuperheater/condenser and after condensation is returned to the mixer by means of a small pump. The heat extracted in the desuperheater can be regenerated or used for industrial and domestic purposes. The liquid metal is returned from the separator back to the mixer through the heat source, either by means of a mechanical pump, an MHD pump, or a diverging nozzle converting the kinetic energy of the flow into pressure. If a heavy liquid metal is used, these devices may become unnecessary, provided the system is arranged vertically and the pressure needed in the mixer can be created by gravity. After completion of the cycle the liquid metal is only a few degrees cooler than in the heat source (since the specific heat of the vapor which extracts heat from the liquid metal is very low compared to the specific heat of the metal), and therefore the necessary temperature rise in the heat source is also only a few degrees.

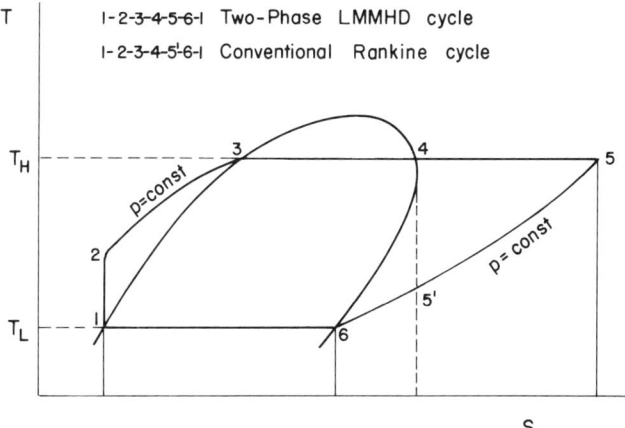

Fig. 2 Modified (LMMHD) cycle and a conventional (turbine) organic Rankine cycle.

As follows from the above description, the liquid metal has two main functions in this system. First, it serves as the electroconductive medium through which the electrical currents pass. Second, the liquid metal serves as a heat source for the expanding gas bubbles. Therefore, unlike a turbine cycle, the gas remains at practically a constant temperature throughout the entire expansion process, which provides a higher conversion efficiency (similar to a cycle with infinite reheat). A modified Rankine cycle of the type described for an organic fluid is presented in Fig. 2, and compared with a conventional turbine Rankine cycle. As seen from this figure, an additional advantage of the Rankine cycle with almost isothermal expansion is that after the performance of the mechanical work, the vapor is still at a high temperature, T_H, so that the heat extracted (process 5-6 in Fig. 2) will still be usable.

There are a limited number of metals whose physical properties make them more or less acceptable for use in the described systems within the given temperature range. Of these, the following should be mentioned: mercury (melting point -39°C), NaK eutectic (22% Na/78% K, melting point -11°C), gallium (melting point +29°C), tin (melting point +232°C), and Wood's metals of various compositions (melting point depending on composition).

Regarding the volatile fluid, there is a rather large choice of different organic liquids with widely varying boiling points. However, the need for compatibility with a liquid metal often narrows the choice drastically. For instance, NaK is incompatible with many organic fluids, including all Freons, and only a few thermodynamic fluids can be used in systems with NaK.

III. The Experimental Facilities

There were several purposes for which the models were built and which influenced their design. These were:

1) To demonstrate the performance of a complete two-phase liquid-metal MHD system.

2) To prove experimentally the basic principles of liquid-metal MHD power generation.

3) To observe the compatibility of the two working fluids over a long period of system operation.

4) To study the two-phase flow pattern, mixing, and separation.

5) To measure the characteristics of the systems.

To aid in the demonstration and visualization, many parts of the systems were made of glass without any thermal insulation, which, of course, greatly limited the maximum acceptable pressure and decreased the efficiency. The absence of any insulation resulted in a rapid drop in the temperature of the liquid metal after it left the heat source. However, since these facilities were by no means designed to be optimal, the obtained level of efficiency was not important.

The main special features of the systems are:

1) The systems are mounted vertically; thus, since a heavy metal (mercury) is used, there is no need for a liquid-metal pump. The metal is returned to the heat source and mixer by gravity.

2) Hydrocyclonic gravitational separators are used, which insure a more perfect separation than purely gravitational separators (though in the case of mercury, separation does not present a serious problem) and also conserve - at least partially - the kinetic energy of the flow.

Mercury/Refrigerant-113 Facility

Mercury was chosen as the liquid metal and Refrigerant-113 as the volatile fluid. The working temperature in the mixer was kept at approximately 80°C, which corresponds to a possible coupling of this system with flat-plate (water or mercury cooled) solar collectors. In these experiments the heat was actually generated by a resistance-type electrical heater. A general schematic diagram of the facility is shown in Fig. 3. A permanent magnet was used for the MHD generator. (At this stage a very weak magnet with a field of only 0.08 T was used; however, the second stage of testing, with a 1.1 T magnet, is now being prepared.)

A general view of the facility is presented in Fig. 4. The cross section of the MHD channel was 2.2 x 0.5 cm^2 and the copper electrodes were 3.5 cm long. The electrical load of the generator was a shunt with a resistance of 2 x 10^{-3}Ω. Since, as will be seen below, the induced electrical current exhibited strong pulsations (±50%) of mean measured value, a transformer could be used to amplify the initially very low voltage so that a miniature bulb could

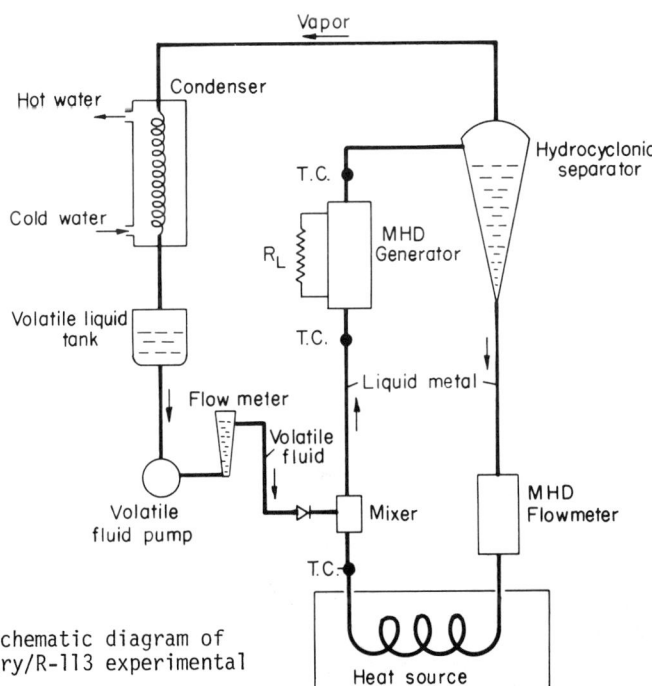

Fig. 3 Schematic diagram of the mercury/R-113 experimental facility.

be lit. The system was equipped with a number of thermocouples and pressure taps. Flowmeters for the liquid metal
and the thermodynamic fluid were also provided.

Mercury/n-Pentane Facility (MP-1)

The schematic of the mercury/n-pentane facility is presented in Fig. 5. It is in general similar to the mercury/Refrigerant-113 facility but there is no condenser and the n-pentane vapor is exhausted to the atmosphere. The main characteristics of the channel were as follows: cross section 0.3 x 5.0 cm^2, electrode length 8.0 cm, magnetic field

Fig. 4 General view of the mercury/R-113 experimental facility.

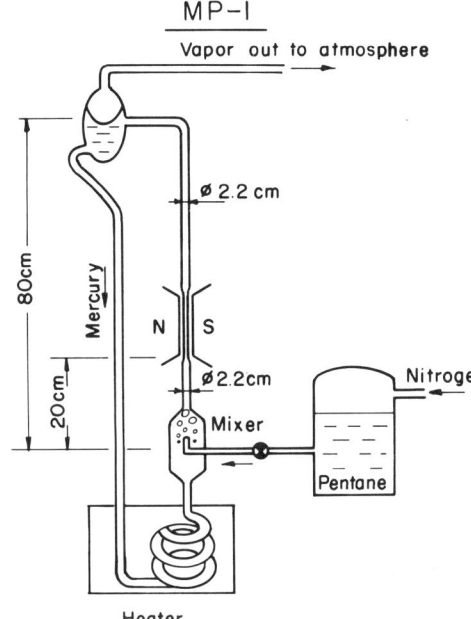

Fig. 5 Schematic diagram of the mercury/n-pentane experimental facility.

length 11 cm. In this system the big electromagnet of the Institute of Physical Chemistry of Uppsala was used and the magnetic field was 2.90 T.

IV. Results and Discussion

Preliminary Tests of Mercury/n-Pentane Facility

The tests were performed with the thermodynamic cycle between 65 and 36°C. The flow rate of n-pentane was 2.16 cm^3/s. In the open-circuit test the rms of strongly pulsating voltage was 150 mV while the maximal voltage approached 250 mV (Fig. 6). With a $2.6 \times 10^{-4} \Omega$ load the rms voltage was 40 mV and the current 154 A, the electrical power being 6.16 W. The n-pentane pumping power output was 5.31 W. The thermal efficiency of the cycle was 6.26% (Carnot efficiency being 8.5%) and the conversion efficiency was 16.7%. The latter number represents the ratio between electrical net power output and total heat input minus total amount of heat rejected to the atmosphere. The total system's efficiency, defined as the ratio of net power output and the total heat input, was 1.05%.

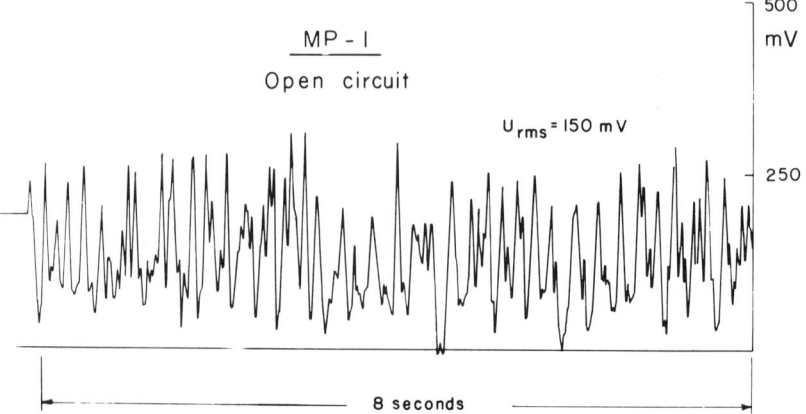

Fig. 6 Oscillogram of open-circuit voltage output of mercury/n-pentane facility.

Fig. 7 MHD generator with 0.08 T permanent magnet of mercury/R-113 facility.

Tests of Mercury/R-113 Facility

The main test runs were performed as follows. The liquid-metal loop was completely filled with mercury (including the separator) to a little above the generator outlet pipe. Then the heater was switched on and the mercury was heated to 85°C in the mixer. Next, the injection of Refrigerant-113 began, which caused circulation. For each flow rate of R-113, the generator's open-circuit voltage and the temperatures and pressures along the liquid-metal and vapor loops were measured. In some of these tests, the generator voltage under load was measured, while in the other tests the fluctuating component of the current was used to operate a transformer connected to a light-emitting diode (LED), see Fig. 7. Because of the very low intensity of the magnetic field (0.08 T) and the small distance between the electrodes (2.2 cm), the highest open-circuit voltage did not exceed 4×10^{-3} V. Since the two-phase flow was essentially a slug-type flow, and since the elec-

Fig. 8 Hydrocyclonic gravitational separator of mercury/R-113 facility.

trode was only 3.5 cm, the output exhibited strong pulsations up to 50% of mean measured voltage. In view of the above, it is understandable that most of the mechanical energy in this system is used to overcome the hydraulic resistance, while in the case of a system with a much stronger and bigger magnet most of the mechanical energy would be used to overcome the electromotive force in the generator and would thus eventually be converted into electrical power.

The main result of the present work is the fact that the complete two-phase liquid-metal MHD system operated as designed. So far, the system has been run for 1000 h and no serious failures have occurred. Since no changes in the mercury or the Refrigerant-113 were observed, it can be concluded that these two fluids are compatible in a MHD system operating at temperatures up to approximately 80°C. The gravitational hydrocyclonic separator (Fig. 8) performed properly.

Let us now consider the main quantitative results regarding the operation of the system. The two-phase flow

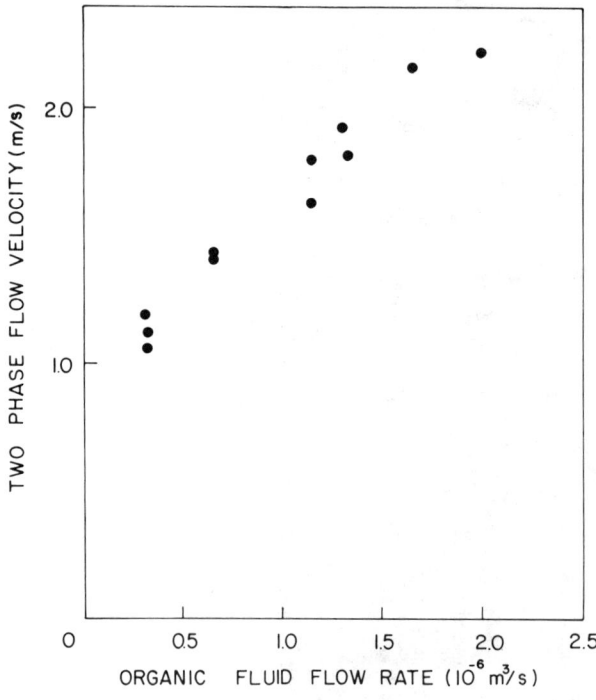

Fig. 9 Two-phase flow velocity vs flow rate of volatile liquid.

velocity (measured directly in the generator) is shown in Fig. 9 (in the open-circuit tests the generator was actually a two-phase flow velocity meter) vs the volatile liquid flow rate.

The mechanical net power was defined as follows:

Mechanical net power = kinetic energy change + hydraulic losses in the loop - volatile fluid pump power

The mechanical net power (calculated on the basis of velocity measurements) vs the flow rate of the volatile thermodynamic fluid is plotted in Fig. 10.

The total system's efficiency defined as the ratio of mechanical net power and heat input seems to depend little on the volatile liquid flow rate (Fig. 11). The total efficiency is rather poor, being only about 15% of the cal-

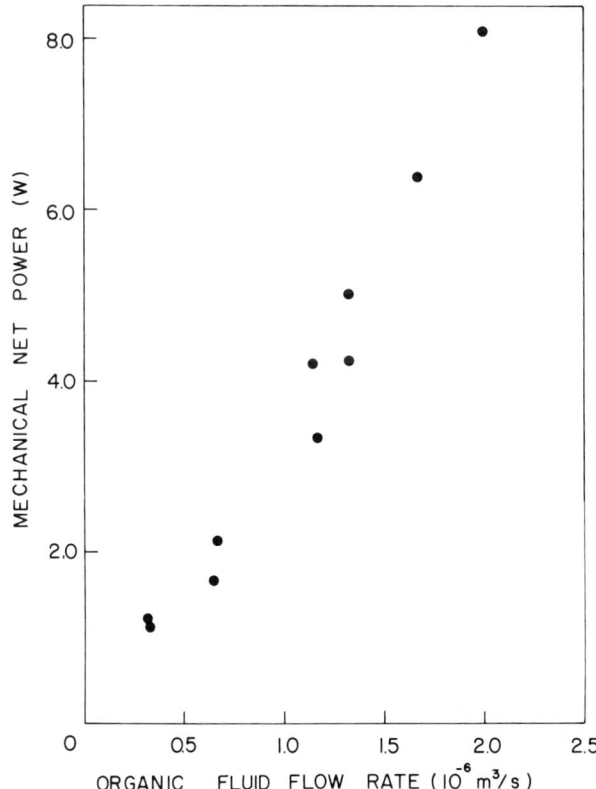

Fig. 10 Mechanical net power vs flow rate of volatile liquid.

culated efficiency of the thermodynamic cycle, η_{th}. This is mainly due to the very high slip ratio between the flow phases. As already mentioned the flow was a slug-type flow, and large velocity differences between the bubbles and the mercury could be observed visually. Typical two-phase flow patterns are shown in Fig. 12. It can be seen clearly that with the increase of the distance (indicated in Fig. 12) from the mixer the flow becomes more and more "torn." The ratio of the experimentally measured kinetic energy and calculated (assuming no slip) kinetic energy is presented in Fig. 13. Here the calculated two-phase (TP) velocity $(V_{TP})_{th} = V_\ell/(1-\alpha)$. The void fraction α was calculated as $\alpha = Q_g/(Q_g + Q_\ell)$ and Q_g, Q_ℓ are volumetric flow rates of gas and liquid, respectively. This ratio for higher volatile fluid flow rates is only about 0.25, while in our previous experiments with more homogeneous water and Refrigerant-113 flows it easily attained the value of 0.95 (Branover et al. 1980b).

It is hoped that the use of surface-active additives (cesium for mercury) will make the two-phase flow much more homogeneous. It should also be taken into account that, in an electrically loaded generator, the very strong longi-

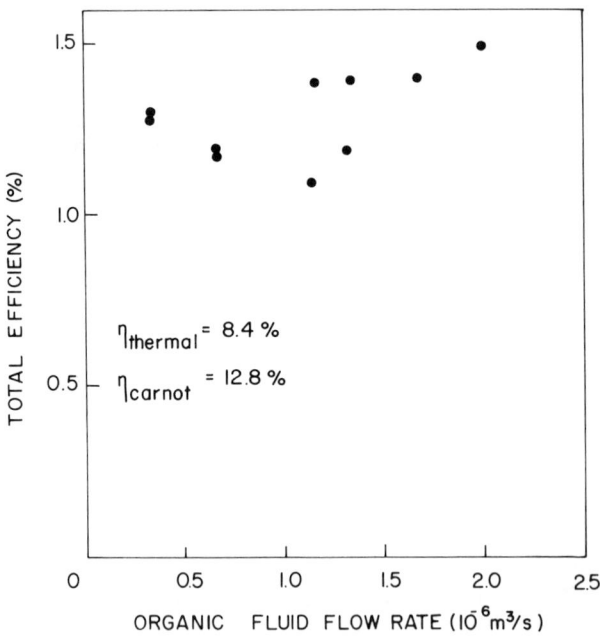

Fig. 11 Total efficiency vs flow rate of volatile liquid.

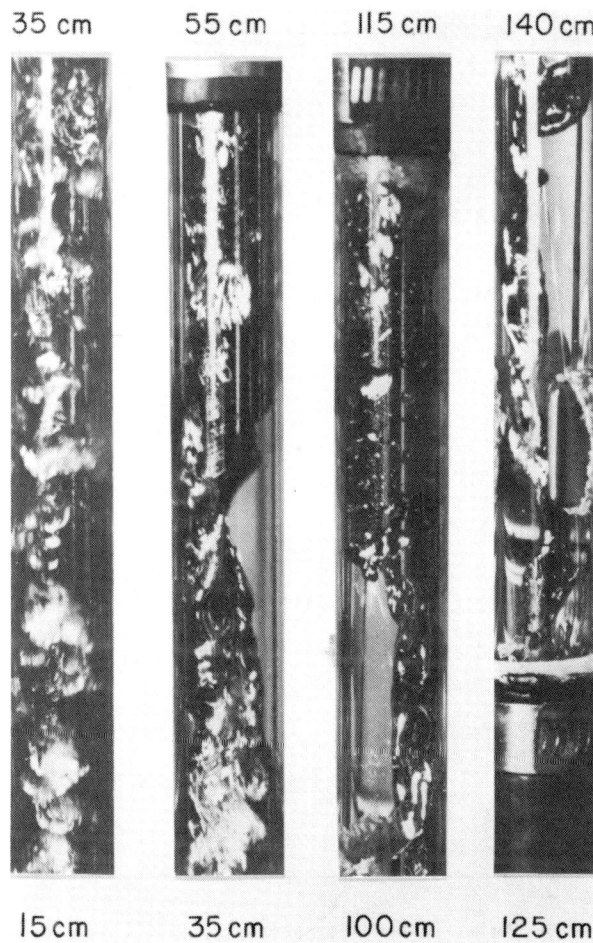

Fig. 12 Two-phase flow pattern.

tudinal pressure gradient breaks big bubbles, so that the slip is reduced considerably (Fabris and Hantman 1981).

One more comment should be made regarding the value of the thermal efficiency of the cycle η_{th}. The characteristics of the systems under consideration depend on the quality of the boiling process. Boiling of the volatile liquid should start and be completed at a pressure corresponding to T_H, the high temperature of the cycle. However, in our system, where the mixer has the same diameter as the rest of the vertical mercury pipe, drops of volatile liquid which get into the hot mercury are lifted by the high-velocity flow into the section with a gradually decreasing pres-

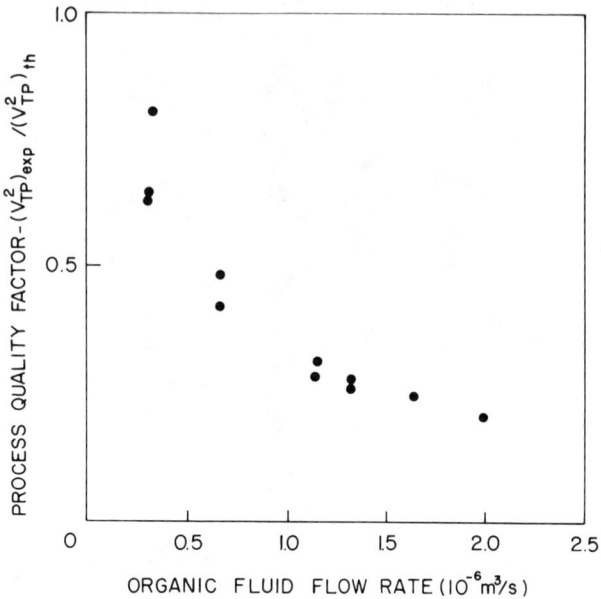

Fig. 13 Ratio of measured and calculated kinetic energy vs flow rate of volatile liquid.

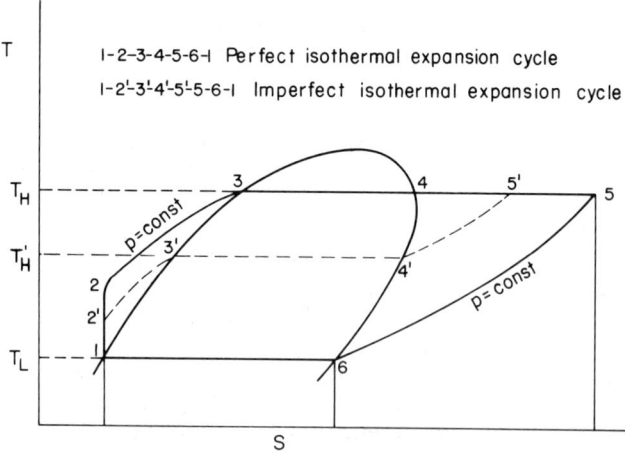

Fig. 14 Modified perfect and imperfect organic Rankine cycles.

sure. Boiling starts when the heated drop attains a height where the pressure corresponds to the vapor pressure at the temperature reached by the drop. This means that boiling occurs at a temperature T_H' which is considerably lower than T_H. The incomplete cycle corresponding to this case (supposing, for simplicity, that after boiling starts the drops stop rising) is shown in Fig. 14, in comparison with a complete cycle. To avoid this undesirable reduction of the thermal efficiency, the present mixer should be replaced by an expanded mixer in which the flow velocity will be reduced. The exact diameter and height of the mixer will have to be so calculated that boiling is completed before the bubble leaves the mixer.

Acknowledgments

This work was supported by Solmecs Corporation. The authors wish to express their sincerest thanks to M. Poisner, A. Ratner, and I. Zilberman, who actually built the facility and who put a great deal of creativity into this project.

References

Amend W.E. and Petrick M. (1971) "Performance of an efficiency 'low temperature' liquid-metal MHD power cycle suitable for large scale power generation," Proc. 5th Int. Conf. on MHD Electr. Power Generation, Munich.

Branover H. (1977) "Method and system for converting solar energy into electricity," U.S. Patent 4 191 901.

Branover H., Borde I., El-Boher A., and Lietner A. (1980a) "On the possible use of MHD generators in solar energy systems," MHD-Flows and Turbulence II, Israel University Press, Jerusalem, p. 147.

Branover H., Yakhot A., and El-Boher A. (1980b) "Solar powered liquid-metal MHD generators and some peculiarities of the performance of two-phase generators," Proc. 7th Int. Conf. MHD Electr. Power Generation, Cambridge, Mass., p. 165.

Fabris G. and Hantman R.G. (1981) "Interaction of fluid dynamic phenomena and generator efficiency in two-phase liquid-metal gas magnetohydrodynamic power generators," Energy Convers. and Manage. Vol. 21, pp. 49-60.

An Analytical Model of a Two-Phase Liquid-Metal Magnetohydrodynamic Generator*

Alexander Yakhot† and Herman Branover‡
Ben-Gurion University of the Negev, Beer-Sheva, Israel

Abstract

An analytical model of a two-phase liquid-metal magnetohydrodynamic generator is developed. The model is based on the averaging equations governing the two-phase flow in the channel of a liquid-metal MHD generator. The inhomogeneity of the distribution of the gas phase over the channel cross section is taken into account by introducing the correlation coefficients β_1, β_2, and β_3: $\langle\alpha u\rangle = \beta_1 \langle\alpha\rangle\langle u\rangle$, $\langle\sigma u\rangle = \beta_2 \langle\sigma\rangle\langle u\rangle$, and $\langle\rho u\rangle = \beta_3 \langle\rho\rangle\langle u\rangle$. Expressions for these coefficents in terms of the correlation coefficient β_1 are obtained. The calculated characteristics of the generator were compared with experimental data obtained at the Argonne National Laboratory. The comparison showed that the calculated and measured characteristics are in fairly good (±15%) agreement.

I. Introduction

The characteristics of a two-phase liquid-metal magnetohydrodynamic generator should be obtained by solving the problem of the flow of a two-phase (liquid-gas) mixture in a channel placed in a magnetic field. The analytical description of such a flow involves great difficulties, and it is

---

   Paper presented at Third Beer-Sheva International Seminar on Magnetohydrodynamic Flows and Turbulence, Ben-Gurion University of the Negev, Beer-Sheva, Israel, March 23-27, 1981. Copyright © 1982 by the American Institute of Physics. Published by the American Institute of Aeronautics and Astronautics, Inc., with permission.
   *This paper was published in Phys. Fluids, Vol. 25, No. 3, pp. 446-451 (1982).
   †Senior Lecturer, Department of Mechanical Engineering.
   ‡Professor, Department of Mechanical Engineering.

doubtful whether at present it is possible to definitely justify (or disprove) some of the assumptions underlying the analytical model. This pertains, for instance, to assumptions about the effect of the magnetic field on the distribution of the gas phase in the mixture, on the dependence of the conductivity on the void fraction α, on the effect of the magnetic field on the slip ratio, on the effect of electric currents on the two-phase flow pattern, etc. All the aforementioned questions require model assumptions, the validity of which can be clarified by comparing the experimental and analytical results.

The analytical model considered in the present paper is based on the assumption of local nonslip flow, employed by Bankoff (1960). The inhomogeneity of the distribution of the gas phase is included in this model by introducing the correlation coefficient β_1: $<\alpha u> = \beta_1 <\alpha><u>$. Bankoff investigated nonmagnetohydrodynamic two-phase flow, while the two-phase liquid-metal magnetohydrodynamic flow is predominantly affected by the electromagnetic force. Saito et al. (1978) used Bankoff's assumption to investigate two-phase MHD flow and obtained satisfactory agreement with the experimental results. They assumed that the conductivity of the NaK-N_2 mixture is a linear function of the void fraction and arrived at the conclusion that the value of $\beta_1 = 1.3$ depends little on the system's parameters (mass flow rates of the liquid and gas phases, magnetic field induction, and channel geometry). Our calculations and their comparison with experimental data obtained at the Argonne National Laboratory do not bear out the universality of the value $\beta_1 = 1.3$. Both the conductivity of the mixture and the correlation coefficient β_1 are functions of the two-phase flow pattern.

II. General Equations

Let us consider the flow of a two-phase liquid-metal gas mixture in the channel of an MHD generator (Fig. 1). We assume that:

1) The flow is fully developed (end effects are neglected).

2) The pressure is constant over the channel cross section, $p = p(x)$.

3) The flow temperature is constant and the expansion of the gas is described by the equation of state of an ideal gas.

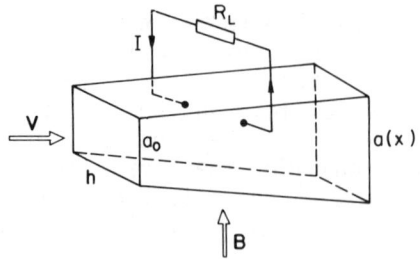

Fig. 1 Schematic of magnetohydrodynamic channel.

4) The channel height (electrode space) is constant (h = const) and channel width $a(x)$ varies in the direction of the flow.

5) Magnetic field B = const and is directed in the z direction.

6) Electric field E_y = const over the cross section.

The two-phase flow is governed by the following equations averaged over the channel cross section:

Equations of motion

$$<\rho u \frac{du}{dx}> + \frac{dp}{dx} = \tau_w + <j_y> B \tag{1}$$

The continuity equations for the liquid and gas phases

$$\rho_\ell <(1-\alpha)u> a(x) h = m_\ell \tag{2}$$

$$\rho_g <\alpha u> a(x) h = m_g \tag{3}$$

Ohm's law equation of current density

$$<j_y> = - <\sigma(uB - E_y)> \tag{4}$$

Equation of state for the gas

$$p = \rho_g RT \tag{5}$$

The density of the two-phase mixture is defined as

$$<\rho> = \rho_\ell (1 - <\alpha>) + \rho_g <\alpha> \tag{6}$$

The symbol $<A>$ in Eqs. (1-6) denotes averaging of $A(x,y,z)$ over the cross section.

Equation (6) was derived on the basis of Eqs. (2) and (3) using the definition of the local density of the two-phase mixture $\rho = \rho_\ell(1 - \alpha) + \rho_g \alpha$.

As mentioned in Sec. I the major difficulty in transition to the one-dimensional description of the flow is the averaging of the products of two or more quantities. Equations (1-4) include averages of the products $<\alpha u>$, $<\sigma u>$, $<\rho u>$, and $<\rho u du/dx>$. At this stage, introducing the so-called correlation coefficients β_1, β_2, and β_3, we shall write these quantities in terms of the products of their averages

$$<\alpha u> = \beta_1 <\alpha><u>, \quad <\sigma u> = \beta_2 <\sigma><u>, \quad <\rho u> = \beta_3 <\rho><u> \quad (7)$$

The meaning of the correlation coefficients will be clarified later. Also we assume that $<\rho u du/dx> = <\rho u> d<u>/dx$ which is equivalent to an assumption of small or constant longitudinal gradient of the velocity.

The system of equations for the averaged quantities becomes (the averaging sign $<>$ is dropped)

$$\beta_3 \rho u \frac{du}{dx} + \frac{dp}{dx} = \tau_w + j_y B \quad (8)$$

$$\rho_\ell (1 - \beta_1 \alpha) u a(x) h = m_\ell \quad (9)$$

$$\rho_g \beta_1 \alpha u a(x) h = m_g \quad (10)$$

$$j_y = -\sigma u B[\beta_2 - k(x)] \quad (11)$$

$$p = \rho_g (p_0/\rho_{go}) \quad (12)$$

$$\rho = \rho_\ell (1 - \alpha) + \rho_g \alpha \quad (13)$$

Here, $k(x) = E_y/uB$ is the load factor, p_0 and ρ_{go} the pressure and the gas density at the channel inlet, $\tau_w = -\lambda \rho u^2/2D$ the shear stress, λ the friction factor, and $D = 2ah/(a + h)$ the hydraulic diameter.

The system of Eqs. (8-13) describes the flow in the channel of the generator. Further transformation of Eqs. (8-13) can be performed only for a specific problem. Some typical examples are:

1) The channel geometry is specified, as are the inlet pressure, liquid- and gas-phase mass flow rates, magnetic field induction, and load resistance; it is required to find the distribution of flow parameters along the channel and the integral characteristics of the generator.

2) It is required to select the channel geometry [its width $a(x)$, for example] in such a manner that the flow velocity would be constant.

3) The channel inlet and exit pressures are specified; it is required to select the channel length or (if this is specified) the magnetic field induction.

Numerical implementation of any of these tasks does not significantly differ from one specific case to another, and hence we shall consider problem (1) in more detail.

III. Generator Model

Let us transform the system of Eqs. (8-13) as follows. From Eqs. (9), (10), and (12) we obtain an expression for the void fraction averaged over the cross section

$$\alpha = \frac{1}{\beta_1}\left(1 + \frac{1-\chi}{\chi}\frac{\rho_{go}}{\rho_\ell}\frac{p}{p_0}\right)^{-1} \qquad (14)$$

where $\chi = m_g/(m_\ell + m_g)$ is the mixture quality.

In order to eliminate the inertia term (with du/dx) in the equation of motion (8), Eq. (9) can be reduced to the form

$$\frac{1}{u}\frac{du}{dx} + \frac{1}{a}\frac{da}{dx} - \frac{\beta_1}{1-\beta_1\alpha}\frac{d\alpha}{dx} = 0 \qquad (15)$$

Using Eqs. (14) and (15) and performing simple manipulations, one can transform Eq. (8) to the form

$$(1-u^2\frac{\alpha^2}{\chi}\beta_1^2\frac{\rho_{go}}{p_0})\frac{dp}{dx} = \rho_\ell u^2 \frac{(1-\beta_1\alpha)}{\beta_3(1-\chi)}\left(-\frac{\lambda}{2D} + \frac{\beta_3}{a}\frac{da}{dx}\right)$$
$$- \sigma u B^2[\beta_2 - k(x)] \qquad (16)$$

The second term in the parentheses on the left-hand side of this equation is the square of the Mach number in the two-phase mixture and is related to the compressibility of the mixture. The first two terms on the right-hand side describe the effects of friction and of the divergence of the channel, while the last term takes into account the electromagnetic effect on the flow. Its contribution is dominant, since the ratio of the electromagnetic and friction forces is of the order of the Hartmann number which is typically higher than 1000.

Let us supplement Eq. (16) by the output current equation

$$dI/dx = a(x)\sigma uB[\beta_2 - k(x)] \qquad (17)$$

We shall now consider the parameters involved in Eqs. (16) and (17), separately.

Friction Factor λ

As previously noted, the effect of friction forces is negligible as compared with the electromagnetic force, and thus λ need not be specified with particular accuracy. The two-phase flow in the model under study is treated as a single-phase flow, whose physical properties are expressed in terms of the void fraction α.

Correlation Coefficients and Conductivity of the Two-Phase Mixture

The correlation coefficients β_1, β_2, and β_3 take care of nonuniformity in the distribution of the gas phase and, as a result, of the nonuniformity of the conductivity and density of the mixture over the channel cross section. If the linear relationship governing the local dependence of the conductivity or density on α is known from theoretical or empirical considerations, then the coefficients β_2 and β_3 can be expressed in terms of β_1. Thus, from the expression $\rho = \rho_\ell(1 - \alpha) + \rho_g \alpha$ and Eqs. (2) and (3) we find that

$$\beta_3 = \frac{\beta_1(1-\beta_1\alpha)}{\beta_1(1-\alpha)-\chi(\beta_1-1)} \qquad (18)$$

Saito et al. (1978) using a linear relationship for the local conductivity of the NaK + N_2 two-phase mixture as a function of α [Tanatugu et al. (1972), Fujii-e et al. (1975)] obtained an expression for β_2 in terms of β_1.

The conductivity of the two-phase mixture is a function of the flow pattern. Investigations by Bouman et al. (1974)

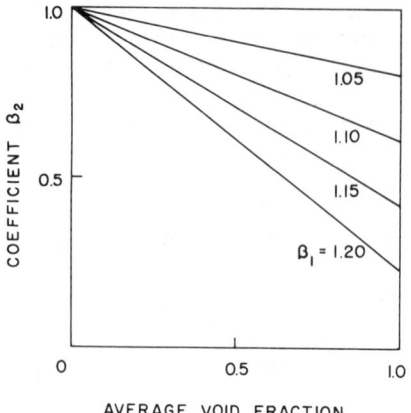

Fig. 2 Correlation coefficient β_2 vs average void fraction.

show that the linear dependence of the conductivity on α is typical for a bubble or slug flow. For annular flow this dependence is close to exponential. Our calculations based on Saito et al.'s (1978) results (linear conductivity and $\beta_1 = 1.3$) led to a serious disagreement with the experimental data of Petrick et al. (1978).

In the case of arbitrary dependence of the local conductivity on α we proceed as follows. Assuming that inhomogeneity over the channel cross section is weak ($\alpha - <\alpha>$ is small), let us expand $\sigma(\alpha)$ in a power series in ($\alpha - <\alpha>$), keeping only the first two terms

$$\sigma(\alpha) = \sigma(<\alpha>) + \sigma'(<\alpha>)(\alpha - <\alpha>) \qquad (19)$$

The meaning of the term "weak" can be clarified only by measuring α over the entire cross section and by comparing the measured flow and generator characteristics with their calculated values obtained in accordance with Eq. (19). It follows from Eqs. (19) and (7) that

$$\beta_2 = 1 + \frac{\sigma'(<\alpha>)}{\sigma(<\alpha>)} (\beta_1 - 1)<\alpha> \qquad (20)$$

The expression for $\sigma(<\alpha>)$ used in our calculations was suggested by Petrick and Lee (1964), $\sigma(<\alpha>) = \sigma_\ell \exp(-3.8<\alpha>)$.

Figure 2 shows the correlation coefficient β_2 as a function of $<\alpha>$ for different values of β_1. The value of β_1 depends on the flow pattern and should be determined experimentally. In the case of a uniform distribution of the gas phase $\beta_1 = 1$ and, as follows from Eqs. (18) and (20), $\beta_2 = \beta_3 = 1$.

Load Factor

In the generator model under consideration the electric field E_y is assumed to be constant and the load factor $k(x)$ can be expressed in terms of load voltage U_ℓ or output current I and load resistance R_ℓ

$$k(x) = U_\ell/uBh = IR_\ell/ubh = k_0 u_0/u \qquad (21)$$

Here we have introduced the load factor k_0 at the channel inlet

$$k_0 = IR_\ell/u_0 Bh \qquad (22)$$

The load factor k_0 and the coefficient β_1 complete the model of the generator and if the inlet pressure p_0 is specified, one can calculate the pressure distribution $p(x)$ from Eq.

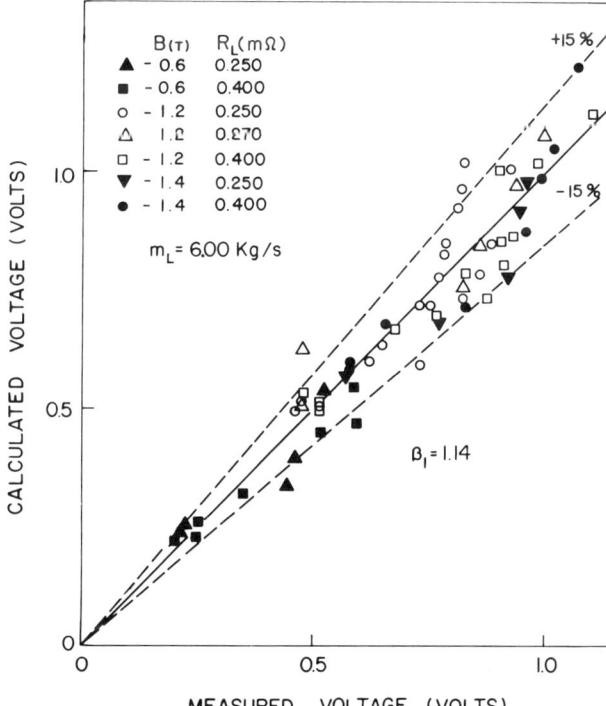

Fig. 3 Calculated load voltage vs measured load voltage (m_ℓ = 6.00 kg/s, β_1 = 1.14).

(16), the void fraction $\alpha(x)$ from Eq. (14), the flow velocity from Eq. (9), the output current I from Eq. (17), the load voltage $U_\ell = IR_\ell$, and the output electric power $P_e = IU_\ell$.

The load factor k_o is a function of load resistance R_ℓ and hence it cannot be taken arbitrarily. In the case when R_ℓ is specified, k_o should be selected in such a manner that the current I obtained from Eq. (22) and the value of I derived from integrating Eq. (17) over a generator length L should be equal. In our calculations k_o for a given R_ℓ was selected by trial and error.

IV. Argonne National Laboratory Experimental Data

In order to compare our calculated results with those obtained from experiments, we used the experimental data of Petrick et al. (1978). The experiments were carried out with a two-phase (NaK-N_2) liquid-metal magnetohydrodynamic generator at ambient temperature. A channel (LT-3) with variable cross section (a_{in} = 0.278 in, a_{ex} = 0.375 in, L = 15.25 in, h = 4.0 in) was investigated. The measured quantities were the load voltage and the pressure distribution along the channel. In addition, the void fraction, averaged in the direction of the magnetic field, was measured by the γ-ray technique. For m_ℓ = 6.00, 9.00, and 12.00 kg/s we processed the data of all the experimental runs listed in Petrick et al. (1978). These values of mass flow rates are the even values. In our calculations, the exact values of m_ℓ as well as m_g listed in Petrick et al. (1978) were used.

Load Voltage

Our calculations showed that for m_ℓ = 6.00 kg/s the best agreement with the measured characteristics is obtained at β_1 = 1.14. Figure 3 shows plots of the calculated and measured load voltage. The rise in the voltage corresponds to increasing the gas mass flow rate. The results show a fairly good agreement (deviation of about ±15%) with experimental data.

The effect of inhomogeneity of the cross-sectional distribution of the gas phase is taken into consideration within the suggested analytical model by the correlation coefficient β_1. In order to estimate the degree of void fraction inhomogeneity, we calculated the generator characteristics at $\beta_1 = 1.00$ and compared them with experimental results. The results are depicted in Fig. 4 and show that the calculated load voltage systematically exceeds (up to 45%) the measured values.

Note that the load voltage, in a series of experimental runs with similar magnetic field induction and load resistance increases with increasing gas mass flow rate and, as is seen from Fig. 4, the calculated points "run away" from the measured points. This indicates that the assumption of the homogeneity of the cross-sectional distribution of the gas ($\beta_1 = 1.0$) can lead to a serious error in the calculations, particularly at high void fractions.

Our calculations showed that β_1 increases with the increase in the liquid mass flow rate. For $m_\ell = 9.00$ and 12.00 kg we found that at $\beta_1 = 1.00$ the deviation of the calculated load voltage from the measured is rather small (about 10%). These results are shown in Fig. 5.

Let us now clarify the fact that $\beta_1 = 1.00$ for relatively high m_ℓ in more detail.

The analytical model under study is based on the local nonslip flow assumption. Bankoff (1960) came to the conclusion that the bulk gas velocity is higher than the liquid-phase velocity when the gas is concentrated in the higher velocity region. Therefore, one can introduce an average slip ratio s, which in this case reflects the nonequality of the bulk liquid and bulk gas velocities. Let us write the conservation laws for the liquid and gas phases as

$$\rho_\ell (1 - <\alpha>) <u_\ell> a(x) h = m_\ell \tag{23}$$

$$\rho_g <\alpha> s <u_\ell> a(x) h = m_g \tag{24}$$

From Eqs. (9), (10), (23), and (24) the expression for the average slip ratio can be obtained in the form

$$s = \beta_1 (1 - \alpha)/(1 - \beta_1 \alpha) \tag{25}$$

(here the averaging symbol < > is dropped).

It was concluded on the basis of the extensive experimental data obtained in Petrick et al. (1978) that the slip ratio s tends to unity with increasing m_ℓ. It was noted

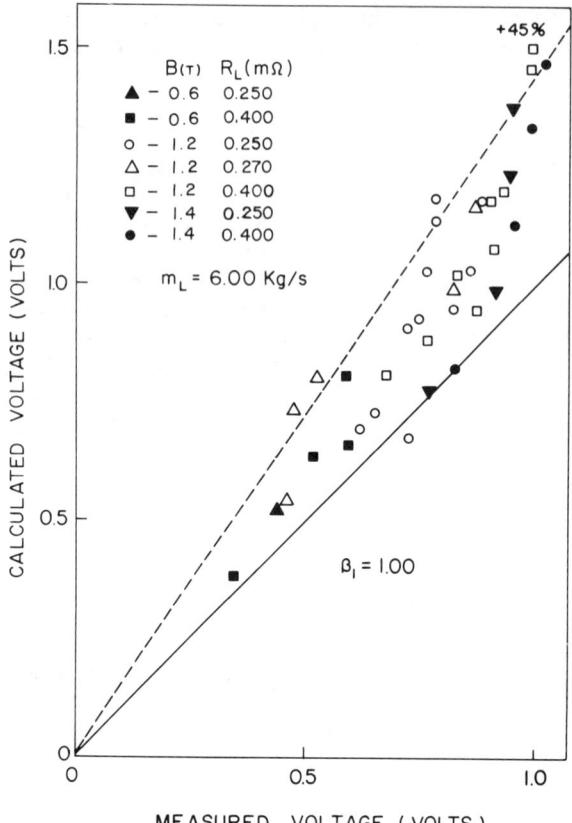

Fig. 4 Calculated load voltage vs measured load voltage (m_ℓ = 6.00 kg/s, β_1 = 1.00).

in these experiments that $s \simeq 1.0$ for m_ℓ = 9.00 and 12.00 kg/s. As can be seen from Eq. (25), the value of β_1 = 1.0, used in the calculations for m_ℓ = 9.00 and 12.00 kg/s, also corresponds to $s = 1$.

It should be emphasized that satisfactory agreement between the calculated and measured load voltages (Figs. 3 and 5) was obtained for a wide range of magnetic field induction and load resistance. This allows the conclusion that β_1 is independent of the magnetic field and the external load resistance.

Pressure Distribution

It would be excessively optimistic to expect that the pressure calculated from a one-dimensional model would agree with high accuracy with the measured pressure. It is

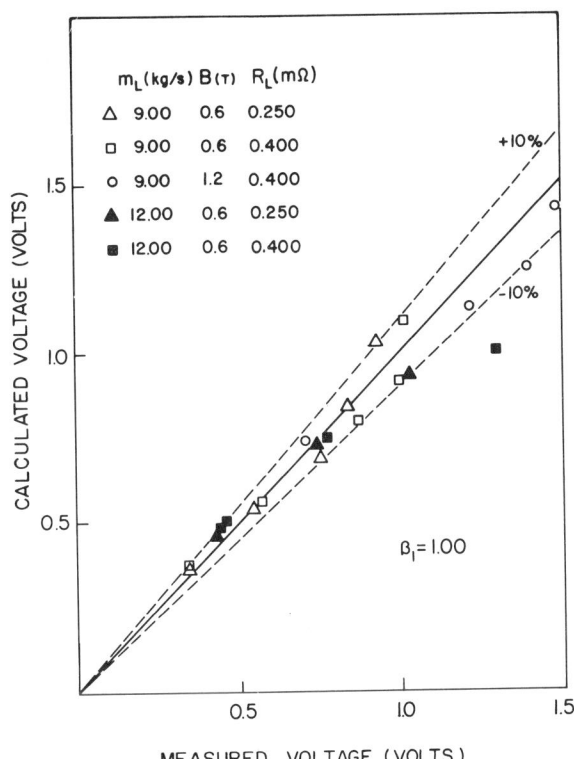

Fig. 5 Calculated load voltage vs measured load voltage (m_ℓ = 9.00 and 12.00 kg/s, β_1 = 1.00).

obvious that a one-dimensional model is incapable of the exact description of three-dimensional effects such as, for example, end effects and the locally measured characteristics (pressure, velocity, etc.) should differ from the calculated. Had this difference been too significant (quantitatively as well as qualitatively), then we would have "felt" this in comparing the integral characteristics (load voltage, for example). As was shown, the agreement between the measured load voltage and that calculated from the one-dimensional model is quite good.

Figure 6 shows the longitudinal pressure distribution. The last experimental points were measured past the generator channel, while the pressure was calculated to the channel exit (L = 0.387 m). The values of pressure at the channel exit, which are higher than those measured, are attributed to the presence of the end effect, which is not included in the one-dimensional model. The rise in pressure in the down part of the channel at m_g = 0 (and also

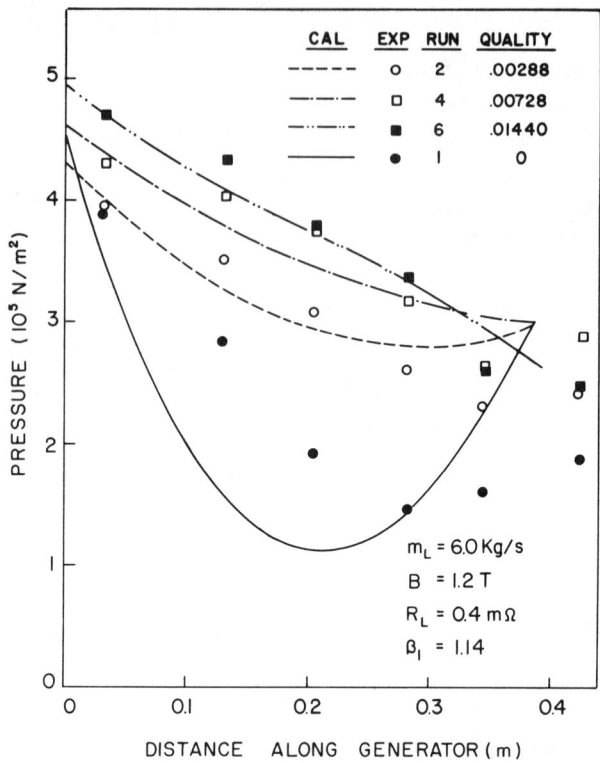

Fig. 6 Calculated and measured pressure distribution along the channel.

sometimes at low m_g) is attributed to the change in the direction of the current, which results in the pumping of the flow.

Void Fraction

The void fraction was measured in the experiments of Petrick et al. (1978) at several points along the channel axis by the γ-ray technique. At each point, the source and detector were traversed parallel to the magnetic field to yield the profile averaged in the direction parallel to the current and the profile was then averaged. The void fraction averaged along the channel was also calculated. In our study we calculated the longitudinal distribution of the cross-sectional average void fraction (Fig. 7) and also their lengthwise averaged values (Figs. 8 and 9). As seen from Figs. 7-9, the calculated values are somewhat lower than those measured at high α.

TWO-PHASE LIQUID-METAL MHD GENERATOR

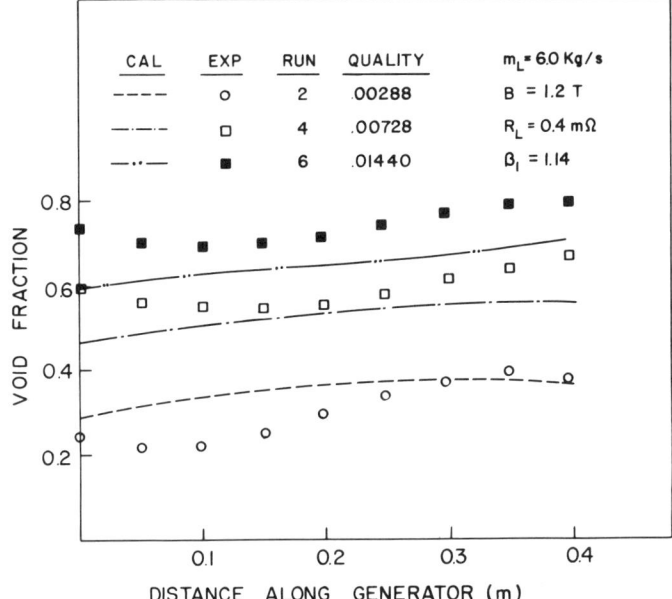

Fig. 7 Calculated and measured void fraction distribution along the channel.

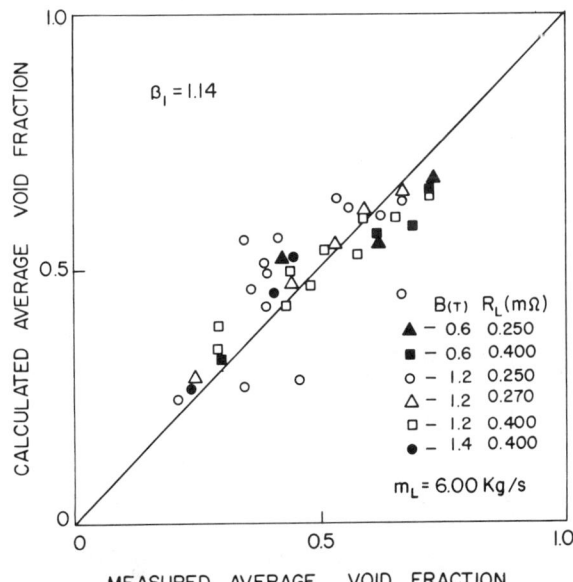

Fig. 8 Calculated void fraction vs measured void fraction (m_ℓ = 6.00 kg/s, β_1 = 1.14).

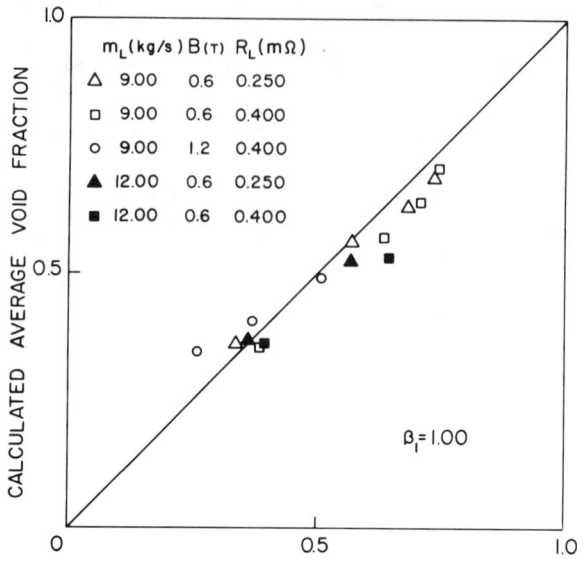

Fig. 9 Calculated void fraction vs measured void fraction (m_ℓ = 9.00 and 12.00 kg/s, β_1 = 1.00).

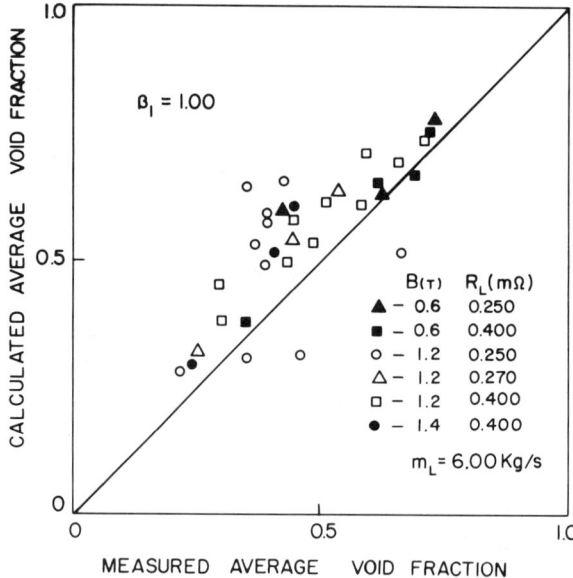

Fig. 10 Calculated void fraction vs measured void fraction (m_ℓ = 6.00 kg/s, β_1 = 1.00).

Calculation of α without allowing for inhomogeneity in the cross-sectional gas-phase distribution ($\beta_1 = 1.00$) yields systematically excessive values of α (Fig. 10).

V. Summary

1) An analytical model of a two-phase liquid-metal MHD generator is developed. The model is based on the averaging equations governing the two-phase flow in the channel of the liquid-metal MHD generator.

2) The inhomogeneity of the distribution of the gas phase over the channel cross section was taken into account by introducing the correlation coefficients. Expressions for these coefficients in terms of the correlation coefficient β_1 were obtained.

3) The calculated characteristics of the generator were compared with experimental data obtained at the Argonne National Laboratory [Petrick et al. (1978)]. The comparison showed that the calculated and measured characteristics are in fairly good (±15%) agreement.

Acknowledgments

This work was supported by the Solmecs Corporation, the Israel Ministry of Energy, and the Office of Naval Research.

References

Bankoff S.G. (1960) "A variable density single-fluid model for two-phase flow with particular reference to steam-water flow," J. Heat Transfer, Trans. ASME, Ser. C., Vol. 82, p. 265.

Bouman K., Van Koppen C.W.J., and Raas L.Y. (1974) "Some investigations on the influence of the heat flux on flow pattern in vertical boiler tubes," European Two-Phase Flow Group Mtg., Harwell, England, Paper A2, KTFS 18297.

Fujii-e Y., Miyazaki K., Inoue S., Saito M., and Suita T. (1975) "Experimental study on two-phase induction power generation using NaK-N_2 mixture," Proc. 6th Int. Conf. MHD Electric. Power Generation, Washington, D.C., Vol. III, p. 247.

Petrick M. and Lee K.Y. (1964) "Performance characteristics of a liquid metal MHD generator," Proc. 2nd Int. Conf. MHD Electric. Power Generation, Paris, Vol. II, p. 953.

Petrick M., Fabris G., Pierson E.S., Fisher A.K., and Johnson C.E. (1978) "Experimental two-phase liquid-metal magnetohydrodynamic generator program," Argonne National Lab. Ann. Rept., ANL/MHD-78-2.

Saito M., Inoue S., and Fujii-e Y. (1978) "Gas-liquid slip ratio and MHD pressure drop in two-phase liquid metal flow in strong magnetic field," J. Nuclear Sci. Technol., Vol. 15, No. 7, p. 476.

Tanatugu N., Fujii-e Y., and Suita T. (1972) "Electrical conductivity of liquid metal two-phase mixture in bubbly and slug flow regime," J. Nuclear Sci. Technol., Vol. 9, No. 12, p. 753.

Critique of MHD Electrical Power Generation

William D. Jackson*
Energy Consultant, Inc., Washington, D.C.

Abstract

The current status of magnetohydrodynamic electrical power generation for the situation in which the working fluid is an ionized gas is reviewed and the remaining steps required to establish a viable design basis are identified. The paper begins with a description of the MHD technology and the current position of the research, engineering development, and the demonstration scale as it progresses to commercialization. This is followed by a definition of the particular requirements which have to be included in an overall MHD development program. Major achievements in the program to date are reviewed in both the United States and other countries, especially the USSR. Future objectives and program emphasis are then set forth and the paper concludes with comments on the institutional arrangements needed to insure the orderly continuation of MHD development to the goal of significant penetration of the central station electricity generation market.

I. Introduction

Magnetohydrodynamic (MHD) electrical power generation is one of the very few advanced systems with the potential of significantly increasing the thermal efficiency of central powerplants while markedly reducing the level of environmental intrusion. While MHD is inherently an energy

---

Paper presented at Third Beer-Sheva International Seminar on Magnetohydrodynamic Flows and Turbulence, Ben-Gurion University of the Negev, Beer-Sheva, Israel, March 23-27, 1981. Copyright © 1982 by Energy Consultant, Inc. Published by the American Institute of Aeronautics and Astronautics with permission.
*President.

conversion process capable of utilizing a variety of both renewable and nonrenewable fuel resources, it is particularly suited to coal. When developed, MHD will offer greatly improved prospects for the utilization of coal, the most abundant of the world's fossil fuels, in the generation of electricity (Jackson 1976).

Based on these considerations, a significant effort on a worldwide basis has been, and continues to be, undertaken to carry MHD through the various phases of research, engineering development, and demonstration. The origins of this effort can be traced back to the late 1950s, especially in those countries with large fossil fuel resources. While much of the early work was directed toward the utilization of clean fuels (natural gas and distillates), coal has been considered throughout the development process. The international aspects of MHD development have been very fully covered in the Status Reports periodically issued by the International Liaison Group on MHD Electrical Power Generation (1977).

This paper deals with the program being conducted in the United States and draws on the activities in other countries where they significantly impact the technology or design base required by American conditions. The United States national effort has been conducted on a major scale for the past eight years but is based on earlier activities, including the clean fuel and special applications work carried out prior to 1970 (Rosa 1969).

The program in the United States has been organized and led by the Department of Energy (DOE) with important assistance from the Electric Power Research Institute (EPRI). Support and participation has come from the electric utility industry, suppliers of equipment to that industry, research and development organizations specializing in energy research and development, and universities. Working within the framework established by DOE, this talented group has made important progress toward the goal of commercialization. It is the objective of this paper to provide an independent assessment of the results obtained and their significance with respect to the ultimate development of MHD.

II. Background

The principle of MHD--the extraction of electrical energy directly from a flowing ionized gas or plasma in a

strong magnetic field oriented perpendicular to the flow--
has been described often and will not be dealt with in any
detail here (Rosa 1968). From the engineering viewpoint
and the application to electric utility systems, the requirement is to combine this "electromagnetic turbine" end
to end with a conventional steam turbine powerplant in a
manner which results in favorable overall plant economics.
In the context of electricity generation in the United
States, this implies the achievement of improved utilization of coal relative to conventional fossil powerplants,
as well as nuclear-fueled plants.

The year 1982 included the date of the 150th anniver-
sary of the first recorded MHD experiment--that by the British experimentalist Michael Faraday from Waterloo Bridge,
London, using the River Thames as the flowing, electrically
conducting "working fluid" (Faraday 1832). Preliminary,
and generally premature, efforts to apply MHD to large-
scale electrical power generation have been undertaken
since the mid-1930s (Karlowitz and Halasz 1964), but it is
important to stress that a serious development effort has
been under way for only 8 out of the 11 years that the
current program has existed. While organized by the DOE
and its predecessor agencies, the significant funding needed for a program of this type has largely resulted from
interest and support in the U.S. Congress. A further factor was the increased emphasis on coal-based electricity
generation as a result of the 1973 oil embargo.

Development of new electrical power generation systems
typically takes 25 years or more and involves large-scale
facilities. The example of nuclear power, which took from
1942 when the feasibility of a heat-producing, steady-state
reactor was first established to 1975 when a 10% penetration of the market had been realized, is illustrative in
this regard. MHD is perhaps one-third of the way to significant commercialization, as will be established later in
the paper.

A further point to be recognized in the assessment of
MHD is that the research, development, and demonstration
process passes easily through recognized stages with de-
finable checkpoints between them. Following a practice
originally developed by the U.S. Department of Defense,
these stages are self-explanatorily defined as: applied research, exploratory development or preliminary engineering,
engineering development, system demonstration, and finally
commercialization.

In the case of MHD, essentially all of the work conducted prior to 1968 with respect to coal was in the first category, and from 1971 to 1977-1978 exploratory development with coal as the preferred fuel progressed to the point where the initiation of engineering development activities could be contemplated. After appropriate reviews, this was set in motion and preparations for the engineering development phase were well in hand when the 1980 change of Administrations occurred. In the accepted sense of the term, engineering development is not yet under way but results from the exploratory phase continue to be obtained (Rudins 1981).

Based on this situation, together with prior experience with the overall development of power generation systems, significant commercialization of MHD cannot now occur until at least the first decade of the 21st century. Accordingly, MHD is the category of longer term engineering projects for which continuity of direction and consistency of support are essential to insure timely and orderly progress.

III. Program Characteristics and Requirements

A typical schematic for an MHD powerplant appears in Fig. 1. From this it can be observed that there are three key activities: 1) system integration, 2) development of components novel to MHD, and 3) adaptation of existing power engineering components to the requirement of an MHD system. In rather more detail, this establishes five major work categories involving an unusually wide range of scientific and engineering skills as follows:

1) Analysis of MHD systems to establish performance, system configuration, and component requirements.

2) Development of novel or first-of-a-kind components unique to MHD.

3) Adaptation of novel components from other advanced technologies to MHD conditions.

4) Development of components which provide the transition from the MHD system itself to the rest of the powerplant and the required environmental performance. This chiefly requires imaginative application of state-of-the-art engineering.

MHD ELECTRICAL POWER GENERATION

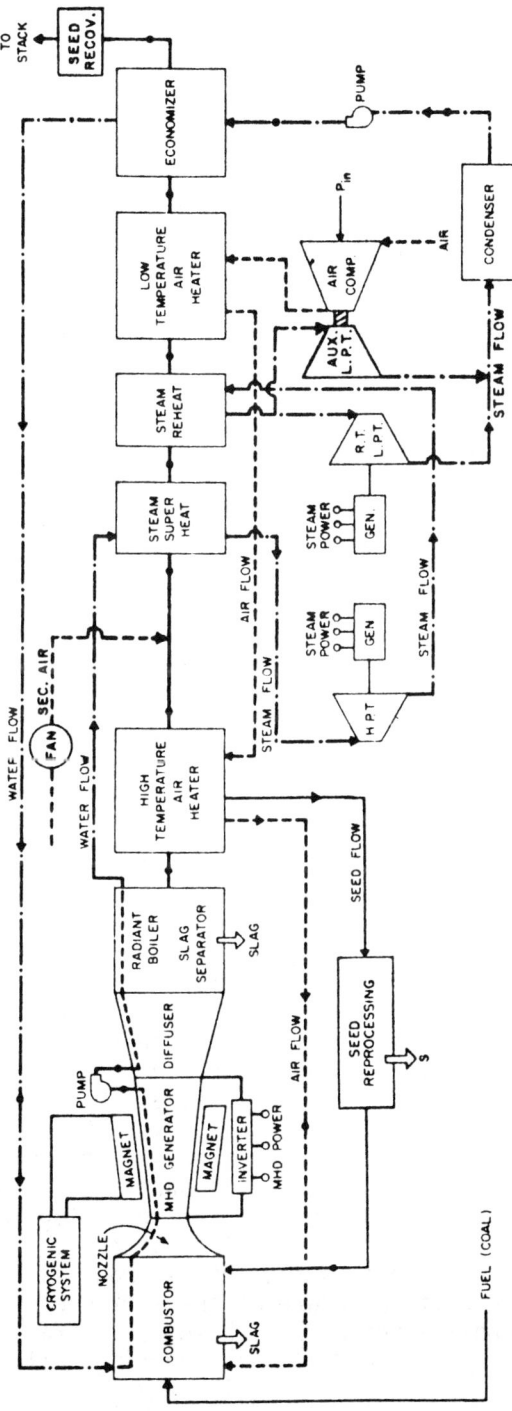

Fig. 1 Simplified schematic of directly fired air preheater MHD system configuration.

5) Integration of key subsystems and overall plant integration.

Examples can be cited as follows: 2) channel and power consolidation circuits; 3) superconducting magnet, combustor, diffuser, and nozzle; 4) heat and seed recovery systems, coal feed, slag removal, and inverters; 5) MHD power train, transition subsystem, seed recycling subsystem, gas path integration, and steam and water integration. Category 1 is self-explanatory.

Each of these work categories involves a different skill base and program success is critically determined by their being brought together in a manner controlled by realistic scheduling. With respect to overall development, this scheduling is controlled by the three major phases:

1) Development of actual subscale components and systems (completion of exploratory development and execution of engineering development).

2) Demonstration of operation of a pilot-scale engineering facility (system demonstration).

3) Development and operations of a commercial-scale plant on a utility system (commercialization).

Several points arise from this phasing and the types of activities covered. First, the scale of activities increases as the phases progress, being in the ranges 10-250, 250-500, and above 500 MW, respectively, the power in each case being the thermal input. However, it is not necessary, especially in phase 1, that all activities proceed at the same scale. Considerable advantage, in fact, accrues from separating performance and duration testing. Thus phase 1 requires several facilities of differing characteristics and scale, the number being governed by the pace of the program and the degree to which parallel efforts can be justified as a risk-reducing procedure. The implementation of this second point is governed by the available budget. If too large a proportion of fiscal resources are devoted to facility operation, then the critical technology development is inhibited. A third point is that the character of the participants changes as the phases progress. The entrepreneurial scientific nature of the work preceding phase 1 (mainly applied research with some exploratory development) has to be replaced with engineering skills appropriate to predominantly first-of-a-kind technology and later powerplant engineers must take over and lead the effort. A final point is that utility participation is essential throughout the phase sequence with the role in phase 1 being guidance, phase 2 being at least equal partnership with the government, and phase 3 being major responsibility.

These issues are often neglected but have been treated at some length here to clarify the strategy to be followed and the roles of the various participants in the overall development process.

IV. Major Program Achievements

During the 1960s, work on clean fuel systems embraced the exploratory development phase and was highlighted by the operation during 1963-1066 of a short-duration (several seconds), self-excited generator with an electrical output of 33 MW. The Avco-Everett Research Laboratory, responsible for this result, also constructed and operated a 4 T model superconducting magnet in 1966. Illustrations of these key early achievements appear in Figs. 2a and 2b, respectively. Together they confirmed the feasibility of energy extraction from an ionized gas in accordance with the predicted performance of MHD generators. Work in the United Kingdom (Heywood and Womack 1969) during the same period made a good start to the combustion of coal and creating a suitable plasma with the combustion products. The actual operation of a MHD generator with direct coal combustion and the use of coal slag was first achieved at the University of Tennessee Space Institute in 1973 using the apparatus shown in Fig. 3.

These initial steps were followed by the careful analysis of MHD powerplants, initially by NASA as part of its Energy Conversion Alternatives Study (ECAS). This study (Seikel et al. 1976) definitively established that MHD plants fueled with coal could be expected to yield overall thermal efficiencies (coal pile to busbar) in the range 46-48% at costs more favorable than those of conventional plants with exhaust scrubbers. The study also found that the MHD coal plants were superior to other advanced systems.

With these major component and system results as a basis, the development of the key components of the MHD power train (combustor, nozzle, channel, diffuser, superconduc-

Fig. 2 Major early achievements: a) Avco Mark V self-excited 33 MW output generator and b) Avco 4 T superconducting magnet.

ting magnet, power consolidation circuits, and inverter) was started in 1975 with the following principal results to date:

1) Operation of two complete superconducting magnet systems by the Argonne National Laboratory (ANL). The first of these (Fig. 4a) was installed in the U-25B loop of the MHD facilities operated by the USSR Institute for High Temperatures (IVTAN) as part of the USA-USSR Cooperative Program and has operated continuously since 1976 under cryogenic conditions in a field of 5 T when energized for generator experiments (Kirillin 1978). The second ANL magnet is sized for a 50 MW (thermal) scale facility [the Coal-Fired Flow Facility (CFFF) in Tullahoma, Tenn.] and has an overall weight of 200 tons. Shown in Fig. 4b, it was successfully operated during 1981 at its rated field of 6 T. However, no channel operation is as yet definitely scheduled.

2) The Avco-Everett Research Laboratory (AERL) has operated several channels with slag deposition at the electrical, thermal, and mechanical stress levels required by the ECAS study. In particular, a 1000 h duration test of electrodes has provided a solid engineering basis for the

Fig. 3 University of Tennessee Space Institute diagonal generator used in pioneering direct coal-fired experiments.

Fig. 4 Superconducting magnet systems constructed by the Argonne National Laboratory: a) 5 T magnet for USA-USSR joint test program on Institute for High Temperature U25B loop and b) 6 T magnet designed for the Tennessee Coal-Fired Flow Facility.

Fig. 5 Avco channel of type used for 1000 h duration test.

design and construction of channels with several thousand hours operational life. A typical AERL channel design and electrode assembly is shown in Fig. 5.

3) At the Arnold Engineering Development Center, also in Tullahoma, Tenn. a large, short-duration generator facility used by the U.S. Air Force until 1968 for special applications MHD development has been modified to enable commerical-type MHD channels to be tested for 15 s operating time in fields up to 6 T. A view of this impressive 250 MW (thermal) facility appears in Fig. 6. The most recent results have yielded over 35 MW of electrical output at a field of 3.6 T. The importance of this result is not only that the scale is impressive but that very good agreement with predicted performance has been achieved. The predictive methods have been largely developed by STD Research Corporation (Vetter et al. 1980).

4) A 20 MW (thermal) combustor of the two-stage type has been operated with the correct coal and oxidizer conditions for about 200 h and has produced a plasma with a conductivity of 7 S, again the value required by the ECAS study. This combustor was designed, constructed, and tested by TRW Inc. at its Capistrano, Calif. test facility shown in Fig. 7.

5) A clean fuel (natural gas) fired MHD power train designed and built by IVTAN has been successfully integrat-

Fig. 6 View of 250 MW thermal short duration test facility at the Arnold Engineering Development Center, Tullahoma, Tenn., showing channel withdrawn from magnet.

Fig. 7 Layout view of TRW 20 MW thermal coal combustor.

ed and operated with the superconducting magnet described in item 1 as part of the USA-USSR Cooperative Program established in 1973. Seven joint power train tests were conducted during 1977-1979 and provided valuable operational data under high field conditions (Rudins et al. 1980).

6) Several system studies subsequent to ECAS have been conducted by EPRI, DOE, and some utility organizations. They have analyzed MHD systems in much more detail than was possible in ECAS, particularly those for near-term applications. In addition to confirming the efficiency range of 46-48% for mature plants and the possibility of significantly exceeding 50% with advanced technology, two important concepts for near-term applications were identified and analyzed. One of these involves the use of oxygen enrichment and the other the "retrofitting" or "repowering" of existing plants, especially in those cases where conversion to coal is required (Griswold 1979). Efficiencies for these conditions are in the range 40-44% and represent the first step in the efficiency improvement made possible by MHD.

These major results have been supported by other contributions in materials, advanced heat exchangers, diffusers, heat and seed recovery systems, NO_x control, seed regeneration, and inverter/consolidation circuits. Documentation of these can be found in the comprehensive status report prepared as part of the USA-USSR Cooperation (Petrick and Shumyatsky 1978) and also in the Proceedings of the Annual Symposia for the Engineering Aspects of Magnetohydrodynamics and the periodic International Conference on MHD Electrical Power Generation. A convenient method of displaying results is the Rosa diagram of Fig. 8. In this, power levels are shown on the y axis and duration times on the x axis, logarithmic scales being used in both cases. Figure 8 summarizes major results on this basis and gives a good indication of the progress made relative to the goal of full-scale commercial operation.

One result of major importance to the American program has been the successful operation of the Soviet U-25 natural gas-fired pilot plant shown in Figs. 9 and 10. Short-duration operation up to 20 MW electrical output into the Moscow grid and continuous plant operation for 250 h have been the major achievements of a project which has made an important contribution to the demonstration of the engineering feasibility of MHD systems. Using this plant as a test facility, a channel (designed by MEPPSCO Inc. and ANL and constructed by Westinghouse Corporation) was to have been shipped early in 1980 to Moscow as another element of the Cooperative Program (Kuczen et al. 1980). The Soviet invasion of Afghanistan led to this activity being indefinitely postponed and the valuable data which would have resultted from the expected 10 MW electrical output for 100 h has yet to be obtained. The completion of the channel did

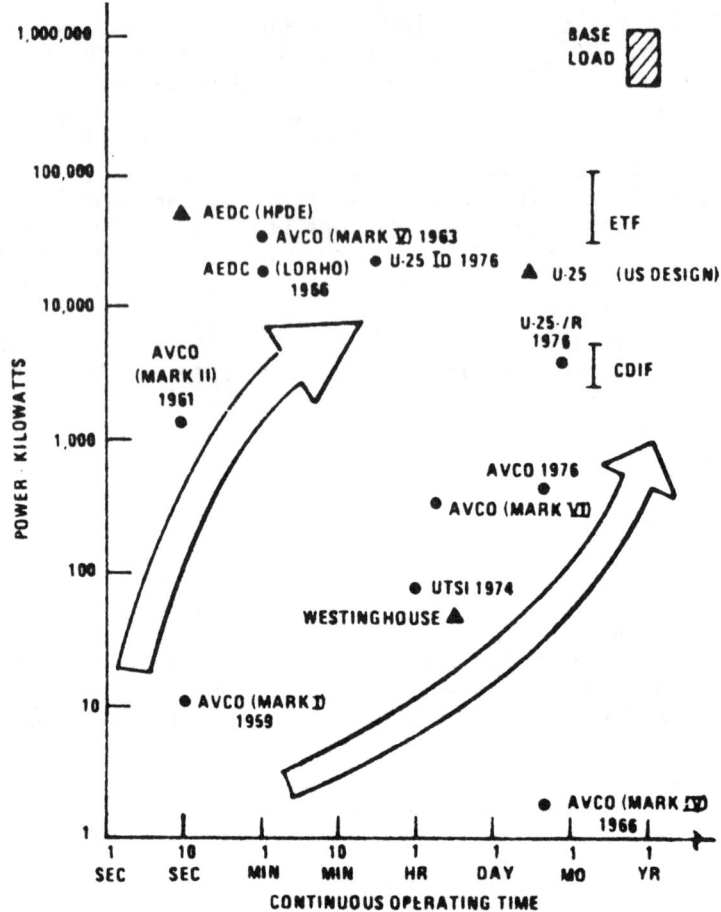

Fig. 8 Selected MHD generator performance data.

yield important advances in understanding the construction of large channels and the possibility of testing at the Arnold Engineering Development Center (AEDC) may be realized as this channel is compatible with the AEDC magnet.

V. Future Objectives

Future objectives can be established from the phases outlined earlier and require that a number of major issues be addressed.

MHD ELECTRICAL POWER GENERATION

Fig. 9 Exterior view of USSR Institute for High Temperatures U-25 pilot plant facility showing 1200°C air preheaters.

In phase 1:

1) Demonstration of operation of a complete coal-fired (not simulated) MHD power train (combustor, channel, diffuser, superconducting magnet, inverter, and consolidation circuits) coupled to the electricity grid at a power level of at least 20 MW (thermal) and for hundreds of hours, of which 100 h must be uninterrupted.

2) Verification of generator scaling laws by operating the AEDC facility up to fields of 5 T with the objectives of attaining enthalpy extraction and turbine efficiency values of 15 and 65%, respectively.

3) Determination in depth of the performance, system configuration, and component requirements for the major facility of phase 2 (known as the Engineering Test Facility or ETF). Associated with this issue is the determination of the merit of the retrofit concept for this unit as an alternative to the construction of a complete test facility.

4) Establishment of an engineering design base for the MHD power train for the ETF. An adequate treatment of this issue will involve both the testing of subscale (probably 50 MW thermal) components in integrated operation and the use of the AEDC facility for short-duration tests of ETF-scale channels.

5) Analysis of heat and seed recovery systems, supported as needed by subscale testing, to determine operating conditions and parameters for ETF-scale components.

6) Continuation for diagnostic development and advanced technology work (e.g., disk generators, high-temperature air heaters) to provide assurance that system and component performance can be adequately monitored and improved toward the goal of overall thermal efficiencies above 50%.

In phase 2 and 3:

1) Focusing of program on the design, construction, and operation of the required plants.

2) Continuation of advanced technology development.

The engineering work which must be undertaken to resolve these issues evidently requires appropriate facili-

Fig. 10 Generator hall of U-25 facility showing 20 MW electrical output flow train with 2 T conventional magnet.

Fig. 11 Schematic of Tullahoma, Tenn., Coal-Fired Flow Facility.

ties. It can be observed that the program now has all of the facilities required for the effective implementation of phase 1. These are indicated in Fig. 8. The recent (1980) initial operation of the Coal-Fired Flow Facility in Tullahoma, Tenn. and the Component Development and Integration Facility (CDIF) in Butte, Mont. are important program milestones from the facility viewpoint. General views of these facilities are shown in Figs. 11 and 12. It is essential that they now be used to the fullest possible extent to generate engineering data and that major modifications (e.g., increasing the CDIF to a 100 MW thermal rating) be deferred until such data are forthcoming. Then, modifications can be made intelligently.

VI. Future Program Emphasis

From the analysis and data presented in this paper, the major conclusions are:

1) The decision to move into engineering development was soundly based and the program is yielding engineering data which support the viability of the process for electrical power generation.

2) Commercialization of the technology is in the longer range category with significant market penetration possible only after the year 2000.

3) The program is in the process of executing phase 1 and major facility investments have just been completed. The next logical point for program review is some years (4-6) hence when the results of the remaining part of phase 1 will be available, mainly from the CFFF and CDIF facilities.

No basis can be established to justify review of the program on a go/no-go basis at the present time, especially in view of the encouraging progress outlined in Sec. IV. In recognition of the national need to control expenditures and the longer range nature of MHD development, it is, however, appropriate to consider and establish a viable "core program" for the next 3-4 years. The following proposals are advanced as a starting point for what must necessarily be a detailed program planning exercise.

Constraints:

1) Focus on open-cycle, coal-fired systems.

2) Limit activities to the development of an engineering base for "first-generation" systems.

Fig. 12 a) General view of Butte, Mont., Component Development and Integration Facility.

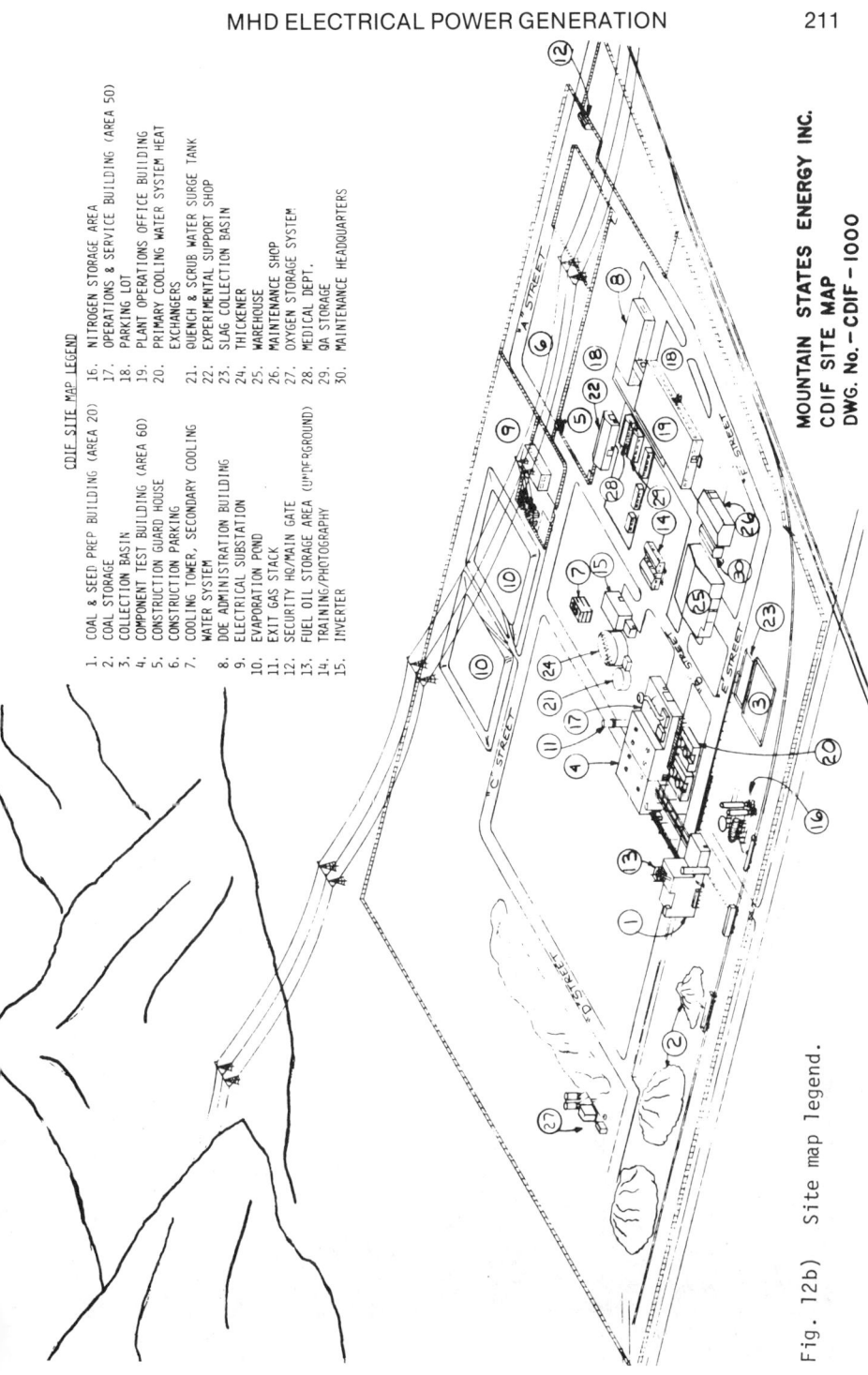

Fig. 12b) Site map legend.

Goals:

1) Demonstrate coal-fired power train operation using existing combustor, channel, and magnet technology.

2) Complete AEDC facility program of channel performance testing.

3) Develop heat and seed recovery component date base using existing facilities.

4) Focus analytical effort on the ETF and particularly the ETF power train.

5) Make every effort to insure that the manufacturing sector of industry is in a position to develop its own design philosophies and manufacturing approaches.

In conjunction with the foregoing, it would be appropriate to encourage other agencies (e.g., the National Science Foundation and Basic Energy Sciences within DOE) to support work on advanced concepts. Other steps could include the shifting of technical emphasis at major facilities toward the manufacturing sector and increasingly in-

Fig. 13 Preliminary model of USSR 500 MW electrical output commercial demonstration plant.

volving the utility industry in the planning and management of the program.

VII. Conclusions

This critique has developed the position that MHD electrical power generation has a solid basis of achievement to date and strong justification for continuation. This is based on the combination of the results of system analysis and the technical results obtained. In the USSR, where somewhat different energy policies have been pursued and the support of MHD has been continuous in nature, the construction of a 500 MW electrical plant is currently in progress at Riazan with initial operation expected in the mid-1980s. This gas-fired unit, shown in preliminary model form in Fig. 13 is the best possible evidence of the reality of MHD.

In the United States, with its logical commitment to coal-fired operation, MHD is at about the midpoint of its overall development and surely qualifies as a project for which "government support should be focused on longer-term, high-risk R&D with the potential for high pay-off" (this quotation is taken from a list of criteria promulgated by the U.S. Office of Management and Budget early in 1981). Industry support for demonstration and commercialization can be expected on a return-on-investment basis only when the engineering issues have been fully resolved and an adequate design basis has been developed for a coal-fired demonstration facility

The abundance of coal relative to other fossil fuels and the uncertain future of nuclear power on cost and public acceptance grounds virtually guarantee the dominant role of coal for electrical power generation in the 21st century. However, serious questions arise concerning the viability of existing low-efficiency technologies, especially when environmental considerations are taken into account. This includes the global effects of carbon dioxide production to which conventional coal technologies are particularly prone. Real opportunities exist for new technologies not just to add a few efficiency points, important though these are, but to enable a major energy resource to be utilized in an environmentally acceptable manner. This is the real importance of MHD, because it emerged from the ECAS study as the electrical power generation technology best qualified to meet these requirements.

It remains only to observe that MHD development is at a critical stage in which the required facilities have been constructed and engineering development of critical components and subsystems can be conducted in accordance with an orderly plan (Jackson and Levi 1978). In this period of fiscal stringency, a core program for this development can be defined and carried through within the framework of the plan of the present program. Only when this has been done will it be possible to reach a realistic assessment of MHD technology and its prospects for a major role in 21st century electricity generation.

References

Faraday M. (1832) "Experimental researches in electricity," Philos. Trans. R. Soc. London, Vol. 15, p. 175.

Griswold J., Moyer J., and Wehrey M. (1979) "The retrofit approach to MHD demonstration and commercialization," Proc. 18th Symp. on Eng. Aspects of MHD, Montana College of Mineral Sci. and Tech., Butte.

Heywood J.B. and Womack G.J. (1969) Open Cycle MHD Power Generation, Pergamon Press, London.

Jackson W.D. (1976) "MHD electrical power generation: prospects and issues," AIAA Paper 76-309.

Jackson W.D. and Levi E. (1978) "Magnetohydrodynamic power generation: program planning and status," Paper F78-658-7 presented at IEEE PES Summer Meeting, Los Angeles.

Joint NEA/IAEA Liaison Group on MHD ELectrical Power Generation (1977) "MHD electrical power generation: 1976 status report," NEA, Paris. (Earlier reports issued in 1969 and 1972, next report expected in 1982.)

Karlowitz B. and Halasz D. (1964) "History of the K and H generator and conclusions drawn from the experimental results," Proc. 3rd. Symp. on Eng. Aspects of MHD, Gordon and Breach Science Publishers, New York.

Kirillin V.A. et al. (1978) "The U-25B facility for studies in strong magnetohydrodynamic interaction," Proc. 17th Symp. on Eng. Aspects of MHD, Stanford Univ., Stanford, Calif.

Kuczen K., Killpatrick D., Pereira A.J., Leavens W.M., Clark R., and Turner A. (1980) "Fabrication of the U-25 MHD generator," Proc. 7th Int. Conf. on MHD Elec. Power Gen., Mass. Inst. of Tech., Cambridge.

Petrick M. and Shumyatsky B.Ya. (1978) "Open-cycle magnetohydrodynamic electrical power generation," A Joint U.S.A./U.S.S.R. Publication, Nauka Press, Moscow and Argonne Natl. Lab., Argonne, Ill.

Rosa R.J. (1968) Magnetohydrodynamic Energy Conversion, McGraw-Hill Book Co., New York.

Rudins G. (1981) "Coal fired MHD programs in the USA," Paper presented at Specialists Meeting on Coal Fired MHD Power Generation, Sydney, Australia.

Rudins G., Sheindlin A.E. et al. (1980) "A study of the U-25B facility MHD generator under conditions of strong electric-magnet fields," Proc. 7th Int. Conf. on MHD Elec. Power Gen., Mass. Inst. of Tech., Cambridge.

Seikel G.R., Sovie R.V. et al. (1976) "A summary of the ECAS performance and cost results for MHD systems," Proc. 15th Symp. on Eng. Aspects of MHD, Univ. of Pennsylvania, Philadelphia.

Vetter A.A., Maxwell C.D., and Demetriades S.T. (1980) "The STD/MHD codes: comparison of analysis with experiment," Proc. 7th Int. Conf. on MHD Elec. Power Gen., Mass. Inst. of Tech., Cambridge.

Formulation of the Slip Loss in a Two-Phase Liquid-Metal Magnetohydrodynamic Generator

G. Fabris*
Argonne National Laboratory, Argonne, Ill.

Abstract

Nicklin's derivation of the slip loss in a vertical two-phase flow has been extended to a two-phase LMMHD generator flow, where that loss can be very important. The slip loss was derived by considering energy exchange between liquid and gas. The energy exchange approach indicates somewhat larger slip losses. Comparison of two expressions shows that the energy exchange approach yields a more universal formulation of the slip loss. Loss of LMMHD generator efficiency is discussed in more detail.

I. Introduction

Although there are many experimental and theoretical studies of two-phase flows, only a few of them treat energy loss resulting from slip. This is probably because the presence of several interrelated loss mechanisms in two-phase flow makes it difficult to isolate and formulate the loss of energy caused by slip. Often researchers are primarily interested in heat transfer, in which case slip loss is not significant. However, because of the existence of pressure drops two to three orders of magnitude higher in a liquid-metal MHD (LMMHD) generator than in ordinary two-phase flows, the loss of energy (or generator efficiency) that re-

---

Paper presented at Third Beer-Sheva International Seminar on Magnetohydrodynamic Flows and Turbulence, Ben-Gurion University of the Negev, Beer-Sheva, Israel, March 23-27, 1981. This paper is declared a work of the U.S. Government and therefore is in the public domain.
 *Senior Scientist. (Present address: Combustion Dynamics and Propulsion Technology Division, Science Applications, Inc., Canoga Park, Calif.)

sults from slip could be much more important. In addition, the slip itself could be greater because of the much higher pressure gradient [Fabris and Hantman (1976)], although the most recent data show very small slips [Fabris et al. (1979)]. Thus, it is necessary to analyze and define clearly the slip loss mechanism in two-phase LMMHD flow. The only existing expression for the LMMHD slip loss is given in Cutting and Amend (1974) based on the derivation of Petrick et al. (1970).

Among the few papers which treat the energy balance in non-MHD vertical two-phase flow in some detail, two can be singled out. The first [Richardson and Higson (1962)] presents extensive experimental data on efficiencies and losses in an air-lift pump. However, it simply states that the slip loss is proportional to the additional airflow rate required by the existence of a slip velocity; no other physical reasoning about the slip loss is given. The second paper, by Nicklin (1962), presents an enlightening discussion on the decomposition of the pressure drop in a vertical two-phase flow and derives an expression for the pressure drop resulting from slip. In this paper, Nicklin's derivation will be used as a starting point and extended to MHD flow and the results will be compared with another derivation based on direct mechanical energy exchange between the two phases.

II. Slip Loss in Vertical Two-Phase Flow after Nicklin

Nicklin (1962) divides pressure drop in a vertical two-phase flow into reversible and irreversible parts,

$$\Delta p = \Delta p_r + \Delta p_i \tag{1}$$

This assumes negligible acceleration effects and no associated pressure change. The pressure drop due to irreversibilities is further divided into those due to wall friction and those due to slip,

$$\Delta p_i = \Delta p_w + \Delta p_s \tag{2}$$

The balance of external forces on the fluid in a vertical two-phase flow column is

$$\Delta p = \Delta p_g + \Delta p_w \tag{3}$$

Eliminating Δp and Δp_w from Eqs. (1-3) yields

$$\Delta p_s = \Delta p_g - \Delta p_r \qquad (4)$$

The gravitational pressure drop is equal to the weight of a two-phase flow column

$$\Delta p_g = (1 - \alpha)\rho_L g \Delta x + \alpha \rho_G g \Delta x \qquad (5)$$

Reversible pressure drop lifts liquid and gas and increases their potential energy. Accordingly, the pressure drop due to slip is

$$\Delta p_s = [(1 - \alpha)\rho_L + \alpha \rho_G] g \Delta x - [\frac{L}{G+L}\rho_L + \frac{G}{G+L}\rho_G] g \Delta x \qquad (6)$$

where L and G are the volumetric flow rates of liquid and gas. Finally, according to Nicklin, the energy loss due to slip is

$$W_s = (G + L)\Delta p_s \qquad (7)$$

III. Slip Loss in Two-Phase LMMHD Flow Using Nicklin's Approach

An infinitesimal element of an MHD channel is shown in Fig. 1. Assume that the channel is horizontal to cancel out gravity effects and that the velocity of the liquid and the magnetic flux density are constants as needed for optimum generator efficiency [Petrick et al. (1970)]. In analogy to the derivation given in the previous section.

$$\Delta p = \Delta p_r + \Delta p_w + \Delta p_s \qquad (8)$$

The momentum balance yields

$$\Delta p = JB\Delta x + \Delta p_w \qquad (9)$$

The interaction of the magnetic field and the liquid-metal flow creates the Lorentz force $(\vec{J} \times \vec{B})_x = JB$ when $\vec{J}$ and $\vec{B}$ are mutually perpendicular and x is the flow direction in the liquid metal. The Lorentz force, in turn, creates useful (reversible) electrical energy in the liquid at the rate of

$$JB \, A \, \overline{U}_L \, \Delta x \qquad (10)$$

where A is the cross section of the generator transverse to the flow. The electrical energy in Eq. (10) includes the

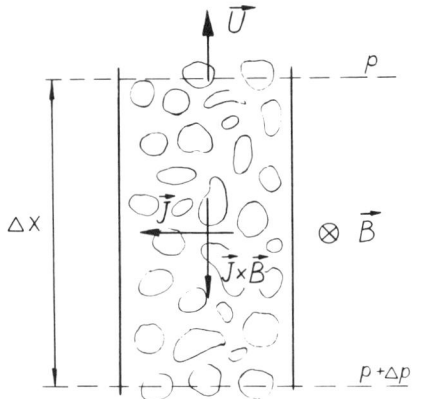

Fig. 1 Infinitesimal element of an MHD channel length.

energy to the external load and the ohmic losses in the generator, end regions, and busbars; they are electrical and are not caused by the slip.

The energy balance equation is

$$(G + L)\Delta p = JB \ A \ \overline{U}_L \ \Delta x + (G + L)(\Delta p_w + \Delta p_s) \quad (11)$$

yielding

$$\Delta p_r = JB \ A \ \overline{U}_L/(G + L)\Delta x \quad (12)$$

Note that the reversible pressure drop is not equal to the electromagnetic pressure drop, JB Δx. From Eqs. (8), (9), and (12),

$$\Delta p_s = JB \ \Delta x - \frac{JB \ A \ \overline{U}_L}{G + L} \Delta x \quad (13)$$

The pressure gradient loss due to the slip is

$$\nabla p_s = \frac{\Delta p_s}{\Delta x} = JB - \frac{JB \ A \ \overline{U}_L}{G + L} \quad (14)$$

Using

$$G = \overline{U}_G \ \alpha \ A \quad (15)$$

$$L = \overline{U}_L (1 - \alpha) A \quad (16)$$

and

$$S = \overline{U}_G/\overline{U}_L \quad (17)$$

in Eq. (14) results in

$$\nabla p_s = JB\frac{\alpha(S-1)}{\alpha(S-1)+1} \tag{18}$$

Figure 2 shows different ways of subdividing the pressure drops according to the foregoing considerations.

From Eq. (14), the energy loss per unit length due to slip is

$$W_s = (G + L)\nabla p_s = \alpha A\ JB(\overline{U}_G - \overline{U}_L) \tag{19}$$

or

$$w_s = \alpha JB(\overline{U}_G - \overline{U}_L) \tag{20}$$

per unit volume, that is, directly proportional to the Lorentz force, local average void fraction, and average local slip velocity.

IV. Energy Approach Derivation of the LMMHD Slip Loss

The mechanical energy exchange between gas and liquid can be analyzed directly. (There is also significant thermal energy exchange [Fabris and Pierson (1978)], but it is not directly associated with the slip.) There are two types of work transfer: 1) displacement of the liquid due to the expansion of the gas bubbles, and 2) interfacial drag force due to slip of bubbles through the liquid. Expansion of the bubbles occurs at quite low velocities, and causes practically negligible viscous motion in and around the bubbles. Accordingly, all indications are that the

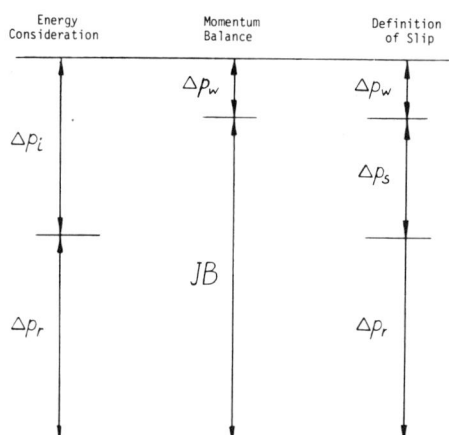

Fig. 2 Balance of pressure drops.

displacement work is performed practically without loss. (This can be compared with the assumed reversibility of piston work in equilibrium thermodynamics.) On the other hand, the gas bubbles do not use their own energy (enthalpy) to slip through the liquid. The pressure distribution around the bubbles forms a resultant force that propels them relative to the liquid, or fluid mechanical energy is expanded to slip the bubbles. Only a very minor fraction of this energy is likely to be dissipated within the bubbles as a result of gas motion; most of it is transmitted back to the liquid through interfacial drag. In the liquid it is mostly dissipated by viscosity and turbulence around the bubbles; the remainder is converted into an increase in the kinetic energy of the liquid immediately surrounding the bubbles ("associated mass"). These spatial inhomogeneities in the liquid kinetic energy are not converted into useful electrical energy but, analogously to the kinetic energy of turbulence, are eventually dissipated.

The force acting on the bubbles because of the total pressure gradient ∇p is $\alpha \nabla p$ per unit volume of mixture (this is a buoyancy-like force). According to the foregoing discussion, the slip loss per unit volume is

$$w_s = \alpha \nabla p (\overline{U}_G - \overline{U}_L) \qquad (21)$$

V. Comparison of Results and Discussion

Comparing Eqs. (20) and (21) for the energy loss per unit volume due to slip, they are identical except that Eq. (20) contains the electromagnetic pressure gradient JB, whereas Eq. (21) contains the total static pressure gradient ∇p. As shown in Fig. 2, Eq. (21) yields somewhat more slip loss, although the difference is small for practical LMMHD generators.

It should be noted that Eq. (20) was derived using ∇p_s, which is not an external force acting on the two-phase flow; it is defined by Eq. (18) only to account for the increased energy loss when the slip is present that cannot be accounted for by the wall friction pressure drop. It is exactly in these terms that the difference between the two expressions can be explained. The total specific (per unit

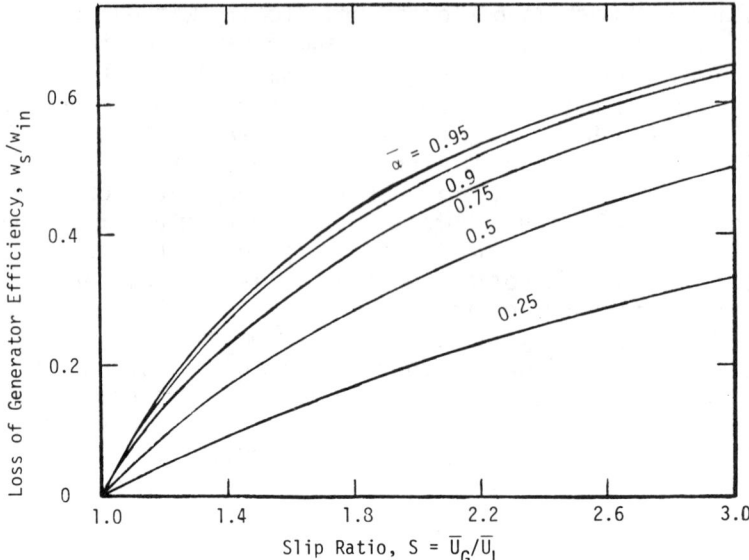

Fig. 3 Loss of generator efficiency due to slip.

volume) work exerted by the fluid is

$$(G + L)\nabla p = (\alpha \overline{U}_G + (1 - \alpha)U_L)(JB + \nabla p_w)$$

$$= \alpha[\overline{U}_L + (\overline{U}_G - \overline{U}_L)] + (1 - \alpha)\overline{U}_L(JB + \nabla p_w)$$

$$= [\alpha \overline{U}_L + (1 - \alpha)\overline{U}_L]JB + \alpha JB(\overline{U}_G - \overline{U}_L)$$

  (reversible electri- (loss due to slip with
  cal work)   electromagnetic force)

    (i)   (ii)

$$+ [\alpha \overline{U}_L + (1 - \alpha_L)\overline{U}_L]\nabla p_w + \alpha \nabla p_w(\overline{U}_G - \overline{U}_L)$$

  (loss due to wall   (loss due to wall
  friction at no slip) friction with slip)

    (iii)   (iv) (22)

It can be readily seen that term ii in Eq. (22) is identical to the expression in Eq. (20), and that if term iv from Eq. (22) is added the expression in Eq. (21) will be obtained.

Term iv is a cross term because of the existence of both wall friction and slip; but, by the same token, term ii is also a cross term because of the presence of the Lorentz force and slip. Accordingly, Eq. (21) more completely describes the energy loss due to slip in a two-phase MHD flow. It also expressed more universally the nature of the slip loss, i.e., as a consequence of the overall pressure gradient force and the slip velocity. For example, if there were a pressure gradient due to acceleration in a two-phase MHD generator, Eq. (21) could be readily used to calculate the slip loss, whereas Eq. (20) could not.

Cutting and Amend (1974) give

$$w_s = C_d' \rho_\ell (1-\alpha)^3 (\overline{U}_G - \overline{U}_L)^2 |\overline{U}_G - \overline{U}_L| \quad (23)$$

as the LMMHD slip loss. The derivation of Eq. (23) is based [Petrick et al. (1970)] on the correct concept that the bubble drag force multiplied by the slip velocity causes the slip loss. However, Eq. (23) contains the empirical coefficient C_d', and the conclusions drawn by Cutting and Amend (1974) are incorrect, e.g., the slip loss increases with the cube of the slip velocity.

VI. Loss in Generator Efficiency Due to Slip

To obtain an expression for the decrease in the LMMHD generator efficiency due to the slip, Eq. (21) should be divided by the insentropic power input per unit volume or

$$\frac{w_s}{w_{in}} = \frac{\alpha \nabla p (\overline{U}_G - \overline{U}_L)}{\nabla p [\alpha \overline{U}_G + (1-\alpha) \overline{U}_L]} = \frac{\alpha(S-1)}{\alpha(S-1)+1} \quad (24)$$

The loss increases with increasing slip and void fraction and can be very substantial, as shown in Fig. 3. Efficient conversion of thermal into electrical energy imposes the need to operate the generator at high void fractions [Fabris and Pierson (1978)]. Accordingly, the best performance of an LMMHD system can be obtained by maintaining the slip ratio as close as possible to unity. The recent high-efficiency generator data did indeed yield slip ratios of approximately unity [Fabris et al. (1979)]. The suggestion [Fabris and Hantman (1976)] of using a surface active additive to create foam flow in the LMMHD generator is very promising.

Acknowledgment

The work reported here was performed under the auspices of the U.S. Department of Energy and supported by the Office of Naval Research.

References

Cutting J.C. and Amend W.E. (1974) "Status of the two-phase liquid-metal MHD program at the Argonne National Laboratory," Proc. 9th IECEC, San Francisco, Calif., pp. 1058-1063.

Fabris G. and Hantman R.G. (1976) "Fluid dynamic aspects of liquid-metal gas two-phase magnetohydrodynamic power generators," Proc. 1976 Heat Transfer Fluid Mech. Inst., Stanford University Press, Stanford, Calif., pp. 92-113.

Fabris G. and Pierson E.S. (1978) "The role of interfacial heat and mechanical energy transfers in a liquid-metal MHD generator," Paper 78-WA/HT-33, ASME 1978 Winter Annual Mtg., San Francisco, Calif. (also Energy Convers., Vol. 19, 1979, pp. 111-118.)

Fabris G., Pierson E.S., Pollack I., Dauzvardis P.V., and Ellis W. (1979) "High-power density liquid-metal MHD generator results," Proc. 18th Symp. on Eng. Aspects of MHD, Butte, Mont., pp. D-2.2.1-D.2.2.6.

Nicklin D.J. (1962) "Two-phase bubble flow," Chem. Eng. Sci., Vol. 17, pp. 693-702.

Petrick M., Amend W.E., Pierson E.S., and Hsu C. (1970) "Investigation of liquid-metal magnetohydrodynamic power systems," Argonne National Lab., Argonne, Ill., ANL/ETD-70-12.

Richardson J.F. and Higson D.J. (1962) "A study of the energy losses associated with the operation of an air-lift pump," Trans. Inst. Chem. Eng., Vol. 40, pp. 169-182.

A Review of Recent Developments in Electromagnetic Flow Measurement

Roger C. Baker*
Cranfield Institute of Technology, Cranfield, England

Abstract

A brief historical summary is followed by a classification of flowmeters by the types of liquid to highlight the important differences. The basic equations are reviewed briefly for each class of operating liquid and an attempt is made to explain the basic physics. The paper reviews some recent published work on design, with emphasis on the most common type of electromagnetic flowmeter, that used for aqueous-based liquids. This is followed by a consideration of the various types of electromagnetic probe, noting the renewed interest in this device. The most evident unsolved problem areas, in particular zero drift, installation effects, standards, and the increasing range of difficult fluid types, are discussed and future trends suggested.

Nomenclature

a	= radius of tube, radius of probe
B	= typical magnetic flux density
$\overline{B}$	= vector flux density
B_y, B_r, B_θ, B_z	= components of vector flux density
$\overline{E}$	= vector electric field
I	= current turns in flowmeter coils
$\overline{J}$	= vector electric current density
$\overline{J}_v$	= vector virtual current density

Paper presented at Third Beer-Sheva International Seminar on Magnetohydrodynamic Flows and Turbulence, Ben-Gurion University of the Negev, Beer-Sheva, Israel, March 23-27, 1981. Copyright © American Institute of Aeronautics and Astronautics, Inc., 1982. All rights reserved.

*Professor of Fluid Engineering and Head, Fluid Engineering Unit.

M	=	Hartmann number = $Ba(\sigma/\eta)^{1/2}$
$\overline{M}$	=	vector magnetization of a fluid
$\overline{P}$	=	vector polarization of a dielectric
Q	=	volumetric flow rate
R	=	Reynolds number = $\rho a v_m/\eta$
R_m	=	magnetic Reynolds number = $\mu\sigma a v_m$
r	=	radial coordinate
S	=	skin effect number
t	=	liner thickness
U	=	electric potential
$\overline{v}$	=	vector velocity
v_z	=	velocity component along pipe
v_m	=	mean velocity in the pipe
$\overline{W}$	=	Bevir weight vector, $W = \int_{-\infty}^{\infty} W_z \, dz$
x,y,z	=	coordinate directions
γ	=	magnetic stream function
ΔU_{EE}	=	potential difference between electrodes
ε	=	fluid permittivity
ε_0	=	permittivity of free space
η	=	fluid viscosity
θ	=	polar coordinate
μ	=	fluid permeability
μ_0	=	permeability of free space
ρ	=	fluid density
σ	=	fluid conductivity
τ	=	variable of integration, pulsed flowmeter elapsed time
$\overline{\omega}$	=	vorticity
ω	=	angular frequency

I. Historical Review

"I made experiments therefore (by favour) at Waterloo Bridge, extending a copper wire nine hundred and sixty feet in length upon the parapet of the bridge, and dropping from its extremities other wires with extensive plates of metal attached to them to complete contact with the water. The wire therefore and the water made one conducting circuit; and as the water ebbed or flowed with the tide, I hoped to obtain currents analogous to those of the brass ball."

"I constantly obtained deflections at the galvanometer, but they were very irregular, and were in succession referred to other causes than that sought for. The different condition of the water as to purity on the two sides of the river; the difference in temperature; slight differences in the plates, in the solder used, in the more or less perfect contact made by twisting or otherwise; all produced effects in turn; and though I experimented on the water passing through the middle arches only; used platina plates instead of copper; and took every other precaution, I could not after three days obtain any satisfactory results."

With these intriguing words Faraday (1832) recorded his attempts to obtain a measurable voltage across the Thames at Waterloo Bridge, as it flowed through the vertical component of the Earth's magnetic field. Wollaston (1881) succeeded where Faraday failed by using the newly laid cross-channel telegraph cable and the flow in the English Channel. Smith and Slepian (1917) were the first to suggest a use for the phenomenon which Faraday had predicted--namely, the induction of a voltage due to the flow of a fluid in a magnetic field. Their device was a ship's log. Williams (1930) was the first to experiment with flow in a pipe with an imposed magnetic field.

However, medical researchers were the first to make use of the device and Kolin (1936, 1941) in particular suggested several features which are now established--reference coil, nonpolarizable electrodes, cuff-type flowmeter for medical use, measurement from outside the blood vessel, ac excitation, coil to provide a compensation for the transformer signal due to the ac induction in coil loops. Medical researchers have continued this tradition with the square-wave excitation proposed by Denison et al. (1955) and the thorough experimental work of Wyatt (1961).

The application of the electromagnetic flowmeter in industry was stimulated by the introduction of the Tobiflux and Foxboro meters, partly in response to the needs of the Dutch dredging industry; by the nuclear reactor developments; and by the work of Shercliff in developing the theoretical understanding of the subject. His work is collected in his book (1962) which still provides the standard work on the subject.

Other applications were also developing. Guelke and Schoute-Vanneck (1947) built a current meter to sit on the seabed. Remeniéras and Hermant (1954) built an improved current meter and a flow-through current meter.

The brief interest of Faraday in obtaining experimental results did little to foreshadow the present interest in the electromagnetic flowmeter as a commercial and scientific tool. It can be found today in fast breeder reactors for measuring sodium flow, in medical laboratories for blood and other flows, in naval and merchant ships as a probe log, in pleasure craft as a flush log, and in other configurations for special needs. But above all, it now accounts for about 14% by value of Western Europe's flowmeter products which represents a total market of about 120 million lb (Medlock 1980), and it is manufactured commercially to meet general industrial needs by about 10 manufacturers.

II. Classification of Types of Electromagnetic Flowmeter†

Classification by Liquid

Electrolytes. The most common type of electromagnetic flowmeter operates with liquids which are electrolytes. Electrolytic conduction is due to the movement of ions in the liquid. Under normal conditions there is dissociation in a water-based solution and the resulting ions move randomly within the body of the fluid until an external emf is applied. The ions then migrate to one or another electrode depending on their charge, and here chemical combination occurs resulting in an electrical current. However, the chemical combination may result in the release of gas at the electrodes. This gas layer partially insulates the electrodes from the liquid and causes change of the apparent resistance between the electrodes. This process, known as polarization of electrolytes, may be greatly reduced by using an alternating voltage across the electrolyte, and because of this the electromagnetic flowmeters for use with water-based fluids have usually used alternating excitation for the field coils.

Although the introduction of alternating excitation for these flowmeter applications resolves the major problems associated with polarization, we are still left with minor effects which, in the context of high-accuracy flow measurement, can become important. The liquid which is polarizable exhibits capacitance-like effects. These are accentuated at

---

†This section is taken from Sec. 7.2 of Developments in Flow Measurement, edited by R.W.W. Scott, Applied Science Publishers Ltd., Barking, England, 1982, and is included by permission of the publisher.

higher frequencies and may vary depending on the physical state of the liquid/solid interface. Thus, we shall have cause to mention later the problem of zero drift which may be caused by this effect.

Electronic Conductors. Liquid metals, like solid metals, conduct by electron movement so that the problems associated with polarization of electrolytes do not exist and a steady magnetic field may be used. However, the conductivity of the liquid metals is 10^5 or so higher than that of the electrolytes and the currents generated by motion through the magnetic field are correspondingly higher. Three dimensionless quantities determine the effects of this high conductivity. The magnetic Reynolds number, R_m, compares the size of magnetic field due to the motion-induced currents with the applied magnetic field. M^2/R (M is the Hartmann number and R the Reynolds number) gives the size of the secondary velocities induced by the interaction of the flow with the magnetic field. S^2, a skin-effect parameter, gives the distortion of an alternating magnetic field due to the eddy currents induced. These three parameters are given by

$$R_m = \mu \sigma v_m a$$

$$M^2/R = \sigma B^2 a / \rho v_m$$

$$S^2 = \mu \sigma \omega a^2$$

Figure 1 shows the various regions of importance. For sodium at 100°C, B = 0.1 Wb/m², $\mu = 4\pi \times 10^{-7}$, a = 0.01 m, v_m = 1 m/s, and ω = 1 kHz, their values are R_m = 0.13, M = 123, R = 13,500, M^2/R = 1.11, and S^2 = 1.3. The size of M^2/R can be reduced by reducing the magnetic field, insuring only that output signals from the flowmeter are adequate in size. Before discussing the other axes, it is necessary to distinguish between the two types of flowmeter: 1) induced voltage and 2) induced magnetic field.

So far we have discussed type 1 and for linear operation R_m must be approximately zero. Type 2 depends for its operation on a nonzero value for R_m and thus two regions exist. For low values of R_m, the induced magnetic field flowmeters will be linear, but at higher values they will become nonlinear. The effect of high values of R_m is known as "field sweeping" since the magnetic field appears to be stretched downstream by the flow. This effect will cause nonlinearity for type 1 unless the axial extent of the field is sufficient to insure that the "sweeping" will cause negligible change in strength at the electrodes.

Increase in S^2 will change the response of a meter to varying flow conditions, but will not normally introduce nonlinearity if S^2 and the velocity profile remain constant.

<u>Dielectrics</u>. A dielectric medium will generate a voltage due to motion in a magnetic field, but the problem of measuring this voltage becomes very difficult. The effect of the motion of a dielectric in a magnetic field is to align the electric dipoles from a random orientation and this effect is known as polarization of the dielectric (to be distinguished from polarization of electrolytes).

Not only is the impedance of the source very high but the signal is beset by high spurious signals from transformer effects and the embedded charge in the dielectric; thus, the optimum signal level is not greatly above the fundamental physical noise in the electronic circuit components. Thus, while the first two categories of electromagnetic flowmeter are commercially viable, it is still uncertain whether this type could ever be a commercial success.

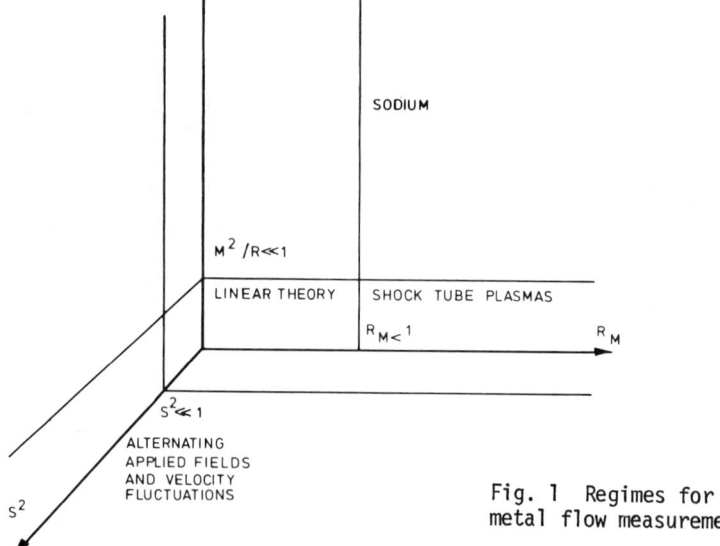

Fig. 1 Regimes for liquid-metal flow measurement.

Classification by Design

We will not consider this in any detail here as the designs are naturally discussed in succeeding sections. Here it is sufficient to point out that as well as the basic flowmeters, probes have been designed for measurement in blood, sodium, large water ducts, estuaries, and as ship's logs. Such probes may operate as an annular passage or in an unconfined flow. In addition, flush-fitting logs have been proposed and open-channel flowmeters installed for river flows.

III. Basic Equations

The theory of operation of the electromagnetic flowmeter has been fully discussed by Shercliff (1962), Bevir (1969, 1970), Al-Rabeh et al. (1978), and others. In this section an attempt is made to clarify the physical effects by using simplified forms of the general equations. First the full-field equations and constitutive relations are set down (Al-Rabeh et al. 1978).

$$\nabla \times \overline{B} = \mu_0 (\overline{J} + \dot{\overline{P}} + \varepsilon_0 \dot{\overline{E}} + \nabla \times \overline{M}) \tag{1}$$

$$\nabla \cdot \overline{B} = 0 \tag{2}$$

$$\nabla \cdot (\varepsilon_0 \overline{E} + \overline{P}) = \rho \quad \text{(where } \rho \text{ is a volume change)} \tag{3}$$

$$\nabla \times \overline{E} = -\dot{\overline{B}} \tag{4}$$

$$\overline{J} = \sigma(\overline{E} + \overline{v} \times \overline{B}) + \rho \overline{v} \tag{5}$$

$$\overline{P} = (\varepsilon - \varepsilon_0)\overline{E} + \frac{\varepsilon\mu - \varepsilon_0\mu_0}{\mu} \overline{v} \times \overline{B} \tag{6}$$

Normal Flowmeter Operating Conditions

$$(\varepsilon = \varepsilon_0, \overline{P} = 0, \rho = 0)$$

Assuming that the induced E field due to B is negligible, Eqs. (1), (4), and (5) give

$$\overline{E} = -\nabla U \tag{7}$$

$$\nabla \cdot (-\nabla U + \overline{v} \times \overline{B}) = 0 \tag{8}$$

$$\nabla^2 U = \nabla \cdot (\bar{v} \times \bar{B}) \qquad (9)$$

These equations lead to Bevir's (1970) weight vector

$$\bar{W} = \bar{B} \times \bar{J}_v \qquad (10)$$

where $\bar{J}_v$ is the distribution of current density within the flowmeter tube due to a unit current entering through one electrode and leaving through the other.

The signal is then given by

$$\Delta U_{EE} = \int_{\text{volume of flowmeter}} \bar{v} \cdot \bar{W} \, d\tau \qquad (11)$$

where

$$\nabla \times \bar{W} = 0 \qquad (12)$$

for an ideal flowmeter, one which measures the mean velocity regardless of flow profile (Bevir 1970).

The physical explanation of this regime of operation can be understood from a consideration of a two-dimensional rectangular channel with a uniform profile and a uniform imposed magnetic field.

The $(\bar{v} \times \bar{B})$ term causes a surface charge on each side of the channel which, in turn, causes an electric field within the fluid. The charge separation will allow the voltage to be measured.

Pure Dielectric Fluid Flowmeter

$$(\mu = \mu_o, \sigma = 0, \bar{M} = 0, \rho = 0)$$

Equations (3), (4), and (6) with the same assumption give

$$\nabla \cdot \{-\nabla U + [1 - (1/\varepsilon_r)](\bar{v} \times \bar{B})\} = 0 \qquad (13)$$

$$\nabla^2 U = [1 - (1/\varepsilon_r)] \nabla \cdot (\bar{v} \times \bar{B}) \qquad (14)$$

Bevir's weight vector is now modified:

$$\bar{W} = \bar{B} \times \bar{J}_v [1 - (1/\varepsilon_r)] \qquad (15)$$

Using the same simple flowmeter geometry and remembering that we now are dealing with bound charges which must always experience a net field force to remain "displaced," we find that the $(\bar{v} \times \bar{B})$ field is not completely balanced by polarization. The field of the "displaced" dipoles must be less than the $(\bar{v} \times \bar{B})$ field to insure a nonzero polarizing field. Consequently the surface charge will be less than in the previous case by a factor which depends on two quantities in particular: the number of bound charges per unit volume and the "spring" constant of each bound charge.

The Clausius-Mossotti equation relates the following quantities to ε:

$$\frac{\varepsilon - 1}{\varepsilon + 2} = \frac{N\alpha}{3\varepsilon_0}$$

where N is the number of molecules per unit volume and α the molecular polarizability, c.f., Cushing (1971).

Fluid with Embedded Static Charges

$$(\mu = \mu_0, \varepsilon = \varepsilon_0, \bar{M} = 0, \bar{P} = 0, \sigma = 0)$$

Equations (3) and (5) give

$$\varepsilon_0 \nabla \cdot \bar{E} = \rho \qquad (16)$$

$$\bar{J} = \rho \bar{v} \qquad (17)$$

Equation (17) is a statement of the fact that a flowing fluid with embedded charges constitutes an electric current. Equation (16) defines the field of the charges. These charges have been used for flow measurement by correlation techniques. It will be apparent that a two-dimensional geometry will not be useful. The theory of static charge flowmeters has been developed by Al-Rabeh and Hemp (1981).

Flowmeters for Liquid Metals

$$[\mu = \mu_0, \varepsilon = \varepsilon_0, \bar{M} = 0, \bar{P} = 0, \rho = 0, R_m = O(1)]$$

Induced Voltage. Equation (9) may no longer be assumed to have a constant value of B. We therefore need to solve B using Eqs. (1), (4), and (5), and neglecting very small terms.

$$\nabla \times \nabla \times \bar{B} = \mu_0 \sigma [\nabla \times (\bar{v} \times \bar{B}) - \dot{\bar{B}}] \qquad (18)$$

Equation (9) will then be used to obtain the distribution of U for induced voltage flowmeters.

<u>Induced Magnetic Field</u>. In the case of induced field flowmeters, Eq. (17) can be simplified by the axisymmetry of most flowmeters of this type. Thus writing

$$\frac{\partial \gamma}{\partial r} = -rB_z$$

$$\frac{\partial \gamma}{\partial z} = rB_r \qquad (19)$$

Eq. (17) becomes

$$r \frac{\partial}{\partial r} \left(\frac{1}{r} \frac{\partial \gamma}{\partial r} \right) + \frac{\partial^2 \gamma}{\partial z^2} = \mu_0 \sigma \left(\frac{\partial \gamma}{\partial t} + v_z \frac{\partial \gamma}{\partial z} \right) \qquad (20)$$

IV. Recent Developments in Pipe Flowmeter Designs

This section is divided according to the nature of the fluid. Figure 2 gives a typical flowmeter with essential components.

For Dielectric

Cushing (1958, 1965, 1971) pioneered this type of flowmeter and reported the first experiments. Hentschel (1973) more recently wrote a dissertation describing his experiments on a conventional meter modified for oil. Engl (1970, 1972) has also discussed the theory. Recent work by Al-Rabeh et al. (1978) has clarified the theory as already dis-

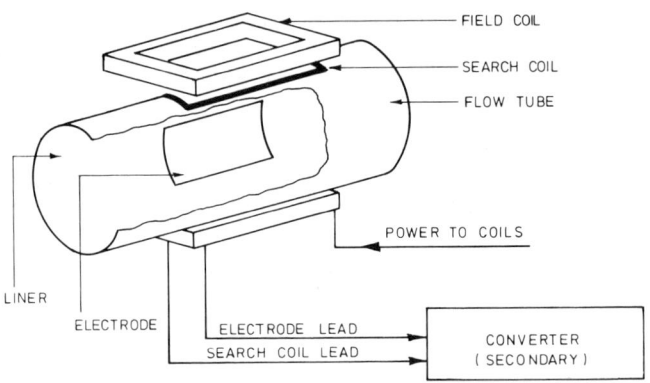

Fig. 2 Typical electromagnetic flowmeter design.

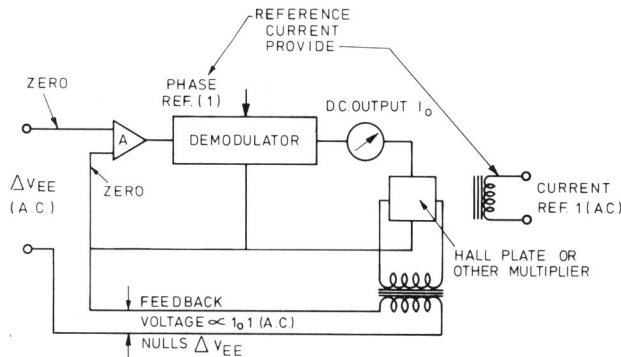

Fig. 3 Converter (secondary) of Hutcheon and Harrison (1965)

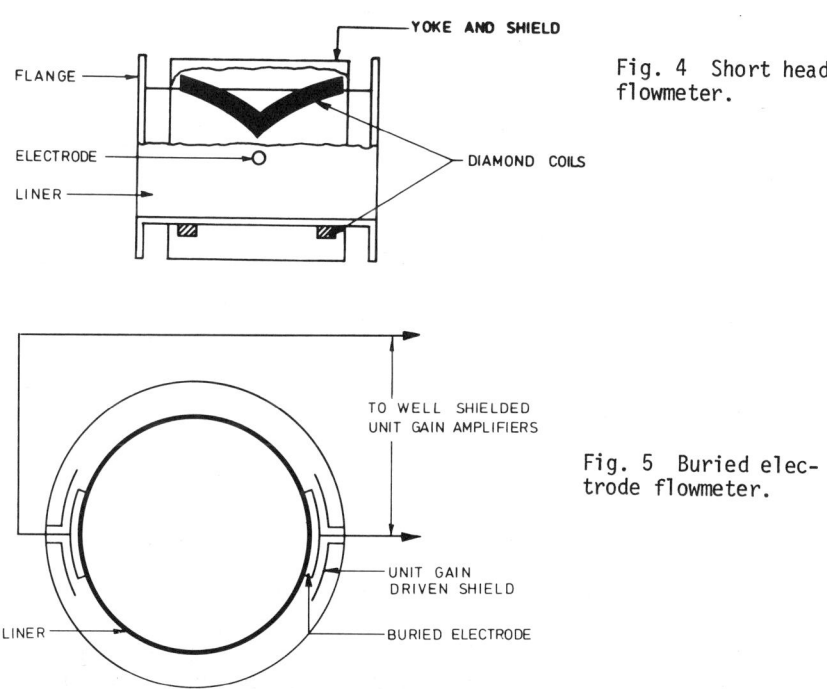

Fig. 4 Short head flowmeter.

Fig. 5 Buried electrode flowmeter.

cussed. Al-Rabeh (1981) has also described various experiments confirming the theory and pointing the way to the successful operation of these devices. His work is a demonstration of the extreme care and skill needed in successfully overcoming the very great problems in such a device.

For Low Conductivity Fluids

This group is by far the most important since it forms the bulk of the commercially available meters. The commercial importance of these meters has also spurred research to improve performance. It is necessary, therefore, briefly to review the recent history of commercial developments.

After the Tobiflux and Foxboro instruments the flowmeter became increasingly accepted in industry, finding a particularly safe niche in slurry flows. By the mid-1960s a second generation was beginning to appear represented by the Kent Veriflux converter (Fig. 3) of Hutcheon and Harrison (1965) and the "short head" design (Fig. 4) by Rummel and Ketelson (1966, 1968). This led to the theoretical work of Bevir (1969) and a general realization that a flowmeter could be made, if not ideal, at least comparable to the "long-head" designs. Kent Instruments made use of the coils analyzed by Bevir (1969) to provide the designer with a series of coil shapes with which to optimize his design. Others followed the trend. The importance of the "short head" was most apparent in the large flowmeter sizes where material savings become appreciable.

Schlumberger produced a large electrode flowmeter (Bourg and Tempe 1972) which showed very good flow profile insensitivity. Then in the mid-1970s, flowmeter manufacturers reverted to the square-wave excitation of Denison et al. (1955), now known by a variety of terms (e.g., Hofman 1978, Kiene 1978), and as a result of good marketing persuaded the users of its superiority in many applications.

The most recent trends are noncontacting or "buried electrodes" in which the potential sensing is via a capacitance link rather than by contact electrodes (Fig. 5), very short heads of about one diameter length (Sen 1978), and flowmeters which operate when partially filled.

Apart from the commercial developments, the medical people have continued to develop designs for their own use and the possibility of "dry calibration" has remained an ideal which has inspired certain work. Another important development is the differential flowmeter.

Point Electrodes. Following Bevir's work Hemp (1975) attempted an optimization to obtain the field which under certain constraints of power and efficiency would approach

ELECTROMAGNETIC FLOW MEASUREMENT

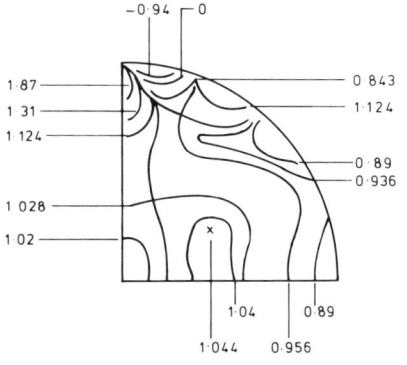

Fig. 6a Hemp's weight function.

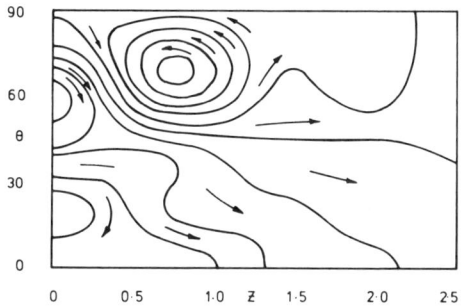

Fig. 6b Hemp's field copper strip pattern.

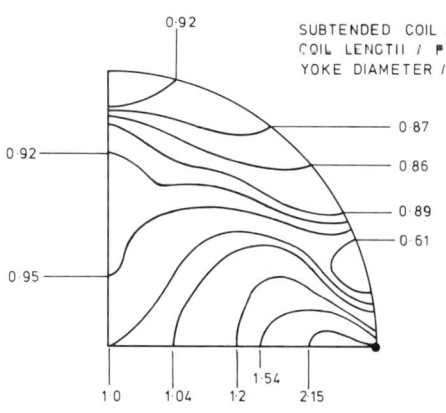

Fig. 7 Al-Rabeh's weight function for optimized diamond coil point electrode flowmeter.

the ideal for rectilinear flows. This is shown in Fig. 6a. The field winding for this (Fig. 6b) is unlikely to be a commercial proposition and so Al-Rabeh and Baker (1979) undertook a further optimization in which they constrained the coil shapes and yoke configurations to those already available commercially. By so doing they were able to demonstrate the optimum shape of the coils and yoke and to provide the designer with a powerful tool in which performance and running costs, etc., could be balanced and a suitable design compromise obtained. The best performing design weight function is shown in Fig. 7.

The ability to dry calibrate electromagnetic flowmeters has been both an ideal and a misleading idea. Haacke (1978) describes a flow simulator which he has used to achieve dry calibration and claims an accuracy of 1.75% or better. Recent work has emphasized the need for extreme skill if "dry calibration" is to be a useful method. Bevir et al. (1981) describe a very elegant piece of work in which magnetic field measurements around the tube wall are used to compute the weight function and very good agreement is obtained between prediction based on the pipe profiles and actual flowmeter response.

Al-Rabeh (1981) has described experiments using one of his standardized coil shapes and a carefully machined flow tube. Despite great care there was a difference in amplitude between the predicted magnetic field and the measured field, leading to the need for at least one magnetic field measurement.

Hemp and Wyatt (1981) have also described a very interesting device for obtaining the "worst flow" to test a flowmeter design. This consists essentially in reversing the flowmeter so that it becomes a pump causing its own worst flow.

Large Electrodes. Bevir (1971a) demonstrated the value of large electrodes for obtaining a uniform W. Al-Khazraji and Baker (1979a) have analyzed the performance of three designs of large electrode flowmeter and have also shown with a two-dimensional analysis the likely effects of electrode fouling (Fig. 8). In a further paper (1979b) they have tested the effect of fouling and showed that unsymmetrical fouling can introduce zero drift of very serious sizes. Al-Khazraji et al. (1978), in a study comprising various flowmeter types, showed that errors due to an eccentric orifice of D/2, 5.5 D upstream of the midplane, were very small and similar to those computed.

Square-Wave Excitation. The problem of zero drift (Sec. VI) has caused manufacturers to return to the square-wave excitation (Hofman 1978b, Kiene 1978). The systems known to the author work at about 3 Hz and sampling takes

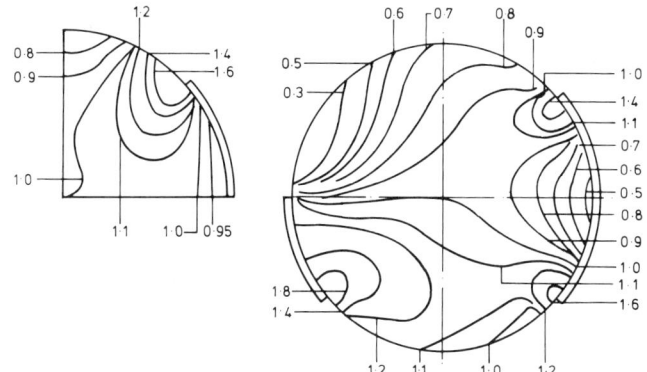

Fig. 8 Change of weight function due to electrode fouling.

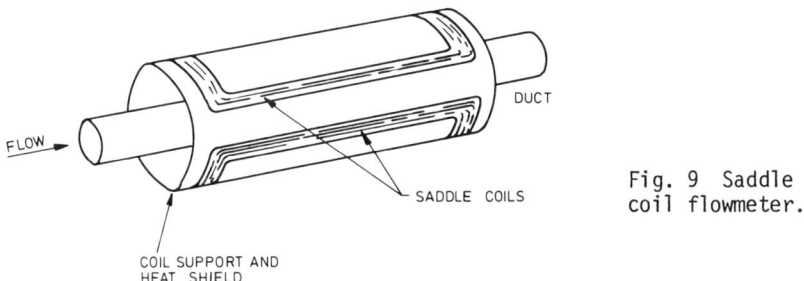

Fig. 9 Saddle coil flowmeter.

place at least three times to eliminate the effects of a constant baseline drift rate. The system is more susceptible to errors due to flow pulsations and it is not clear what effects changes in the electrochemistry of the fluid will have on calibration.

Noncontacting Electrodes. Hofman (1978a) describes such a flowmeter in which the electrodes are bonded to the outside of a Teflon liner. With the liquid they form a capacitor of which the dielectric is the liner. For a 2.5 mm thickness, $\varepsilon \simeq 2$, and electrode diameter of 25 mm for a 40 mm diam pipe, the coupling capacitance is about 2 pF leading to a source impedance at 50 Hz of about 1.5 GΩ. Problems arise from the microphonic effects when the liner is under stress; material selection to obtain high insulation quality and good bonding to electrode; preamplifier design to reduce noise. Operation is possible down to 0.2 µS/cm. It performed with deposits up to 10 mm thick, although no accuracy checks were made for these conditions.

Al-Khazraji and Baker (1979b) have given results for a 3 in. i.d. contactless flowmeter with an optimized integrating design by Al-Khazraji (1979) which has a 2% error for eccentric orifice at 2.5 diam and an effect on sensitivity due to smeared insulating material well below 1%. Al-Khazraji and Hemp (1980) have described methods for designing optimum flowmeters using noncontracting electrodes. This type promises to have a very low sensitivity to generalized flow profile variations as it is possible to approach the Bevir ideal within limitations on axial length. Short axial length will introduce an error which may be assessed.

Partially Filled. Strauss (1978) has discussed data showing the errors resulting from partially filled flowmeters. Hemp (1980) has described a method for handling a flowmeter which will operate correctly despite the fact that it runs only partially filled. Al-Khazraji and Baker (1979b) have given the weight function for a partially filled tube.

Differential Flowmeter. This flowmeter resulted from the needs of the medical community to measure very small differences in flows occurring, for example, in kidney machines. It was developed by Gray and Sanderson (1970) [c.f., Sanderson and Gray (1981)]. They demonstrated extreme skill in controlling the spurious signal sources in order to achieve good differential accuracies. They use an elegant system with the two rectangular channels back-to-back between the magnet poles.

For Liquid Metals

This flowmeter, although offered by a few commercial organizations has seen its main engineering application in nuclear reactor programs. It has been developed by organizations such as UKAEA for use in fast reactors. Its development has been aimed almost exclusively at this engineering field. It seems certain that it will have an increasing use in molten metals handling where continuous casting will require accurate monitoring. The discussion below will, therefore, center on developments from the nuclear industry [more fully reviewed by Baker (1977)].

Induced Voltage. The work of Thatcher et al. (1970) on the saddle coil flowmeter provided the best way to achieve linear response (unaffected by change) by making the field long so that sweeping would be confined to the end effects. The saddle coil was so called due to its support outside a concentric insulating cover and its saddle shape (Fig. 9). The possibility of compensating for the nonlinearity due to field sweeping by displacing the electrodes downstream was suggested by Turner (1960), but this method does not eliminate the conductivity dependence.

Tarabad (1980) has described computations to obtain a design which will not be too greatly affected by variations in the velocity profile, and he has calculated weight function distributions for swept fields. He has also used an analog rig to check some of his calculations.

Tarabad and Baker (1982) discuss axial integration of the electric potential to reduce nonlinearity and conductivity dependence.

The induced voltage flowmeter is also used in the reactor core where it serves to monitor the onset of fault conditions and where accuracy is secondary. It is designed as an insertion device: either a probe around which the flow passes or a tube within the duct through which the flow passes.

Induced Magnetic Field. The Lehde and Lang patent (1948) has been developed into a useful monitoring device for the core duct flows. It is used both in the geometry of a probe in the duct, forming an annulus for the flow, and also as an insert forming a central duct for the flow. It

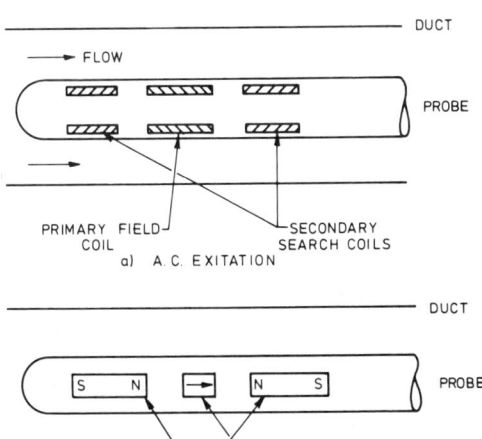

Fig. 10 Induced field flowmeter.

usually operates with ac excitation in the primary field coil and a pair of secondary search coils are used to balance out the no-flow signal. However, it is possible to use permanent magnets and a magnetometer (Weigand 1972), see Fig. 10.

Analysis of these flowmeters has been undertaken by various people. Cowley (1965) analyzed an aerofoil probe using ac excitation. Baker (1968c, 1979, 1970a) analyzed a flowmeter using dc excitation and an axial flux probe. This configuration suggested a flowmeter approaching the ideal of mean velocity measurement. In a later paper (Baker 1977) a three-coil core duct probe was analyzed by two independent computational methods. Thatcher (1978) has used an exact analysis for a flowmeter with no axial variation apart from the magnetic field. Weigand (1967) also analyzed the "flow-through" type.

Due to the axisymmetry of the induced field flowmeter it is possible to replace all conducting material by current loops. The field of a current loop is well known and it is possible to write a matrix describing the interaction of each loop with the others. One advantage of this model is that the solution does not need to be taken to large distances to reduce the boundary condition to zero. The loop fields already do this. It is also possible to use loops of different cross sections according to the accuracy of the results required (Baker 1977).

Modification of the flow due to the presence of the flowmeter has been analyzed for the constant magnetic field case (Baker 1970a).

Pulsed Field Flowmeter. Figure 11 shows a diagram of the pulsed field flowmeter. This type of flowmeter was proposed by Zepgir and Sermons (1965). The current in the field coil is stopped at time $\tau = 0$ and induces a current in the fluid which acts as a marker. The search coil signal increases as this marker is swept toward it and the reciprocal of the time to the peak signal (Figs. 11a, b, c) is related to the flow velocity as shown in Fig. 11c (Tarabad and Baker 1979). This figure shows the effect of a changing profile as would be expected since the marker will be affected by velocity in the wall region. The nonzero value of $1/\tau^*$ for zero V can be explained by the decay of the signal which causes the field distribution to be distorted in a way which moves the peak search coil signal to an earlier point in time.

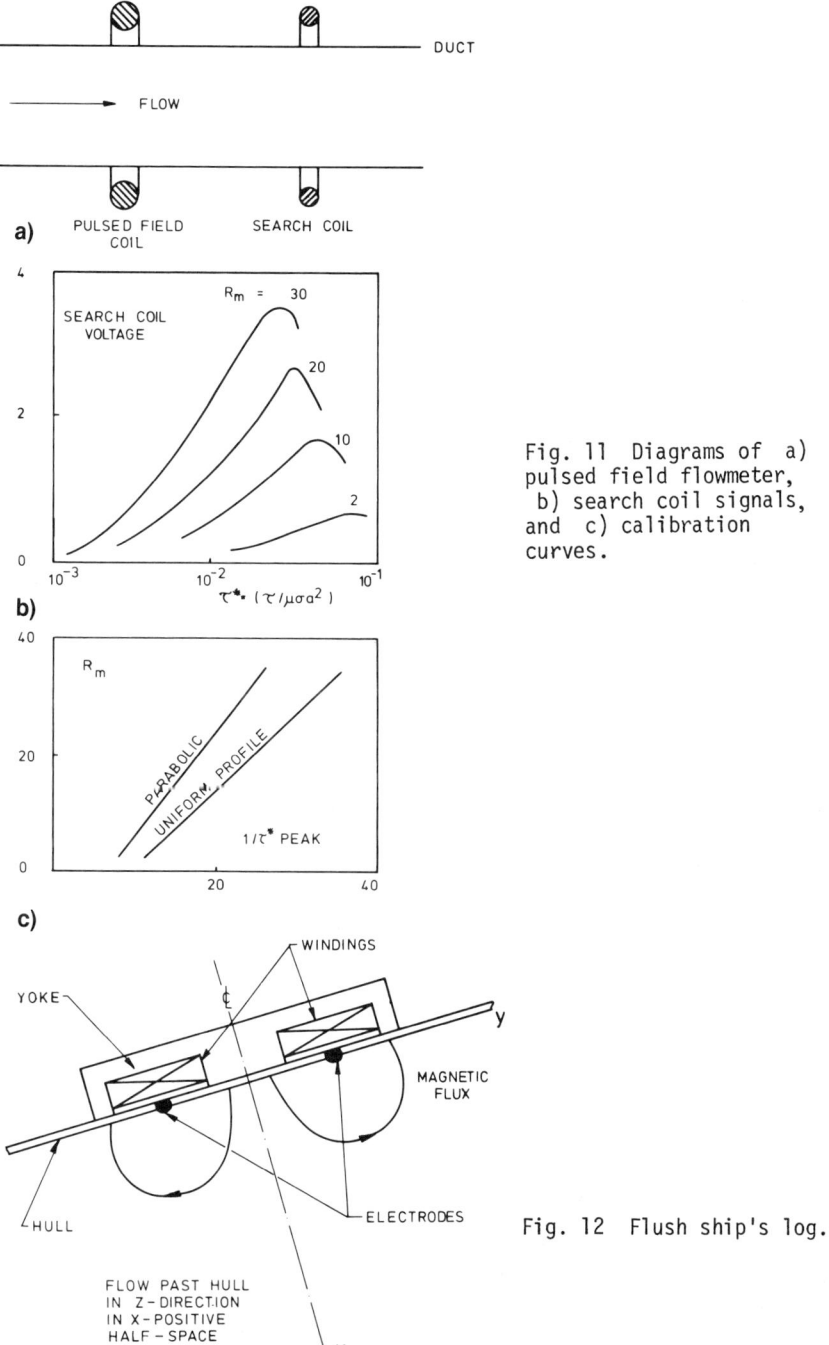

Fig. 11 Diagrams of a) pulsed field flowmeter, b) search coil signals, and c) calibration curves.

Fig. 12 Flush ship's log.

V. Induced Voltage Logs and Probes

Ship's Speed

Flush Log. The Smith and Slepian (1917) patent does not appear to have been exploited. However, the concept was revived in the late 1960s with the application for a patent (Baker 1970b) based on a more complete theory of the flush log.

In Fig. 12 the coordinates are defined which permit an expression to be written for the flush log response as (Baker 1968a)

$$U(o,y,z) = \int_0^\infty V(x/2) \, B_y(x,y,z) \, dx$$

for the potential at the ship's hull. This allows the possibility of improvements in the performance of these flush logs. If the magnetic field is suitably tailored to introduce some cancellation of the boundary layer, then the log will retain its calibration in spite of weed growth. However, flush logs, without such tailoring of the magnetic field have been used successfully for pleasure yachts.

Probe Logs. The fixed probe consists of an aerofoil section sensing head which is bolted to the hull and projects into the water (Fig. 13). A design of this type is in use and is reputed to give extremely high accuracy. The

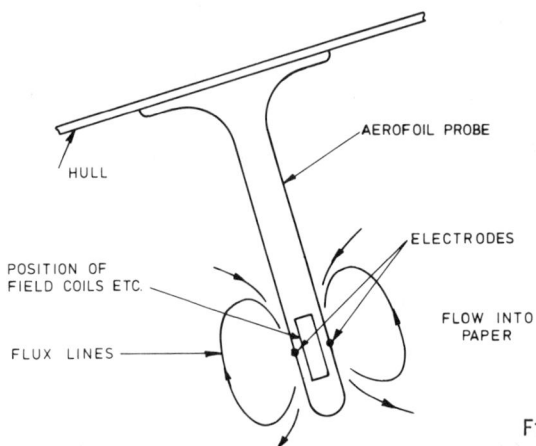

Fig. 13 Fixed probe log.

potential is sensed by electrodes on each face of the aerofoil. The electrodes are far enough outboard to be clear of boundary-layer variation on the ship's hull.

The retractable probe is essentially the same as the fixed probe (in operation) but is mechanically far more complicated since it requires a powerful mechanism to move the probe outboard through the hull fitting. The aerofoil shape may be sacrificed for a circular probe, but this will cause increased drag and uncertainty in the prediction of signal behavior due to variation of flow pattern behind the cylinder.

The Institute of Oceanographic Sciences has developed a two-component probe for its own use which is now available commercially (Tucker et al. 1970). This consists of an ellipsoidal head on an aerofoil stem. The ellipsoid has four electrodes in the top surface and provides a two-component measure of velocity (Fig. 14). This device originated as a current meter and is discussed further below (Bowden and Fairbairn 1956).

Current Meters

Guelke and Schoute-Vanneck (1947) described measurements of tidal movements, first by making use of the Earth's magnetic field and then by building a device with an alternating field. They adopted a circular device positioned on the seabed with axis perpendicular to the flow (and to the channel bed). Four electrodes were used to give directional flow information. Their theoretical assessment of the signal was approximate and to overcome the boundary-layer problems they raised the electrodes above the surface by a small amount.

Remeniéras and Hermant (1954) built two designs of the current meter. The first, ellipsoidal in shape, had a low field strength and suffered from quadrature pickup problems. This was overcome in the second design in which a ferromagnetic core was used to increase the field strength.

Bowden and Fairbairn (1956) described a two-component current meter, which developed into the IOS device already mentioned. The measuring head of the instrument was 100 mm diam and was in the form of an ellipsoid with a minor axis of 38 mm. The axis of the coil was along the minor axis and there were four electrodes equally spaced around the largest diameter. The voltages between diametrically oppos-

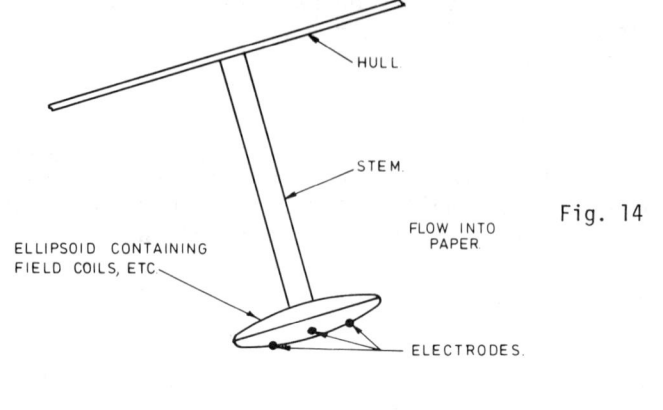

Fig. 14 IOS probe log.

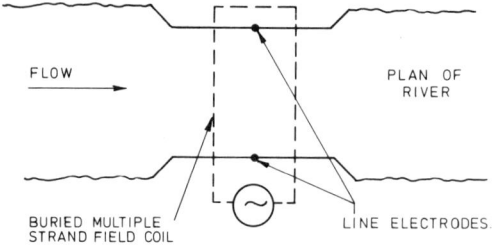

Fig. 15 River flow gaging.

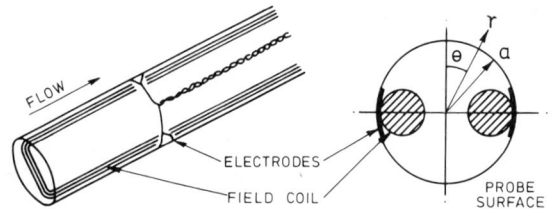

Fig. 16 Mills' catheter tip probe.

ed electrodes were measured, giving outputs proportional to the components of the water velocity.

Robinson (1976) has extended the early work which made use of submarine cables to measure induced voltages. Using a weighted vector technique he has obtained the response measured at two ends of a submarine cable due to the flow of sea water through the Earth's magnetic field.

Important developments of electromagnetic river gaging have resulted in the installation of devices at a few river sites. The magnetic field is produced by a coil consisting of two multistrand cables laid across the river bed perpendicular to the flow (Fig. 15). This loop causes a magnetic field approximately vertical in the measuring region and hence a potential across the river (Herschy and Newman 1974).

Probes

Mills Catheter-Tip Probe. A probe of this type is shown in Fig. 16. Mills (1966) designed it for insertion into an artery on the end of a catheter to measure the blood flow. A coil within the probe causes a field above the electrodes with a component perpendicular to the flow and parallel to the electrode surface. Mills appears to have used the probe with success, developing a more sophisticated version with other sensors built in.

The response of the device is given by (Baker 1968b)

$$\Delta U_{EE} = 2 \int_a^\infty V(\sqrt{ar}) \, B_\theta \, (r, \pi/2) \, dr$$

if the artery radius is large compared with the probe radius.

Bevir (1971b) showed that the weight function is very high over the electrodes. It is surprising that this type of probe is not in more general use for local velocity measurement. One somewhat similar commercial device is available.

Remeniéras and Hermant Probe. The reverse of Mills' probe is a streamlined flow-through type of probe (Remeniéras and Hermant 1954), see Fig. 17. A much greater signal strength is possible with this design, due both to the confined field and the confined flow. The disadvantage will be the danger of blockage of the flow passage occurring in slurries or other solid/liquid flows.

Vortex Probe. This device is shown in Fig. 18. It is based on the electromagnetic flowmeter equation in which $\nabla \times \overline{V}$ is replaced by $\overline{\omega}$ the vorticity

$$\nabla^2 U = \overline{B} \cdot \overline{\omega}$$

The device consists of a five-electrode probe which allows the Laplacian to be measured in finite-difference form. The reading of such a probe will yield the vorticity parallel to the local magnetic field. Some experiments were performed

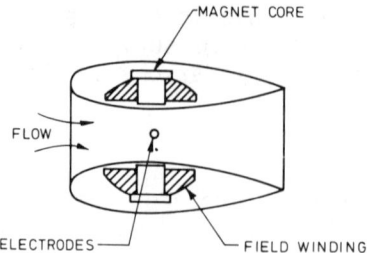

Fig. 17 Reméniéras-Hermant probe.

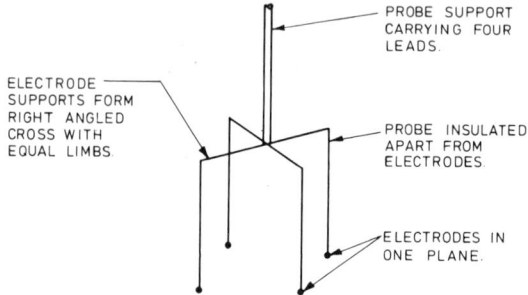

Fig. 18 Vortex probe.

to test such a device and to apply it to a particular situation. However, serious reservations as to its practical value remain (Baker 1971).

Pipe Insertion and Other Probes. Large-sized electromagnetic pipe flowmeters have been particularly expensive as the early designs required stainless steel tubes. For this reason, a few manufacturers have produced a probe as a less accurate alternative to the pipe flowmeter. The probe is inserted far enough into the pipe to sense a point in the flow profile that will have a velocity approximately proportional to the mean velocity in the pipe. The probe design is of circular or aerofoil section with electrodes usually forward of the largest section to insure a stable boundary layer and as little influence from separation behind the probe as possible.

A small insertion probe flush with the pipe inner surface is also available commercially. One would expect this to be highly sensitive to velocity profile changes. Figure 19 shows a possible configuration.

Two recent papers show renewed interest in the probe for local velocity measurement. Cushing (1978) analyzes the boundary-layer effect of the signal of the probe and shows

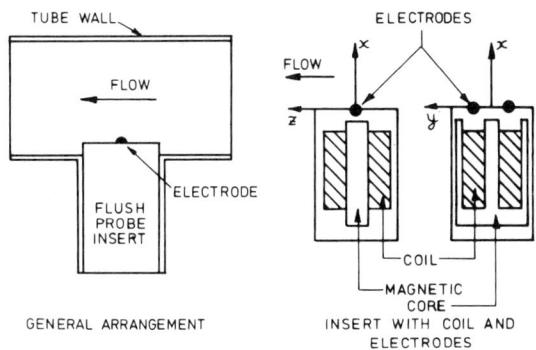

Fig. 19 Flush insertion probe.

that sensitivity can increase or decrease with speed. Nikitin (1978) discusses an optimum design which was 28 mm diam and gave a signal of 250 µV $(m/s)^{-1}$.

VI. Problem Areas and Flowmeter Evaluation

Zero Drift

Zero drift has plagued the flowmeter from its earliest use and increasing accuracy demands have stayed ahead of its solution. It still appears unclear what the causes are, although some probable causes are given below.

Quadrature Rejection. The transformer-induced signal is a major problem. It is reduced at the flow tube either by a final mechanical adjustment of the signal leads during assembly and testing or by using a third signal lead and coupling two leads with a potentiometer which can be adjusted to minimize the signal. The remaining quadrature signal is then removed by a phase-sensitive detector circuit linked either to a reference from the field coils or from a reference coil in the field.

Any change in these corrections due to unequal heating, mechanical vibration, or other unforeseen effects will cause the suppression to be less effective.

Electrode Contamination. This is reduced by "burn-off" --the application of a high current, electrode removal, or ultrasonic electrode cleaning. On point electrodes the effect of localized deposits will be less than on large electrodes where the lack of symmetry can greatly increase the quadrature pickup.

Earthing and Isolation of Signal Leads. Earth loops are a notorious problem and great care is taken to avoid them by separate Earth systems for head and electronics. The pipe work and fluid must also be grounded correctely. Another problem arises when very small capacitive links ($3 \cdot 10^{-12}$F) exist between signal leads and mains power cables. These can inject unwanted signals into the amplifier of order 1 mV (Harrison 1980).

Fields Induced by Eddy Currents. These will cause an in-phase transformer voltage and while this may be eliminated in the initial setup, if any change occurs in the eddy currents during warmup, etc., a drift will occur.

Electrochemical Effects. It appears that charge clouds will form around the areas of voltage maxima and minima. Hogan (1923) noted that such clouds were modified by fluid flow patterns. This is likely to be a particular problem in large-electrode flowmeters where the electrodes form a conducting path. It also seems likely that temperature of the fluid, turbulence level, pH number, etc., will have an effect on these clouds.

Installation Effects

International attention is currently focused on the effect of upstream pipework on calibration. de Jong (1978) has tested a series of commercial flowmeters and demonstrated that for bends and valves 5 diam were sufficient to restrict the calibration change to 1%. Reinhold (1978) reports experiments with meters of 250 mm i.d., indicating that semicircular upstream blockage at 5 diam or less may cause errors greater than 1%.

Apart from the effect of flow profile it is also possible to make estimates of the effect of other parameters. We may calculate, for instance, the effect of expansion of the tube due to temperature increase (Baker 1976). If we assume a search coil to sense the magnetic flux, then

$$\Delta U/B \propto Va$$

and assuming a constant volumetric flow, Q, for a change in temperature

$$V'a'^2 = Va^2$$

ELECTROMAGNETIC FLOW MEASUREMENT

and hence

$$\frac{(\Delta U/B)' - (\Delta U/B)}{(\Delta U/B)} = [(a/a') - 1]$$

Taking a coefficient of expansion of about $10^{-5}/°C$ we may expect a calibration change of 0.1%/100°C which is negligible.

Bevir (1972) has also given the effects of surrounding pipework on the sensitivity of certain electromagnetic flowmeters.

Standards

As well as the International Standards Organization (1980) document, the British Standards Institution (1980) has produced BS 5792 and similar documents may have been produced by other national bodies. These two standards give definitions of the main terms, a guide to the features which should appear for best practice, and a guide to the installation restrictions.

Slurries and Multiphase Flows

These flowmeters offer certain clear advantages for slurries and studies of the effects of local conductivity variation (Baker 1970c, Bevir 1971c) and of the effects of a magnetic fluid (Baker and Tarabad 1978) are relevant.

While the effect of local conductivity variation is likely to be less than 1%, the effect of permeability changes can be more important. The effect of a magnetic slurry is to increase (and probably redistribute) B in the fluid. Referring to Fig. 20 the change in field due to a slurry of very high permeability would be given approximately by

$$B_r a = \mu_0 I \quad \text{for } \mu = \mu_0$$

$$B_r' t = \mu_0 I \quad \text{for high } \mu$$

This would result in an increase in field strength and signal size of a/t.

If, however, a search coil is used then any change in the value of B will be allowed for and errors will be much reduced. However, in addition, it is necessary to insure that the field strength is low enough to avoid attracting deposits of the magnetic solids within the flow tube.

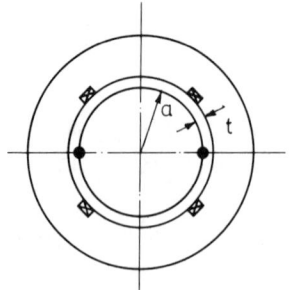

Fig. 20 Geometry for magnetic slurry flowmeter calculations.

VII. Future Trends

Design Trends

A number of new features are under consideration. Work is well advanced on noncontacting flowmeters with "buried" electrodes (made by two manufacturers). The advantages of noncontacting integrating electrodes have not yet been fully exploited and experience on performance with wall deposits is virtually nonexistent.

The need to improve the electronic processing and to introduce the advantages of a microprocessor to interrogate the signal is resulting in a new generation of secondary circuits.

The very short head is also attractive where space is minimal and accuracy is less important. One is reported by Sen (1978).

Self-checking capability to reduce early maintenance calls would be valuable. One way to self-check a flow is to introduce a correlation measurement, one form of which has been suggested by Raptis and Forster (1975). Another form of self-checking is by the use of two pairs of electrodes (Kuromori et al. 1978).

Zero drift has still not been overcome and complete systems which eliminate it will continue to be the object of new developments.

Electromagnetic probes have found uses in the nuclear and medical fields and to some extent in the oceanographic and ship's log fields, but a good probe to measure local industrial flows of slurries, etc., would be useful.

Testing Trends

The need to understand the behavior of existing designs has led to installation tests and is likely to lead to more of these tests to check zero behavior, installation effects, parametric (temperature, etc.) effects, slurry behavior, and two-phase and multiphase flow behavior.

Application Trends

The process industries in general are seeking increased flow accuracy and the electromagnetic flowmeter offers a suitable solution in continuous casting, nuclear waste disposal, and other areas already mentioned. Here, there are increasingly difficult demands made for materials to cope with new liquids and extreme temperatures and pressures.

VIII. Conclusion

In this paper an attempt has been made to review some of the theoretical and experimental developments in electromagnetic flow measurement as well as mentioning some of the outstanding problems.

Much industrial activity remains in the flowmeter development and new applications are still appearing. Theoretical treatment for these more complex flows will be needed in due course to improve design and give operating confidence, while more experimental data will be essential to insure continuing confidence in their use.

Acknowledgments

I am very grateful to the organizing committee of the Beer-Sheva Seminar on MHD Flows and Turbulence for inviting me to give this paper. I am also very much aware of my debt to the many colleagues who have helped me to understand the subject and to keep abreast of the latest technology. My thanks to them all, but particularly to Riadh H. Al-Rabeh who kindly checked this manuscript for me. My thanks are also due to Mrs. L.B. Hampshaw for typing the manuscript in a very short time.

References

Al-Khazraji Y.A. (1979) "Electromagnetic flowmeters with large electrodes," PhD Thesis, London Univ., England.

Al-Khazraji Y.A., Al-Rabeh R.H., Baker R.C., and Hemp J. (1978) "Comparison of the effect of a distorted profile on electromagnetic, ultrasonic and differential pressure flowmeters," Proc. FLOMEKO 1978 - Flow Measurement of Fluids, Groningen, the Netherlands, pp. 215-22.

Al-Khazraji Y.A. and Baker R.C. (1979a) "Analysis of the performance of three large-electrode electromagnetic flowmeters," J. Phys. D, Vol. 12, p. 1423.

Al-Khazraji Y.A. and Baker R.C. (1979b) "New design concepts for electromagnetic flowmeters for the process industries--Some advantages and disadvantages," Proc. IMEKO Symp., Japan, Paper 66-1.

Al-Rabeh R.H. (1981) "The design and performance of electromagnetic flowmeters," PhD Thesis, London Univ., England.

Al-Khazraji Y.A. and Hemp J. (1980) "Electromagnetic flowmeters and methods of measuring flow," Patent application No. 8011624.

Al-Rabeh R.H., Baker R.C., and Hemp J. (1978) "Induction flow-measurement theory for poorly conducting fluids," Proc. R. Soc. London, Ser. A, Vol. 361, p. 93.

Al-Rabeh R.H. and Baker R.C. (1979) "Optimisation of conventional electromagnetic flowmeters," NEL Fluid Mech. Silver Jubilee Conf., Paper 6.1.

Al-Rabeh R.H. and Hemp J. (1981) "A new method for measuring the flow-rates of insulating fluids," Proc. Int. Conf. on Advances in Flow Measurement Techniques, Warwick Univ., England, pp. 267-278.

Baker R.C. (1968a) "On the potential distribution resulting from flow across a magnetic field projecting from a plane wall," J. Fluid Mech., Vol. 33, pp. 73-86.

Baker R.C. (1968b) "Solutions of the electromagnetic flowmeter equation for cylindrical geometries," Br. J. Appl. Phys. (J. Phys. D), Vol. 1, Ser. 2, pp. 895-899.

Baker R.C. (1968c) "The motion induced magnetic field flowmeter for various applications," 6th Symp. on Magnetohydrodynamics, Riga.

Baker R.C. (1969) "Flow measurement with motion induced magnetic field at low magnetic Reynolds numbers," Magnetohydrodynamics (USSR), No. 3, pp. 69-73 (in Russian).

Baker R.C. (1970a) "Linearity of motion-induced-magnetic-field flowmeter," Proc. Inst. Electr. Eng., Vol. 117, pp. 629-633.

Baker R.C. (1970b) "Improvements in or relating to fluid flow velocity measurement," Patent 1 206 463.

Baker R.C. (1970c) "Effects of non-uniform conductivity fluids in electromagnetic flowmeters," J. Phys. D, Vol. 3, pp. 637-639.

Baker R.C. (1971) "On the electromagnetic vortex probe," J. Phys. E, Vol. 4, pp. 99-101.

Baker R.C. (1976) "Some recent developments in ultrasonic and electromagnetic pipe flow measurement at Imperial College," IMEKO VII, Paper BFL/246.

Baker R.C. (1977) "Liquid metal electromagnetic flowmeters in fast reactor technology," Prog. Nucl. Energy, Vol. 1, No. 1, New Series, pp. 41-61.

Baker R.C. and Tarabad M. (1978) "The performance of electromagnetic flowmeters with magnetic slurries," J. Phys. D., Vol. 11, p. 167.

Bevir M.K. (1969) "Induced voltage electromagnetic flowmeters," PhD Thesis, Warwick Univ., England.

Bevir M.K. (1970) "The theory of induced voltage electromagnetic flowmeters," J. Fluid Mech., Vol. 43, p. 577.

Bevir M.K. (1971a) "Long induced voltage electromagnetic flowmeters and the effects of velocity profile," Q. J. Mech. Appl. Math., Vol. 24, p. 347.

Bevir M.K. (1971b) "Sensitivity of electromagnetic velocity probes," Phys. Med. Biol., Vol. 16, p. 229.

Bevir M.K. (1971c) "The predicted effects of red blood cells on electromagnetic flowmeter sensitivity," J. Phys. D, Vol. 4, p. 387.

Bevir M.K. (1972) "The effect of conducting pipe connections and surrounding liquid on the sensitivity of electromagnetic flowmeters," J. Phys. D, Vol. 5, p. 717

Bevir M.K., O'Sullivan V.T., and Wyatt D.G. (1981) "Computation of electromagnetic flowmeter characteristics from magnetic field data," J. Phys. D, Vol. 14, p. 376.

Bourg A. and Tempe P. (1972) "Electromagnetic flowmeter," UK Patent 1 275 137.

Bowden K.F. and Fairbairn L.A. (1956) "Measurements of turbulent fluctuations and Reynolds stresses in a tidal current," Proc. R. Soc. London, Ser. A, Vol. 237, p. 422.

British Standards Institution (1980) "British Standard specification for electromagnetic flowmeters," BS 5792.

Cowley M.D. (1965) "Flowmetering by a motion-induced magnetic field," J. Sci. Instrum., Vol. 42, p. 406.

Cushing V. (1958) "Induction flowmeter," Rev. Sci. Instrum., Vol. 29, p. 692.

Cushing V. (1965) "Electromagnetic flowmeter," Rev. Sci. Instrum., Vol. 36, p. 1142.

Cushing V. (1971) "Electromagnetic flowmeters," Symp. on Flow - Its Measurement and Control in Science and Industry, Pittsburgh, Pa., Paper 2-4-38.

Cushing V. (1978) "Probe type electromagnetic flowmeter," Proc. FLOMEKO 1978 - Flow Measurement of Fluids, Groningen, the Netherlands, p. 163.

de Jong J. (1978) "Comparison of some 500 mm diameter flowmeters," Proc. FLOMEKO 1978 - Flow Measurement of Fluids, Groningen, the Netherlands, p. 565.

Denison A.B., Spencer M.P., and Green H.D. (1955) "A square-wave electromagnetic flowmeter for application to intact blood vessels," Circ. Res., Vol. 3, p. 39.

Engl W.L. (1970) "Der induktive Durchflussmesser mit inhomogenem Magnetfeld: Teil I, Allgemeine Grundlagen und Lösung des ebenen Problems," Arch. Elektrotech. (Berlin), Vol. 53, p. 344.

Engl W.L. (1972) "Der induktive Durchflussmesser mit inhomogenem Magnetfeld: Teil II, Lösung des raumlichen Problems," Arch. Elektrotech. (Berlin), Vol. 54, p. 269.

Faraday M. (1832) "Experimental researches in electricity," Philos. Trans. R. Soc. London, Vol. 15, p. 175.

Gray J.O. and Sanderson M.L. (1970) "Electromagnetic differential flowmeter," Electron. Lett., Vol. 6, No. 7, p. 194.

Guelke R.W. and Schoute-Vanneck C.A. (1947) "The measurement of sea water velocities by electromagnetic induction," J. Inst. Electr. Eng., Vol. 94, Pt. 2, p. 71.

Haacke A.C. (1978) "Calibration of electromagnetic flowmeter using a flow simulator," Proc. FLOMEKO 1978 - Flow Measurement of Fluids, Groningen, the Netherlands, p. 169.

Harrison D.N. (1980) "Design of magnetic flowmeter converters," Short Course on Electromagnetic Flowmeters, Fluid Engineering Unit, Cranfield Inst. of Tech., England.

Hemp J. (1975) "Improved magnetic field for an electromagnetic flowmeter with point electrodes," J. Phys. D, Vol. 8, p. 983.

Hemp J. (1980) "Channel flowmeters," Patent Application 8035858.

Hemp J. and Wyatt D.G. (1981) "A basis for comparing the sensitivities of different electromagnetic flowmeters to velocity distribution," J. Fluid Mech., Vol. 112, pp. 189-201.

Hentschel R. (1973) "Über induktive Durchflussmessung mischleitender und isolierender Flüssigkeiten," Doctoral dissertation, Hanover Tech. Univ., Germany.

Herschy R.W. and Newman J.D. (1974) "Electromagnetic river gauging," Proc. Symp. River Gauging by Ultrasonic and Electromagnetic Methods, Water Research Centre, Medmenham, and Water Data Unit, Reading, England.

Hofman F. (1978a) "Magnetic flowmeters with capacitance signal pick-off," Proc. FLOMEKO 1978 - Flow Measurements of Fluids, Groningen, the Netherlands, p. 493.

Hofman F. (1978b) "New developments in magnetic flow measurement using keyed dc-field," Proc. FLOMEKO 1978 - Flow Measurement of Fluids, Groningen, the Netherlands, p. 499.

Hogan M.A. (1923) "An electrical method of measuring the velocity of moving water," Engineering, Vol. 115, p. 66.

Hutcheon I.C. and Harrison D.N. (1965) "A new magnetic flow converter based on the Hall multiplier," Instrum, Pract., p. 529.

International Standards Organisation (1980) "Measurement of conductive fluid flowrate in closed conduits: method using electromagnetic flowmeters," ISO/TR 6817.

Kiene W. (1978) "Enhanced magmeter performance with pulsed d.c. excitation," Proc. FLOMEKO 1978 - Flow Measurement of Fluids, Groningen, the Netherlands, p. 174.

Kolin A. (1936) "An electromagnetic flowmeter: The principles of the method and its application to blood flow measurement," Proc. Soc. Exp. Biol. N.Y., Vol. 35, p. 53.

Kolin A. (1941) "An A.C. induction flowmeter for measurement of blood flow in intact blood vessles," Proc. Soc. Exp. Biol. N.Y., Vol. 46, p. 235.

Kuromori K., Kobayashi T., and Kanai H. (1978) "A magnetic flowmeter having dual signal detection system," Proc. IMEKO Symp., Japan, Paper bb-4.

Lehde H. and Lang W.T. (1948) "Device for measuring rate of fluid flow," U.S. Patent 2 435 043.

Medlock R.S. (1980) Private communication.

Mills C.J. (1966) "A catheter tip electromagnetic velocity probe," Phys. Med. Biol., Vol. 11, p. 323.

Nikitin B.I. (1978) "Flow rate measurement of conductive liquids by means of electromagnetic velocimeters," Proc. FLOMEKO 1978 - Flow Measurement of Fluids, Groningen, the Netherlands, p. 201.

Raptis A.C. and Forster G.A. (1975) "A signal analysis method using the cross-correlation of turbulence flow signals to determine calibration of permanent magnet sodium flowmeters," Argonne National Laboratory, Argonne, Ill., Rept. ANL-CT-76-8.

Reinhold I. (1978) "Velocity profile influence on electromagnetic flowmeter accuracy," Proc. FLOMEKO 1978 - Flow Measurement of Fluids, Groningen, the Netherlands, p. 181.

Remeniéras G. and Hermant C. (1954) "Electromagnetic measurement of speed in liquids," Houille Blanche, Vol. 9, p. 732.

Robinson J.S. (1976) "Submarine cables as flowmeters," Philos. Trans. Roy. Soc. London, Ser. A, Vol. 280, No. 1297, p. 355.

Rummel T. and Ketelsen B. (1966a) "Inhomogeneous magnetic field enables inductive flow measurement of all practical flow profiles," Regelungstech., Vol. 6, p. 262.

Rummel T. and Ketelsen B. (1966b) "Flowmeter with short head and diamond shaped coils", Patent Application 6 514 384.

Rummel T. and Ketelsen B. (1968) Reglungstech., Vol. 5.4, p. 131.

Sanderson M.L. and Gray J.O. (1981) "Developments in differential fluid flow measurement using electromagnetic techniques," Symp. on Flow - Its Measurement and Control in Science and Industry, p. 177.

Sen J. (1978) "Some properties of magnetic circuits in electromagnetic flowmeters for special use," Proc. FLOMEKO 1978 - Flow Measurement of Fluids, Groningen, the Netherlands, p. 223.

Shercliff J.A. (1962) The Theory of Electromagnetic Flow Measurement, Cambridge Univ. Press, Cambridge, England.

Smith C.G. and Slepian J. (1917) "Electromagnetic ship's log," U.S. Patent 1 249 530.

Strauss K.H. (1978) "On measuring discharge in partly-filled pipes," Proc. FLOMEKO 1978 - Flow Measurement of Fluids, Groningen, the Netherlands, p. 187.

Tarabad M. (1980) "Electromagnetic flowmeters for liquid metals," PhD Thesis, London Univ., England.

Tarabad M. and Baker R.C. (1979) "Electromagnetic flowmeters for sodium cooled nuclear reactors," Proc. IMEKO Symposium, Japan, paper 6b-6.

Tarabad M. and Baker R.C. (1982) "Integrating electromagnetic flowmeters for high magnetic Reynolds numbers," J. Phys. D, Vol. 15, pp. 739-745.

Thatcher G. (1978) "Recent developments in LMFBR coolant flow measurement techniques," Proc. FLOMEKO 1978 - Flow Measurement of Fluids, Groningen, the Netherlands, p. 193.

Thatcher G., Bentley, P.G., and McGonigal G. (1970) "Sodium flow measurement in PFR," Nucl. Eng. Int., Vol. 15, p. 822.

Tucker M.J., Smith N.D., Pierce F.E., and Collins E.P. (1970) "A two component electromagnetic ship's log," J. Inst. Nav., Vol. 23, p. 302.

Turner G.E. (1960) "The non-linear behaviour of large permanent-magnet flowmeters," Atomics International, Rept. NAA-SR-4544.

Weigand D.E. (1967) "Summary of an analysis of the eddy-current flowmeter," 14th Nuclear Science Symposium, Los Angeles, Calif., pp. 28-36.

Weigand D.E. (1972) "Magnetometer flow sensor," Argonne National Laboratory, Argonne, Ill., Rept. ANL-7874.

Williams E.J. (1930) "The induction of emf's in a moving liquid by a magnetic field and its application to an investigation of the flow of liquids," Proc. Phys. Soc. London, Vol. 42, p. 466.

Wollaston C. (1881) "Tidally induced emf's in cables", J. Soc. Tel. Engr., Vol. 10, p. 50.

Wyatt D.G. (1961) "Problems in the measurement of blood flow by magnetic induction," Phys. Med. Biol., Vol. 5, pp. 289 and 369.

Zepgir B.D. and Sermons G.Ya. (1965) "Impulse method of measuring the velocity in electrically conducting liquids," Magnetohydrodynamics (USSR), No. 1, p. 131.

Chapter III Electromagnetic Pumps, Flow Couplers, Fission and Fusion Applications

High Interaction Parameter Studies in a Large NaK Loop

I.R. McNab,* C.C. Alexion,† A.R. Keeton,‡ and P.A. Ciarelli§
Westinghouse Electric Corporation, Pittsburgh, Pa.

I. Test Setup

Experiments have been undertaken in a large (60 ℓ/s) NaK loop having a direct-current electromagnetic pump test duct configuration. The duct had a thin-wall construction, was rectangular in cross section (7.6 by 8.9 cm), and was over 182 cm long with an active length of 23.3 cm. Applied currents of 18 kA were provided by a dc homopolar generator driven by a 50 hp motor. A toroidal coil magnet with cobalt steel pole pieces provided up to 3.5 kG of induction across the 14.0 cm gap. Compensating return conductors, located between the duct and the pole pieces, cancelled most of the magnetic field induced by the applied current.

The loop was mainly 16.1 cm inner diameter stainless steel pipe with flanged reducer and diffuser sections bolted to the end flanges of the duct. An 8.3 cm diameter bypass leg to a NaK cooler had an orifice-plate flowmeter to measure the total flow diverted. The NaK cooler was made from finned

---

Summary of paper presented at Third Beer-Sheva International Seminar on Magnetohydrodynamic Flows and Turbulence, Ben-Gurion University of the Negev, Beer-Sheva, Israel, March 23-27, 1981. Copyright © American Institute of Aeronautics and Astronautics, Inc., 1982. All rights reserved.
 *Manager, Electrodynamics Section, Research and Development Center.
 †Engineer, Electrodynamics Section, Research and Development Center.
 ‡Engineer, Materials Chemistry Section, Research and Development Center.
 §Safety Supervisor, Electronics Division, Research and Development Center.

tubing inside a stainless steel enclosure with a forced airflow. Other components in the main loop included two disk valves (one motorized) and a turbine flowmeter. The loop was vacuum filled with NaK.

During normal operation of the test duct the current, magnetic field, and flow rate were parameters that could be chosen independently. The dependent variables were the pressure, electric potential, and fluid velocity. The pressure and electric potential were measured along the sidewall at 27 locations on one side and 23 on the other. Each measurement station consisted of a 3 mm stainless steel tube within a 6 mm tube, separated by a rubber washer. The inner tube was welded to the inner wall (Inconel) and the outer tube was welded to the outer wall (stainless steel). The inner tube was connected by a gas line to a pressure transducer. Since the inner tube was in contact with the NaK and isolated from the outer wall, it also served as a potential tap. Voltages were referenced to the tap at the middle of each electrode.

Although most data from the pump were taken at the sidewall, an internal measurement probe has recently been used to measure additional parameters within the NaK stream. The measurements included: the fluid velocity (which was calculated from the differential pressure across a Pitot tube at the probe tip), two perpendicular components of magnetic field, the static pressure, and the electric potential at two points.

II. Test Results

The internal probe test results showed that the velocity near the electrode decreased from the inlet to the outlet edge, but outside of the electrode region the velocity was high and a nearly slug-like profile existed. The velocity at the centerline was nearly constant throughout the duct, although with some noticeable acceleration at the inlet and outlet of the active region.

The axial magnetic field profile at centerline was slightly perturbed, being swept downstream approximately 2 cm for a flow rate of 3.25 m/s, showing some influence of the self-induced field even though the magnetic Reynolds number was only 0.5 for this flow rate.

The centerline static pressure fell slightly at the inlet, rose linearly through the active region, and leveled off at the outlet of the pump. This was in marked contrast to the behavior of the pressure at the sidewall where the pressure showed spatial fluctuations at the inlet and outlet. This behavior is as yet unexplained.

Overall efficiency of the pump was within 5% (an average) of the value predicted by the quasi-one-dimensional theory (Hughes and McNab, 1983).

Acknowledgments

The work reported here was aided by National Science Foundation Grant ENG-78-8434, Dr. R.E. Rostenback, program director.

Reference

Hughes W.F. and McNab I.R. (1983) "A quasi-one-dimensional analysis of an electromagnetic pump, including end effects," Liquid-Metal Flows and Magnetohydrodynamics, Progress in Astronautics and Aeronautics, Vol. 84, AIAA, New York, pp. 287-312.

High-Efficiency Direct-Current Electromagnetic Pumps and Flow Couplers for Pool-Type LMFBRs

I.R. McNab* and C.C. Alexion†
Westinghouse Electric Corporation, Pittsburgh, Pa.
and
R.K. Winkleblack‡
Electric Power Research Institute, Palo Alto, Calif.

Abstract

In this study, high-efficiency dc electromagnetic pump concepts for use in pool-type liquid-metal-cooled fast breeder reactors (LMFBRs) were evaluated. Two concepts were examined: 1) the conventional dc electromagnetic pump, and 2) the flow coupler, a device that electromagnetically links the flow in a liquid-metal MHD generator with the flow in an immediately adjacent dc pump. It was concluded that the conventional dc pump would not be feasible for the high-flow, high-pressure requirements of the LMFBR, primarily because of the necessity for a high-current, low-voltage power supply. By contrast, the flow coupler was found to be suitable for the LMFBR because the external electrical power supply is eliminated. The efficiency of the flow coupler was evaluated as a function of the operating parameters and 60% was shown to be attainable in a relatively small volume (3.5 m

---

Paper presented at Third Beer-Sheva International Seminar on Magnetohydrodynamic Flows and Turbulence, Ben-Gurion University of the Negev, Beer-Sheva, Israel, March 23-27, 1981. Copyright © American Institute of Aeronautics and Astronautics, Inc., 1982. All rights reserved.
 *Manager, Electrodynamics Section, Research and Development Center.
 †Engineer, Electrodynamics Section, Research and Development Center.
 ‡Program Manager, LMFBR Projects.

long by 2.5 m diam). Several conceptual flow coupler arrangements are described.

I. Introduction

The objective of this study was to evaluate high-efficiency direct-current electromagnetic (EM) pump concepts for use with pool-type liquid-metal-cooled fast breeder reactors (LMFBRs).

Two basic approaches were considered. The first was the conventional direct-current (dc) EM pump, in which crossed electric and magnetic fields applied to the liquid metal produce a pumping force in a mutually perpendicular direction. This concept has been known for many years and, for example, was used experimentally by Barnes (1953) in one of the early liquid-metal-cooled breeder reactor experiments at the Argonne National Laboratory. The advantage of the dc EM pump, compared with a mechanical pump, is primarily that a high component reliability appears to be inherent as a result of the lack of moving parts other than the liquid itself. Also, the basic pumping section may be made relatively compact compared with a mechanical pump. However, several disadvantages occur with the dc EM pump for LMFBR applications. First, two loss mechanisms are present that can have a significant effect on the pump efficiency:

1) Losses in the entrance and exit regions of the pump where the applied current fringes out into the liquid metal; since the local magnetic field is small, these regions contribute little useful work to the pumping.

2) For many applications, probably including LMFBR, it is unlikely that an electrical insulation material can easily be found for use as, or on, the pump duct sidewalls. Metal walls therefore have to be used. This allows a fraction of the applied current to bypass the liquid metal so that it does not perform useful work. This effect can be minimized by the use of thin walls of high-resistivity material; however, structural and material compatibility considerations limit the extent to which this approach can be followed. The ultimate solution to this problem lies with the development of a high-temperature wall insulation. However, although materials such as high-density, high-purity alumina offer such a prospect, the qualification of such materials for the LMFBR is likely to require great development effort.

Another problem associated with the conventional direct-current electromagnetic pump arises from the inherent characteristics of the pump itself (geometry and material properties) and the above loss mechanisms. These combine to cause the power supply requirements for a large pump to be unusual, with very high currents (> 100 kA) being required at low voltages (< 1 V). Most conventional power supplies are not well suited to this type of operation and are inefficient. The problem is made worse by the need for extremely large busbar assemblies to keep the high-current transmission losses low. In a pool LMFBR, where space is at a premium, this is unattractive.

Considerable progress has been made in recent years in development of high-efficiency, high-current power supplies by Westinghouse and other organizations. However, while such homopolar generators may be more efficient than conventional power supplies, the problem of transmission losses is still present.

In this study an assessment was made of a direct-current EM pump system for LMFBR operation. The loss effects mentioned above (and others) have been included and estimates have been made of the total system efficiency and practical feasibility. This is described in Sec. III.

An alternative which utilizes similar principles and technology but which offers the prospect of overcoming some of the problems raised above has also been evaluated. This is the flow coupler system. In this arrangement the pumping of the primary liquid metal takes place in exactly the same way as in the dc EM pump. The main difference is that the necessary dc current is supplied locally by a companion liquid-metal dc generator which is in the intermediate fluid circuit. Thus, the external low-voltage, high-current power supply is no longer required and the busbars become unnecessary (except for the short generator-to-pump connection). The power input necessary to drive the liquid metal in the generator is provided by a second pump (electromagnetic or mechanical) in the intermediate circuit. The concept title arises from the fact that the two fluid circuits are electromagnetically coupled together through the generator/pump sections. Additional advantages of this arrangement are that the local generation of current enables lower voltages and higher currents to be used than would be possible with an external power supply; the lower voltages reduce end current losses and permit higher duct efficiencies to be attained.

Interest in the flow coupler concept for the LMFBR arises for several reasons. First, the potential reliability is improved compared with the mechanical pump. Second, the smaller pump duct volume may permit inproved reactor layouts, including the possibility that the pump (or flow coupler) could be integrated with the heat exchanger assembly. Third, the primary coolant flow rate could be controlled by the flow rate in the intermediate coolant loop. Fourth, the integrity of the heat exchanger may be improved, in the event of an external pumping failure, as a result of the coupling between the two fluid streams. The flow coupler insures that a flow in one loop will cause flow to occur in the other, even as a result of natural convection. Fifth, compared with induction pump concepts, there is an absence of electrical power leads into the liquid-sodium pool. Finally, although estimates are uncertain at this time, it seems likely that the relative simplicity and easy operation of the direct-current flow coupler should enable a relatively low-cost component to be developed, thereby providing overall plant cost savings compared with the mechanically pumped system.

The study of the flow coupler concept as applied to the pool-type LMFBR is described in Sec. IV of this paper.

II. Quasi-One-Dimensional Theory of a dc EM Pump Duct

In the evaluation of the efficiency and flow conditions in the dc EM pump and flow coupler, use was made of a quasi-one-dimensional theory of flow in EM ducts that was developed by Westinghouse prior to this study. This theory includes the effect of current and magnetic field fringing on pump efficiency, as well as taking into account the development of internal current leakage through wall boundary layers of slow-moving fluid (so-called Hartmann layers). Slug flow is assumed for the fluid and finite duct wall conductivity is included. This theory provides the best, relatively simple, available model to describe flow processes in a practical pump duct geometry. The same theory was also used to describe flow in the generator side of the flow coupler.

The theory outlined here was used as the basis for the evaluation of efficiency and flow conditions in the dc EM pump and in the flow coupler. The theory was developed prior to this contract and is described in detail elsewhere [Hughes and McNab (1983)] although a brief summary is provided here.

The EM pump (or generator, since the theory is equally applicable) is assumed to consist of a center region in which the magnetic field strength and current density are uniform, plus two end regions in which spatial fringing of the current and magnetic field occur. In these end regions it is assumed that the magnetic field decays exponentially from the edge of the electrodes. Finite sidewall conductance is also assumed in the interelectrode region and up and downstream of the electrodes.

The theory describes the total pump flow in terms of six parameters: a dimensionless voltage V^*, the Hartmann number M_0, an axial duct aspect ratio λ, a magnetic field fringing decay parameter τ, a duct cross-section aspect ratio β, and a wall conductance ratio $\bar{\sigma}$. The total pressure rise, total current, and total efficiency are then expressed (in dimensionless terms) as functions of these parameters,

$$\Delta P_T^* = F_1(V^*, M_0, \lambda, \tau, \beta, \bar{\sigma})$$

$$I^* = F_2(V^*, M_0, \lambda, \tau, \beta, \bar{\sigma})$$

$$\eta_T = F_3(V^*, M_0, \lambda, \tau, \beta, \bar{\sigma})$$

Auxiliary equations relate the duct and flow parameters to the above parameters.

III. Direct-Current Electromagnetic Pump

Optimal Duct Size

Small dc EM pumps suffer from inefficiencies caused by hydraulic friction at the duct walls and by internal current leakage through the relatively slow-moving boundary layers near the walls. Increasing the duct size is beneficial in reducing the relative effect of these losses and provides an increase in efficiency. However, this process does not continue indefinitely as the duct dimensions are increased. If the pump pressure head, and hence pump length, is fixed by external constraints, the end current leakage losses increase faster than friction and wall layer losses decrease, so that an optimum duct size is found.

To examine these effects in duct sizes that are relevant to the pool-type LMFBR, the equations outlined in Sec. II were evaluated using a short computer program. A unit pump size of 2.9 m³/s (46,000 gal/min) was assumed for this evaluation and subdivision of this unit size into N ducts, each having a 1/N flow rate, was examined. Three different sidewall arrangements were examined, two of which were electrically conducting (Inconel 706 of 1.5 and 0.5 mm wall thickness, respectively) and one of which, for comparison, was assumed to be electrically insulating (no thickness). Of these, the thicker conducting wall configuration is judged to be the most practical at present. A square duct cross section was assumed for these calculations; other parameter values are shown in Fig. 1, where the results are displayed.

From Fig. 1 it can be seen that: 1) the duct efficiency is strongly dependent on the wall conductance, and 2) for the 1.5 mm wall, the duct efficiency is maximized when the flow per duct is about 0.3 m³/s.

It is clear from Fig. 1 that the use of a single pump duct of 2.9 m³/s (46,000 gal/min) would introduce a significant efficiency penalty compared with the optimal (smaller) duct. Nonetheless, the pump efficiency is still fairly attractive (>76%) so that such an arrangement cannot be dismissed on efficiency considerations.

Table 1 provides a comparison of the current and voltage requirements per duct for various duct sizes. From this it can be seen that the voltage/current characteristics for an optimized single duct of 2.9 m³/s are rather unattractive, requiring 2.15 MA at 0.76 V. Supplying such a high

Table 1 Comparison of current and voltage requirements (per duct) for N pump ducts[a]

N	η, %	I, kA	Vo, V	P, kW	P_{TOT}, kW	2a, cm	B, T
1	76.4	2152	0.761	1638	1638	60.23	0.141
2	78.8	1186	0.669	793	1586	42.49	0.177
4	80.6	664	0.584	387	1548	30.12	0.220
10	81.6	322	0.475	152	1520	19.05	0.284
12	81.6	282	0.454	128	1536	17.39	0.297
32	79.6	145	0.340	49	1578	10.65	0.358
48	77.9	112	0.296	33	1584	8.69	0.378

[a] With V_p = 8 m/s, ξ = 2 m⁻¹, λ = 15, t_w = 1.5 mm, α = 1.

current in an efficient (low loss) manner in a pool-type LMFBR appears to be impractical at present. The single duct approach does not therefore appear practical.

Multiduct Assemblies

One way to improve the voltage/current characteristics for a 2.9 m³/s pump could be to use several smaller ducts, each of which would require a smaller current than the single large duct. For example, Table 1 shows that if 12 ducts were used, in addition to having a higher efficiency, each duct would only require 282 kA rather than several megamps. However, much depends on how such smaller ducts are configured electrically and hydraulically. Various layouts can be considered. First, consider putting 12 optimum ducts all in parallel electrically and hydraulically. This would insure that the efficiency of the total multiduct system would remain as high as that of the single optimum duct.

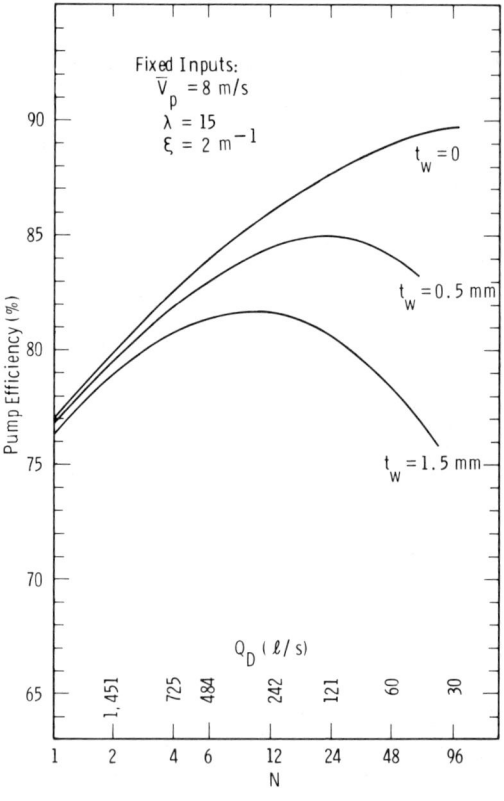

Fig. 1 Pump duct efficiency as function of number of ducts N and volumetric flow per duct Q_D.

However, electrically paralleling the 12 ducts would provide no benefit so far as the power supply problem is concerned. [Note that this multiduct arrangement is equivalent to that of a single rectangular duct having a 12/1 (width/height) cross-sectional aspect ratio.]

Second, consider placing the 12 ducts hydraulically in parallel but electrically in series. At first sight this seems most attractive since it permits each duct module to operate at the peak efficiency while increasing the supply voltage by a factor of 12. However, if all of the ducts were joined by a common supply pipe or plenum, the arrangement would be equivalent to that of a single duct having a 1/12 (width/height) cross-sectional aspect ratio. In this case the end losses caused by current leakage would be much increased (approximately 12 times larger than for the optimum duct because of the changed length-to-width ratio).

To overcome these end losses it is necessary to try to make each duct module electrically independent. This may be done by extending out the thin wall pipework for each pump duct for a significant length upstream and downstream of the pumping section. However, this introduces additional electrical and hydraulic losses. The hydraulic losses occur because of the increased length of small-diameter pipe, compared with a single run of full-diameter pipe. As usual, increasing the length of the small-diameter pipe sections increases the hydraulic losses.

The electrical losses arise because, since the 12 ducts are arranged in series electrically, the electric potential of the fluid (on the centerline) in each duct is at a different value. If these fluid streams were brought together a short distance from the ends of the duct electrodes, electrical short circuiting would take place (see Fig. 2). Increasing the length of the duct extensions reduces the amount of electrical shorting but increases the hydraulic losses. These two losses may be traded off against each other to find the point of minimum losses. This evaluation has been undertaken showing that for 12 ducts in series electrically the required duct length extensions are impractically long ($\sim$ 15 m).

Some improvement in this situation can be achieved by changing the stacking configuration of the duct modules. Thus, rather than having all 12 ducts in series electrically,

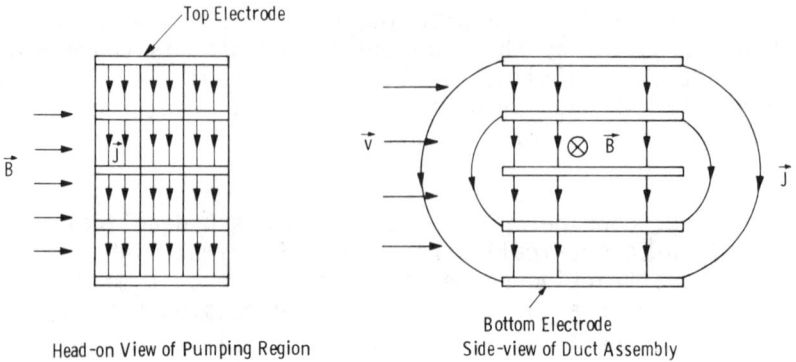

Fig. 2 Electrical shorting in a typical multiple-duct EM pump assembly.

we can have: 6 in series, 2 in parallel; or 4 in series, 3 in parallel, etc. The improvements in end losses that take place in this way have to be offset against the supply voltage and current changes and the impact that this has on the total system efficiency.

This assessment has been undertaken for several duct module geometries. However, it has been found that no truly attractive total system arrangements appear to exist. The one that appears best (a 4 series, 3 parallel arrangement) in terms of yielding reasonably good system efficiencies ($\sim 60\%$), although preferable to the above arrangements, still involves major system problems in two areas. First, the supply current/voltage characteristics are still difficult (850 kA, 1.8 V) so that not only is the generator itself likely to prove very difficult for this application, but associated aspects, such as busbars, become large in cross section (> 1 x 1 m) if low transmission losses are required and gas cooling has to be used.

The second major difficulty is that the duct extension length for minimum losses is still significant (> 3 m at each end). Although a pump configuration using such extensions could possibly be developed, it is clearly not an attractive prospect since it contradicts the original hope of the basic dc EM pump, namely that of a high specific power density (that is, pumping power per unit volume).

It has therefore been concluded that the dc EM pump is probably unsuitable for large-scale in-pool liquid-metal pumping. Large scale in this context is more than a few hundreds of liters per second. This conclusion may not necessarily preclude the use of such pumps for out-of-pool applications where large busbars may be more readily accommodated.

IV. The Flow Coupler

Introduction

Several questions have to be addressed in evaluating whether the flow coupler is a feasible concept for the pool-type LMFBR. First, the basic operation of the flow coupler module has to be evaluated and its performance verified (at this stage, theoretically only). Second, the arrangement of flow couplers in a practical geometry which will fit in with the other pool reactor components has to be devised. Third, the operational requirements of the flow coupler have to be evaluated, e.g., magnet design details, part-load operation, etc.

The first of these three aspects is discussed in this section. Practical considerations are described in Sec. V.

Basic Flow Coupler Operation

The basis of the flow coupler analysis is the EM theory outlined in Sec. II of this paper. This has been applied both to the pump and generator sides of the flow coupler, together with auxiliary equations which couple together the two components. Thus, it has been assumed that the current produced in the generator by the intermediate fluid flow is entirely used in the pump section for pumping the primary fluid, that is, the current in the two sections is the same.§ Because of resistive losses in the busbar electrodes which connect the generator with the pump, it has been assumed that the voltage applied to the pump section is lower than that produced in the generator section by an amount that represents the losses.

Flow Coupler Calculations

Flow Coupler Model. The following calculations are applicable to flow couplers in general, both radial and azimuthal current designs. They pertain only to a flow coupler module, i.e., one rectangular pump duct cor ected electri-

---
§The current supplied to the pump is the "effective" generator current, that is, the current generated by the fluid interacting with the magnetic field minus the current shunted back through the sidewalls.

cally in series to one generator duct. The parallel and antiparallel flow coupler modules shown in Fig. 3 have been assumed. The analysis used in these calculations is the quasi-one-dimensional approach mentioned in Sec. II of this report. For reasons of simplicity and practicality, the following assumptions are made:

1) Both ducts are rectangular.

2) All magnetic parameters, such as the active length, air gap, field strength, and exponential decay constant, are the same for both ducts.

3) The conducting sidewalls are of the same thickness for both ducts.

4) The fluid and sidewall are at the same temperature for each duct.

5) The pump and the generator are interconnected with a low-resistance electrode.

Program Inputs. To simulate this system, a computer program has been developed that performs all of the necessary calculations. The physical inputs to the program are: 1) the total flow rate required per pump, 2) the number of

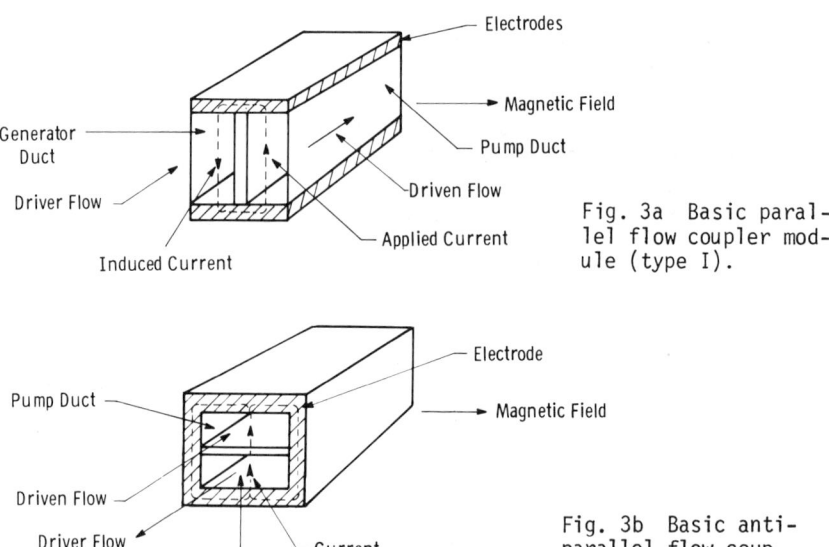

Fig. 3a Basic parallel flow coupler module (type I).

Fig. 3b Basic antiparallel flow coupler module (type II).

individual pump ducts, 3) the pump fluid flow velocity, 4) the total pressure rise required per pump, 5) the number of pumping stages in series, 6) the temperature of the primary fluid, and 7) the temperature of the secondary fluid.

The parametric inputs to the program are: 1) the conducting sidewall thickness, 2) the pump aspect ratio (width/height), 3) the longitudinal aspect ratio (length/half-width) for pump and generator, 4) the magnetic field exponential decay constant, and 5) the loss factor between generator voltage and pump voltage due to busbar connections.

Program Outputs. Using the analysis developed by Hughes and McNab (1982) the following output is obtained: 1) generator height, 2) generator width, 3) generator velocity, 4) generator duct flow, 5) generator pressure drop per stage, 6) generator current, 7) generator voltage and dimensionless voltage, 8) generator efficiency, 9) pump height, 10) pump width, 11) pump flow per duct, 12) pressure rise per pump stage, 13) pump current, 14) pump voltage and dimensionless voltage, 15) pump efficiency, 16) active length, 17) magnetic field, and 18) the flow coupler efficiency.

Parametric Studies. A brief parametric study has been conducted on how some of the inputs affect the total flow coupler efficiency. These relationships are graphed in Figs. 4-8 for a type I flow coupler. It can be seen that, for a practical system, the efficiency can range about 60-75%, depending on the combination of N, v_p, λ, ξ, and t_w that is chosen.

V. Practical Flow Coupler Considerations

Basic Requirements

A number of basic and practical requirements have to be considered in the design of flow coupler geometries. Although the scope of the present study has not permitted a complete practical flow coupler design to be achieved, a number of concepts have been generated for future evaluation and some practical concerns have been identified.

Perhaps the most fundamental requirement for the flow coupler, as for the dc EM pump, is that the magnetic field and applied/generated current have to be applied to the liquid metal in directions that are mutually orthogonal to the desired flow direction. In addition, since there cannot be

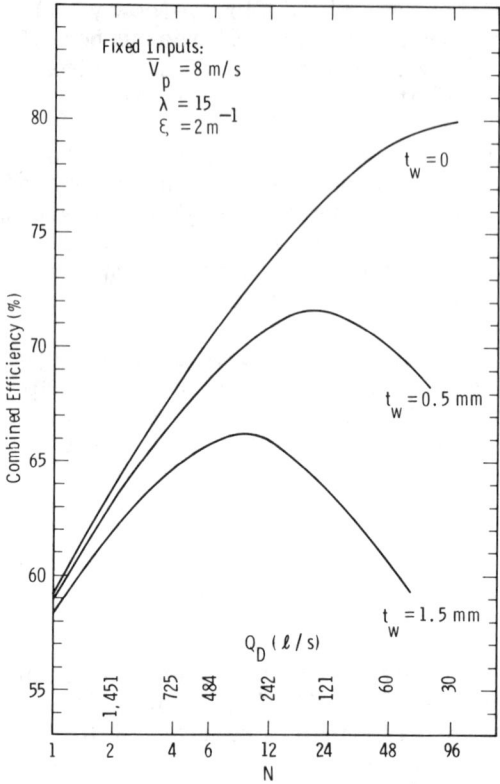

Fig. 4 Flow coupler duct efficiency as a function of number of ducts N and volumetric flow per duct Q_D.

sources or sinks of magnetic field or current, both of these quantities have to be close upon themselves. This causes certain practical topological constraints in the design of flow couplers.

Second, as the evaluations in Sec. IV show, there is a preferred size for the basic flow coupler duct module, based on maximizing the coupler efficiency. The preferred size is a function of several parameters, including particularly the thickness of the conducting sidewalls and the material of which it is manufactured. Since the dependence of optimum efficiency on wall thickness is quite strong, it is necessary to choose the desired thickness in order to advance the conceptual coupler design. For the studies described here, it has been assumed that a sidewall thickness of 1.5 mm (0.060 in.) can be used. A thinner wall would be preferable if it could be practically achieved, since it would permit a higher efficiency. A thinner wall would also enable smaller ducts to be used (for maximum efficiency). The optimum

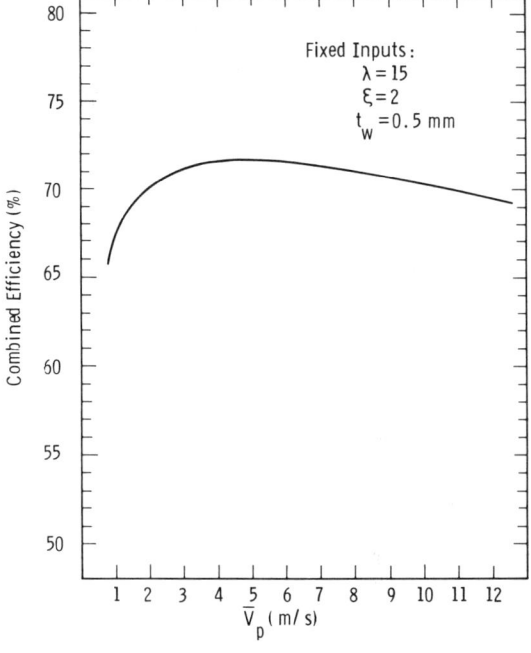

Fig. 5 Flow coupler duct efficiency as a function of flow velocity V_p in the pump duct.

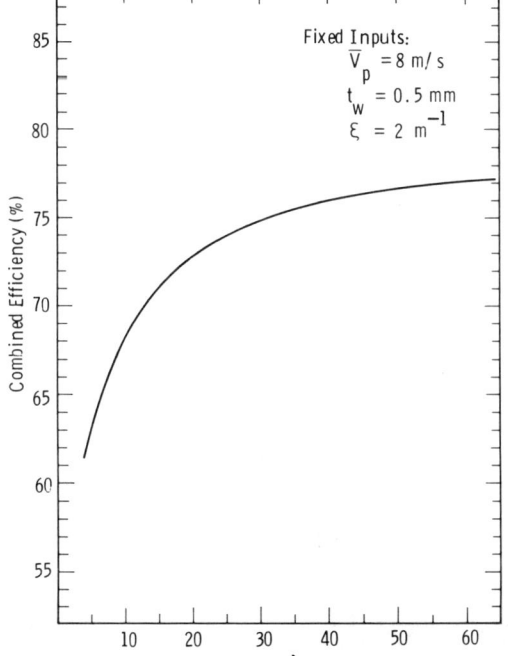

Fig. 6 Flow coupler duct efficiency as a function of length to half-height aspect ratio λ.

for an infinitely thin conducting wall (i.e., an electrically insulated wall) would be a rather small duct having a flow rate of a few tens of liters per second. However, although there may be ultimate prospects for the development of such insulated-wall ducts (e.g., by plasma spray or detonation gun coatings of high-density, high-purity aluminum oxide), we have chosen the more conservative approach of a relatively thick conducting wall for these studies. The chosen wall thickness has, of course, also to be compatible with the requirement to withstand the primary/intermediate fluid pressure differential.

Since an essential requirement of the flow coupler is the presence of an applied magnetic field, the manner in which the field is generated and applied to the working section has to be considered. Generally, two alternatives are available: either permanent magnets or separately excited wound electromagnets may be employed. The choice between these two is not clear. Permanent magnets have the great advantage of not requiring separate power supplies and, once magnetized they can retain their properties for a long time. On the other hand, the materials used are generally expensive and often contain a significant quantity of co-

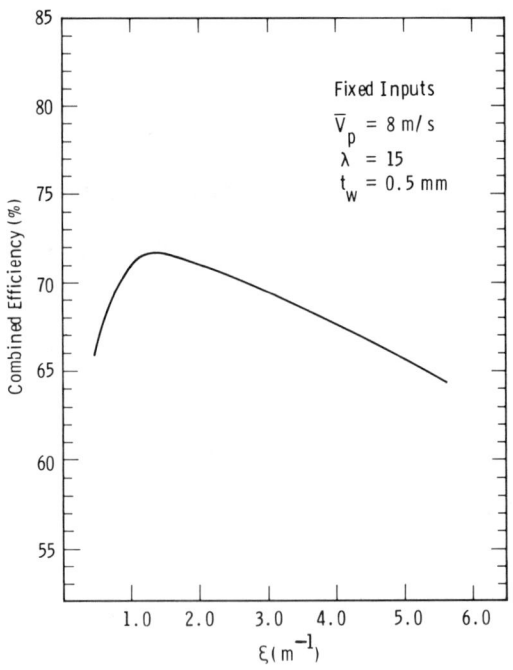

Fig. 7 Flow coupler duct efficiency as a function of magnetic field fringing parameter ξ.

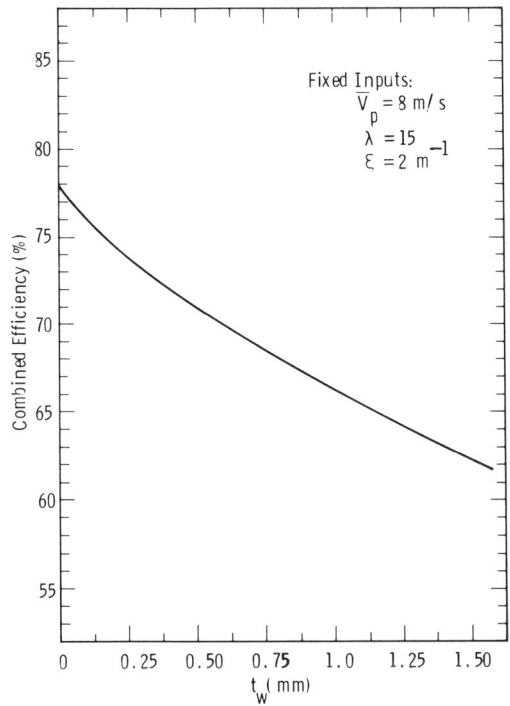

Fig. 8 Flow coupler duct efficiency as a function of conducting sidewall thickness t_w.

balt. If exposed to a high-radiation flux in the reactor (which is not inevitable, however) this may cause the material to become radioactive and/or to suffer some degradation of its magnetic properties. The activity may be alleviated if the permanent magnet material is clad with stainless steel, for example, and such techniques may be needed to protect the permanent magnet material from corrosive attack by the liquid sodium. The performance of permanent magnet materials is also dependent on temperature. It seems likely that several materials may be available for cold-pool operation (< 600°F). However, the safety aspects of a pumping arrangement that could be significantly affected by an overtemperature transient requires careful attention. The possibility that the material may become demagnetized as a result of such an event may require the presence of externally excited remagnetizing coils.

Iron-cored electromagnets, which use separately excited copper wound field coils, are likely to be much less expensive than permanent magnets. However, a separate power supply is required to provide a continuous dc supply for the magnet coils, and this could be deemed either a safety or reliability concern. Methods of constructing a reliable, electrically insulated, high-temperature magnet that must work in a liquid-sodium environment at high temperatures have also to be addressed and will need development.

At present the choice between these two magnet technologies for the flow coupler is not clear. A possibility may be a compromise in which both technologies are utilized, with the excited copper-iron magnet being employed for the primary flow coupler magnetic field, but with some permanent magnet material being present to insure the continued operation of the flow coupler even in the event of an external power failure.

A further choice that has to be made relates to the strategy to be adopted for the use of the flow coupler. One approach centers around the use of the flow coupler to provide only the pressure head necessary to drive fluid through the heat exchanger - this Δp is probably less than 100 kPa. The duty required of this flow coupler is significantly less than the requirement to provide full core pressure rise ($\sim$ 1 MPa). The main advantage for the flow coupler for this "low-risk" application may be the protection of the intermediate heat exchanger (IHX) from any thermal transients that would result if the intermediate loop pumps were to fail. The flow coupler would link the intermediate and primary fluids together, thereby insuring a thermal convection driven flow of the intermediate fluid. However, since present IHX designs cater for such a thermal transient, it is likely that the changes in the heat exchanger design resulting from eliminating this requirement would be minimal. However, further investigations are required.

The approach that has been followed here may be considered a higher risk, higher payoff strategy. Thus, use of the flow coupler to provide the full core plus heat exchanger pressure head has been investigated. The objective is to achieve a substantial improvement in the pool reactor layout through the use of the flow coupler, either by reducing the number of components or by rearranging their layout. In the sections below, several possible flow coupler configurations are discussed. The first of these - the radial-current, azimuthal flow arrangement - has been evaluated in the greatest detail. Other arrangements, while investigated superficially, have not been evaluated in this study in as much detail. This does not imply that such configurations may not ultimately prove practical, but rather that with the limited funds and resources available for this study all possible arrangements could not be studied in the required detail.

Radial-Current, Azimuthal-Field Flow Coupler

The approach evaluated in the greatest detail in this study has been the radial-current, azimuthal-field flow coupler which is connected to (in line with) the intermediate heat exchanger. This arrangement is intended to permit an in-line arrangement of the pump and heat exchanger, thereby reducing (approximately halving) the number of major penetrations of the top cover on the reactor vessel and possibly permitting a concomitant reduction in the vessel diameter and associated cost savings. A major feature of the design is that it assumes no basic changes in the heat exchanger, although it seems likely that a segment-shaped arrangement could be developed, if necessary, to provide a better pool layout. The hydraulic connection between the flow coupler and the heat exchanger would be via a plenum, the hydraulic design of which has not been addressed in this study.

The preferred arrangement is with the flow coupler located below the IHX, thereby providing an improved pump inlet pressure distribution. Location of the flow coupler below the IHX will also enable it to be maintained at essentially the cold-pool temperature conditions ($< 660°F$) and, if required, the incoming intermediate circuit fluid (at $\sim 600°F$) can be utilized to cool critical components.

A further preferred arrangement of the IHX is with the primary fluid in tubes and with the primary and secondary fluid flows in opposite directions. If the self-contained flow coupler units are utilized, this necessitates an electrode return arrangement of the type shown in Fig. 3.

Including the return electrode, each of 10 flow coupler sections will have external dimensions of approximately 0.35 x 0.4 m with a length of approximately 2.5 m. Adding inlet and exit plenums to provide low-loss connections to the heat exchanger is likely to increase the total flow coupler length to approximately 3.5 m.

Subject to the more detailed evaluation of this concept, these flow coupler sizes and efficiencies appear to offer an attractive prospect for a practical unit.

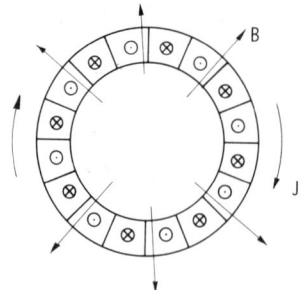

Fig. 9 Radial-field, azimuthal-current flow coupler concept.

Radial-Field, Azimuthal-Current Flow Coupler

The second general group of concepts that may prove to be attractive, but which has not yet been evaluated in detail, uses the radial-field, azimuthal-current flow geometry. This uses the flow coupler arrangement shown in Fig. 3 as the basic module, but has the added advantage that return current conductors are not always required.

A basic layout of this type is shown in Fig. 9. In this arrangement, primary and secondary flow ducts alternate around the periphery of the total unit. For this sketch, the secondary (generator) flow is out of the paper and the primary (pumped) flow is into the paper. This annular arrangement could be used in the form shown, with the annulus diameter being the same as that of the pool. However, this would probably require a substantial change in the total reactor concept compared with the present practice. In addition, a difficulty with this arrangement could be that if a fault were to occur in any one of the generator or pump ducts in such a way as to increase the electrical resistance to current flow, the total pumping power of the entire unit would be affected. In addition, it is not immediately clear how all of the necessary electrical connections for high-current flow could be made and maintained.

A more practical arrangement of the same basic concept may therefore be that shown in Fig. 10, in which the flow coupler unit is divided into smaller annular sections, each of which is associated with a conventional heat exchanger. In this way, if a fault develops in any one of the ducts, only one flow coupler/heat exchanger unit will be lost, leaving the other five units available for reactor operation. In the configuration shown in Fig. 10, the magnetic field path has to be closed at a point below the plane of the sketch.

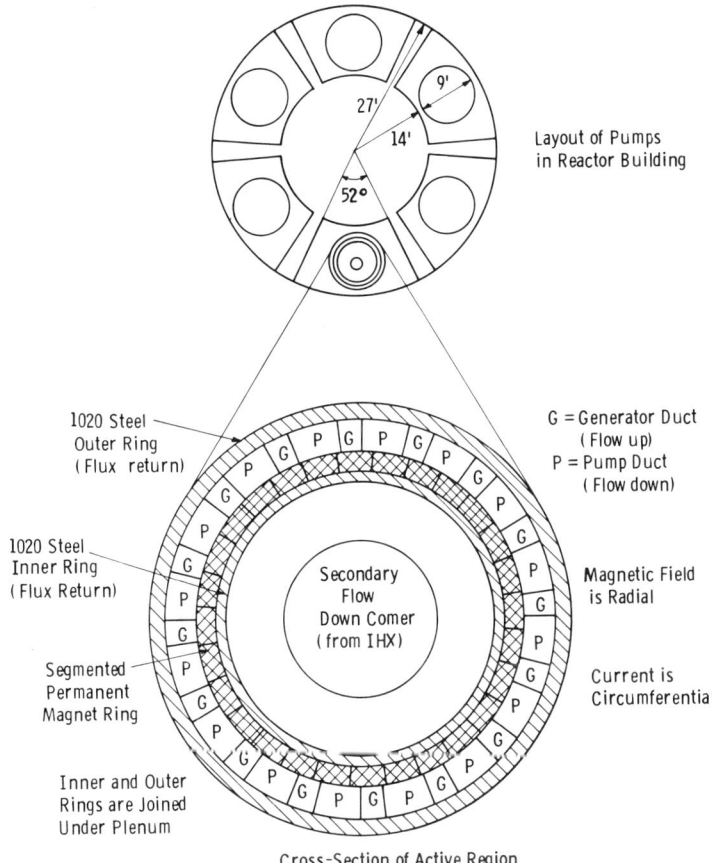

Fig. 10 Single-pass field flow coupler unit within pool.

VI. Conclusions

1) Evaluation of the direct-current EM pump shows that its operating characteristics are not well matched to the pool-type LMFBR operation. This arises primarily because of the necessity to supply very high currents (megamps) at low voltages (less than 1 V).

2) In contrast, the flow coupler offers much better prospects for use in the pool reactor since efficient ($\gtrsim$65%), low-volume (3 m long x 2.5 m diam) systems appear feasible without the requirements for external power supplies.

3) The configurations evaluated to date have utilized either radial-current/azimuthal-field or radial-field/azimuthal-current geometries. It is not yet clear which of these geometries is best.

4) The choice between permanent magnets or separately excited electromagnets is not yet clear and, in part, will depend on the system operating and safety philosophy.

Acknowledgments

The authors wish to acknowledge the many useful ideas generated during this study by our consultant, W.C. Roman. The work reported here was supported by the Electric Power Research Institute under Contract TPS-79-774.

References

Barnes A.H. (1953) "Direct-current electromagnetic pumps," Nucleonics, Vol. 11, pp. 16-21.

Hughes W.F. and McNab I.R. (1983) "A quasi-one-dimensional analysis of an electromagnetic pump including end effects," Liquid-Metal Flows and Magnetohydrodynamics, Progress in Astronautics and Aeronautics, Vol. 84, AIAA, New York, pp. 287-312.

A Quasi-One-Dimensional Analysis of an Electromagnetic Pump Including End Effects

W.F. Hughes*
Carnegie-Mellon University, Pittsburgh, Pa.
and
I.R. McNab†
Westinghouse Electric Corporation, Pittsburgh, Pa.

Abstract

A quasi-one-dimensional theory has been developed to describe the performance of a direct current electromagnetic pump or generator duct. Boundary-layer effects are considered of the order 1/M and wall conductivity effects are included. The efficiency, pressure rise, operating voltage, and current are among the parameters evaluated. The description includes end effects caused by current fringing in regions where magnetic field fringing also occurs. General performance equations are formulated for the particular case where the magnetic field decays exponentially. Examples of The pump efficiency variation with each of the major parameters, including wall conductance for a conducting side wall of finite thickness, are given.

Nomenclature

a = half-channel dimension in the direction of the magnetic field
B_0 = applied magnetic field in center of channel

Paper presented at Third Beer-Sheva International Seminar on Magnetohydrodynamic Flows and Turbulence, Ben-Gurion University of the Negev, Beer-Sheva, Israel, March 23-27, 1981. Copyright © American Institute of Aeronautics and Astronautics, Inc., 1982. All rights reserved.
*Professor of Mechanical Engineering, Department of Mechanical Engineering.
†Manager, Electrodynamics Section, Electrotechnology Department, Research and Development Center.

b	=	half-channel dimension perpendicular to the magnetic field
I	=	current
I_F	=	total fringing current for both ends
I^*	=	dimensionless current = $I/L\bar{v}\sqrt{\sigma\mu}$
i	=	transverse current in fluid per unit z length
i_w	=	transverse current in one side wall per unit z length
i_T	=	total transverse current per unit z length
i_z	=	axial current in the fluid for one-half the duct area
i_{zw}	=	axial current in both side walls for one-half the duct area
i_{zT}	=	total axial current for half the duce area
L	=	center section length
L_{cr}	=	initial length at which $dP/dx = 0$
ℓ	=	length at which the exponentially graded magnetic field is terminated
M	=	Hartmann number = $aB_0\sqrt{\sigma/\mu}$
N	=	number of parallel ducts
ΔP	=	pressure rise
ΔP^*	=	dimensionless pressure rise = $a^2\Delta P/L\mu\bar{v}$
Q	=	volume flow rate
$\bar{Q}$	=	dimensionless flow parameter = $Q\sqrt{\sigma\mu}/2abi$
R_F	=	axial fluid electrical resistance
R_W	=	axial wall electrical resistance
T	=	temperature
t_w	=	side wall thickness (one wall only)
V	=	local voltage between electrode walls
V_0	=	applied interelectrode voltage in center section
V^*	=	reduced voltage = $V_0/2bvB_0$
v	=	mean fluid velocity = $Q/4ab$
w	=	$[1 + (1 + a/b)\bar{\sigma}]/(1 + \bar{\sigma})$
x	=	direction perpendicular both to applied magnetic field and fluid flow direction
z	=	direction of fluid flow
α	=	$ab\sigma[1 + (1 + a/b)\bar{\sigma}]/2$
β	=	$a\sigma(1 + \bar{\sigma})/b$
γ	=	$\sigma Q/2b = 2a\sigma\bar{v}$
η	=	efficiency
λ	=	length aspect ratio = L/b
μ	=	fluid viscosity
ν	=	wall resistance parameter = $\{[1 + (1 + a/b)\bar{\sigma}]/2(1 + \bar{\sigma})\}^{\frac{1}{2}}$
ξ	=	magnetic field spatial decay exponent
σ	=	fluid electrical conductivity
σ_w	=	wall electrical conductivity
$\bar{\sigma}$	=	wall conductivity parameter = $\sigma_w t_w/\sigma a$

τ = magnetic field fringing parameter = ξL
ϕ = electric potential

Superscripts

* = dimensionless parameter

Subscripts

c = pertaining to center section
d = pertaining to duct
E = pertaining to end region
F = pertaining to fluid
T = total (fluid and wall)
w = pertaining to wall

I. Introduction

In recent years attempts have been made at the Westinghouse Research Center to evaluate the performance of direct-current electromagnetic pumps in more detail than has previously been the case. A combined theoretical and experimental program has been undertaken. This paper describes one of the theoretical approaches which was intended to provide a simple yet effective design methodology.

The improved evaluation of dc electromagnetic pump performance is based on a high Hartmann number analysis undertaken by Hunt and Stewartson (1965) for magnetohydrodynamic (MHD) flow in rectangular channels. The first part of the analysis permits the performance of the central part of the pump to be modeled more accurately than previously. This is described in Sec. II. The analysis is based on a Hartmann number expansion for the velocity profile and includes boundary-layer effects to order $1/M$. The velocity profile is essentially slug-like except for the boundary layers. The wall conductivity is also considered since losses in the wall may be important in an actual situation.

Experimentalists who have operated dc pumps have shown that it is necessary to allow for end losses when evaluating or predicting pump performance. Blake (1956) points out that end effects, caused by current bypassing the working section, become increasingly important as the ratio of induced emf to internal voltage drop increases. He specifically mentions the possibility of grading the magnetic field to match the current fringing as a means of minimizing current bypassing and gives typical practical curves for both parameters to serve as a guide. However, he quotes Barnes (1953) in advising that an experimental determination is ultimately necessary in each case to establish the relationship of fringing resistance to field profile, current, and duct geometry.

Sutton et al.(1962) have evaluated the influence of an exponentially graded magnetic field profile on the performance of a gaseous MHD generator, assuming a slug flow with no boundary-layer effects and insulating walls. They show, for their particular case, that the generator efficiency can be doubled by extending the magnetic field with an e-folding distance equal to the electrode spacing. We are not aware of any experimental work which would confirm this prediction. Gherson et al. (1980) have also addressed this problem by using a numerical technique to solve the equation for the potential. However, they assumed slug flow with no boundary-layer effects and that the walls are all insulators. The boundary layers are extremely important in determining viscous losses and electrical leakage paths and are included in the present analysis. Further, the wall conductivity must be considered since in any practical pump an insulating layer inside the duct is probably not feasible.

Hoffman and Carlson (1971) have made a detailed study of end effects in liquid-metal flows in relation to lithium-cooled blankets for fusion reactors. They point out that the complete solution of combined losses is a complex three-dimensional problem. They study linearly fringing magnetic fields, utilizing Shercliff's solution (1962) of Laplace's equation for a step change in the magnetic field, to compute the fluid behavior for multiple small steps. The highly beneficial effect of a slowly changing magnetic field is shown in this way.

In Sec. III attention is focussed on the end effects caused by the interaction of the fringing current with the fringing magnetic field; a quasi-one-dimensional approach is used (i.e., lumped circuit model). No attempt is made to account for distortion of the applied magnetic field in the pump duct. Inlet relaxation effects are also ignored.

As an alternative to grading the magnetic field, several authors [Blake (1956), Barnes (1953), Sutton et al. (1962), Childs (1973), Moszynski (1967)] have suggested the use of nonconducting baffles in the end regions to prevent transverse current flow and hence the associated losses.

However, such baffles may have to be extremely long to be effective and their presence may not, in any event, be acceptable in the primary coolant circuit of a reactor. The approach employed here is, therefore, considered to offer a more practical solution to this problem.

II. Pump Central Region

Hughes (1976) analyzed the central region of a rectangular duct dc electromagnetic (EM) pump with conducting walls, constant applied electric and magnetic fields, and no end losses. The velocity profile to order 1/M is obtained by using the method of Hunt and Stewartson (1965) for high Hartmann number laminar flow. The following parameters were determined:

Dimensionless pressure rise

$$\Delta P_c^* = M^2[V^* - 1 - (1/M_o)] \tag{1}$$

Dimensionless total (fluid plus walls) current

$$I_{T_c}^* = 2M[V^* - 1 + V^*\bar{\sigma}] \tag{2}$$

The pump efficiency is defined as

$$\eta = \frac{\text{workout per unit time}}{\text{power input}} = \frac{Q\Delta P}{V_o I} \tag{3}$$

Using the definitions provided in the nomenclature, this may be written as

$$\eta = 2\Delta P^*/MI^*V^* \tag{4}$$

which becomes

$$\eta_c = \frac{V^* - 1 - 1/M}{V^*[V^* - 1 + V^*\bar{\sigma}]} \tag{5}$$

Using the auxiliary relationship

$$V^* - 1 = 1/\bar{Q}M \tag{6}$$

This may be rewritten as

$$\eta_c = [\frac{\bar{Q}M(1 - \bar{Q})}{1 + \bar{Q}M}][\frac{1}{1 + (1 + \bar{Q}M)\bar{\sigma}}] \tag{7}$$

In Eqs. (5) and (7), the influence of conducting walls can be clearly identified. When insulating side walls are used, $\bar{\sigma} = 0$ and the efficiency becomes

$$\eta_c = \frac{V^* - 1 - 1/M}{V^*(V^* - 1)} = \frac{\bar{Q}M(1 - \bar{Q})}{(1 + \bar{Q}M)} \tag{8}$$

From this expression, it is apparent that increasing M always leads to an increase in efficiency. Differentiating with respect to Q shows that the optimum reduced flow rate for maximum efficiency is (for M >> 1)

$$\bar{Q} = 1/\sqrt{M} \tag{9}$$

at which point the efficiency is

$$(\eta_c)_{opt} = (\sqrt{M} - 1)/(\sqrt{M} + 1) \tag{10}$$

Since M essentially measures the magnetic field strength, a high magnetic field would seem desirable. However, consideration of losses and end effects shows this not to be the case.

III. End Effects

Numerical evaluations, using the approach outlined in Sec. II, showed that relatively short ducts were capable of providing the pressure heads required for typical liquid-metal coolant flows in fast breeder reactors. For example, for 4.4 m^3/s and 1.28 MPa (70,000 gal/min and 500 ft sodium head) the pump duct dimension could typically be 0.65 m wide by 0.65 m high by 1.5 m long (2.13 x 2.13 x 4.92 ft). Under these circumstances, effects which occur at the ends of the electrodes and in the magnetic field decay region become important and require a detailed analysis.

Physically, the problem reduces to that of shaping the magnetic field in the region beyond the electrodes in an optimum way to minimize unusable current fringing.

Within the central portion of the pump duct, the back emf and applied voltage are almost equal (for high efficiency so that the net electric field driving current through the fluid is small. If the applied magnetic field could be terminated at the ends of the electrodes, there would be no back emf and the fluid in these regions would present an apparent short circuit to the applied current. This would lead to a low pump efficiency. Conversely, if the magnetic field were extended out at a constant value well beyond the ends of the electrodes, a point would be reached (relatively quickly) where the back emf would equal the local electrical field strength due to the applied voltage on the electrodes. Out to this point pumping action will occur, but at this point it will cease. Beyond this point the back emf would exceed the local applied electrical field and the fluid would act as a generator, again with an adverse effect upon the pump efficiency.

The required compromise entails a shaping of the magnetic field in the end regions in such a way that as much as possible of the current fringing that inevitably occurs can be utilized to provide pressure head. This solution has been used to evaluate the total fringing current (both ends), the total generated pressure head, and the overall pump duct efficiency. At the same time, several other factors should be considered in an engineering sense: 1) most of the pumping action should take place in the center section, 2) the total applied voltage/current characteristics must match an available power supply having an acceptable efficiency, 3) the magnetic field profile must be physically realizable, 4) the current density in the electrodes must not be unacceptably high, and 5) dimensions must be kept within reasonable limits.

Assumptions

The assumptions made, or implied, in this treatment are the following:

1) The quasi-one-dimensional approach based on high Hartmann number expansions for laminar flow is a valid approximation for treating this problem.

2) MFD flow effects related to corners may be neglected with negligible loss of accuracy for large Hartmann numbers.

3) The flow solutions used in the central region (see Hughes 1976) apply at any axial location in the end regions, that is, the inertial relaxation effects can be ignored.

4) Distortions of the applied magnetic field caused by the current are ignored.

5) Current is permitted to flow in the conducting walls as well as in the fluid.

6) No change in the duct cross-sectional area or shape takes place in the region treated here.

7) The electrodes are assumed to have infinite conductivity.

8) A simple lumped circuit of the fluid is assumed. A more detailed two-dimensional potential solution may easily be effected, but the additional accuracy seems unwarranted. (Such solutions have been carried out by the authors and the overall results are different from the present ones by less than 5% over a wide range of parameters. See Appendix.)

Formulation of Basic Equations

From Figs. 1a or 1b, we consider a simple lumped circuit model. A simple current balance yields

$$di_{zT}/dz = -i_T \qquad (11)$$

Along the centerline of the duct (see Fig. 1a), the electric potential is zero. Hence, in the quasi-one-dimensional approximation, the average voltage in the z direction per half channel will be the mean of that at the electrode wall and in the center of the duct. Approximately, the IR drop in the fluid and in the walls is

$$i_z R_f = i_{zw} R_w = (V/2|_z - V/2|_{z+\Delta_z})/2$$

From Fig. 1c,

$$R_f = \frac{\Delta z}{2ab\sigma} \quad \text{and} \quad R_w = \frac{\Delta z}{2(a+b)\sigma_w t_w}$$

so that Eq. (11) becomes

$$\frac{dV}{dz} = -\frac{2i_z}{\sigma ab} - \frac{2i_{zw}}{\sigma_w t_w (a+b)} \qquad (12)$$

In the x direction, the applied voltage is equal to the back emf plus the internal IR drop in the fluid and is also equal

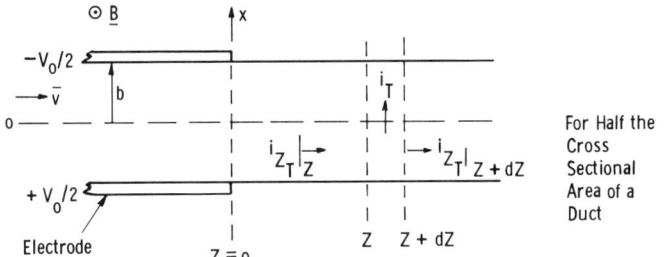

Fig. 1a End region geometry and currents.

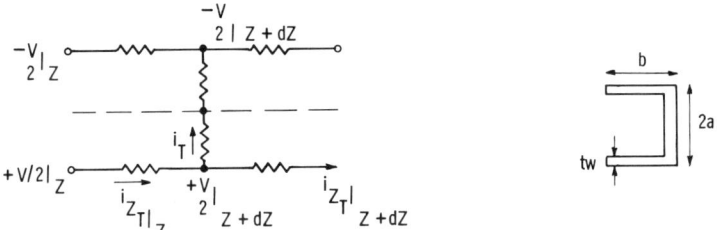

Fig. 1b Equivalent network for current flow.

Fig. 1c Wall geometry.

to the wall drop,

$$V(z) = \frac{bi(z)}{a\sigma} + \frac{B(z)Q}{2a} = \frac{i_w 2b}{\sigma_w t_w} \quad (13)$$

Finally, assuming that the asymptotic solutions presented in Hughes (1976) hold locally,

$$Q = \frac{4a^3 b}{\mu M} [- \frac{dP}{dz} + \frac{B(z)i}{2a}] \quad (14)$$

Combination of the above equations yields

$$-\frac{di_z}{dz} - \frac{di_{zw}}{dz} = \frac{a\sigma}{b}(V - \frac{BQ}{2a}) + \frac{V\sigma_w t_w}{b} \quad (15)$$

$$-\frac{di_z}{dz} = \frac{\sigma ab}{2} \frac{d^2 V}{dz^2} \quad (16)$$

and

$$-\frac{di_{zw}}{dz} = \frac{(a+b)\sigma_w t_w}{2} \frac{d^2 V}{dz^2} \quad (17)$$

Combining the above we obtain

$$\frac{d^2V}{dz^2} - \left(\frac{\beta}{\alpha}\right)V = -\left(\frac{\gamma}{\alpha}\right)B \qquad (18)$$

where α, β, and γ are constants defined in the nomenclature. The solution of Eq. (18) is

$$V = C_1 \exp[z\sqrt{\beta/\alpha}] + C_2 \exp[-z\sqrt{\beta/\alpha}] + F(z) \qquad (19)$$

where C_1 and C_2 are constants and $F(z)$ depends on the form assumed for $B(z)$.

Exponential Magnetic Field Decay

As one example of a practical spatial variation of the centerline magnetic field strength in the region beyond the end of the pump electrodes, an exponential magnetic field decay will be assumed. This approximates the actual decay from an abruptly terminated pole piece (Bewley 1948),

$$B(z) = B_0 \exp(-\xi z) \qquad (20)$$

For this case, the particular integral is

$$F(z) = -\left(\frac{\gamma}{\alpha}\right) \frac{B_0 \exp(-\xi z)}{[\xi^2 - (\beta/\alpha)]} \qquad (21)$$

The required boundary conditions are (see Fig. 1) $V = V_0$ at $z = 0$ and $V = 0$ at $z = \infty$, yielding for the complete solution

$$V = \left\{V_0 + \left(\frac{\gamma}{\alpha}\right)\frac{B_0}{[\xi^2 - (\beta/\alpha)]}\right\} \exp[-z\sqrt{\beta/\alpha}] - \left(\frac{\gamma}{\alpha}\right)\frac{B_0 \exp(-\xi z)}{[\xi^2 - (\beta/\alpha)]} \qquad (22)$$

Use of this equation permits the pump performance in these end regions to be evaluated.

Note that for the special case where $\xi^2 = \beta/\alpha$, a singularity is introduced. This may be overcome by expanding the last term in Eq. (22) as a series (to the first term only) to yield

$$V\big|_{\xi=\sqrt{\beta/\alpha}} = V_0\left[1 + \frac{z\,\beta/\alpha}{2(1+\sigma)V^*}\right] \exp(-z\sqrt{\beta/\alpha})$$

Evaluation of Pump Performance in End Regions

The general expression for the rate of pressure rise is (see Hughes 1976)

$$\frac{dP}{dz} = \sigma v B_0^2 \left(\frac{B}{B_0}\right) \left[V^*\left(\frac{V}{V_0}\right) - \left(\frac{B}{B_0}\right) - \frac{1}{M}\right] \quad (23)$$

Substituting Eqs. (20) and (22) in this yields the pressure rise for the exponentially decaying magnetic field

$$\frac{\exp(\xi z)}{\sigma v B_0^2} \left(\frac{dP}{dz}\right) = V^* \exp(-z\sqrt{\beta/\alpha}) - \exp(-\xi z) - \frac{1}{M}$$

$$+ \frac{\{\exp[-z\sqrt{\beta/\alpha}] - \exp(-\xi z)\}}{(1 + \bar{\sigma})[\xi^2(\alpha/\beta) - 1]} \quad (24)$$

If the exponent ξ is not chosen properly, generator action may occur in the outer fringing region. The location, $z = L_{cr}$, at which pumping action may cease and generator action begin, may be found by setting the left-hand side of this expression equal to zero and solving the resulting equation. In general, the expression obtained is rather complicated, but for many practical cases with small pump ducts, where $\beta/\alpha \gg 1$, the exponential terms on the right-hand side of Eq. (24) which contains $\sqrt{\beta/\alpha}$ may be neglected without appreciable loss of accuracy, and L_{cr} is then found from the resulting equation

$$L_{cr} = \frac{1}{\xi} \ln_e \left\{ M \left[\frac{1}{(1 + \bar{\sigma})[1 - \xi^2/(\beta/\alpha)]} - 1 \right] \right\} \quad (25)$$

Clearly, the exponential decay parameter must be chosen so that pumping action occurs throughout the region of the magnetic field, i.e., the value of the magnetic field becomes negligible at a position denoted by ℓ less than L_{cr}. At this point, the total transverse voltage across the fluid and walls is sufficiently low so that negligible small leakage currents occur beyond that point. The total pressure rise in each end region is then found by integrating Eq. (24) from $z = 0$ to $z = \ell$ to yield

$$\frac{\Delta P_E}{\sigma v B_0^2} = \frac{\Delta P_E^*}{(M^2/L)} = \frac{V^*\{1 - \exp[-(\xi + \beta/\alpha)\ell]\}}{[\xi + \sqrt{\beta/\alpha}]} - \frac{[1 - \exp(-2\xi\ell)]}{2\xi}$$

$$-\frac{[1 - \exp(-\xi\ell_c)]}{M\xi} + \frac{\{1 - \exp[-(\xi + \sqrt{\beta/\alpha})\ell]\}}{(\xi + \sqrt{\beta/\alpha})(1 + \bar{\sigma})[\xi^2(\alpha/\beta) - 1]}$$

$$-\frac{[1 - \exp(-2\xi\ell)]}{2\xi(1 + \bar{\sigma})[\xi^2(\alpha/\beta) - 1]} \quad (26)$$

Provided $\xi\ell$ is reasonably large, the total pressure rise (for <u>both</u> ends) takes the approximate form

$$\Delta P_E^* = 2M^2 [-\frac{V^*}{(\xi L + L/b\sqrt{2/w})} + \frac{1}{(\xi L + L/b\sqrt{2/w})^2 (b/L)^2 (\xi L) w (1 + \bar{\sigma})}$$

$$- \frac{1}{\xi L} \left(\frac{1}{M} + \frac{1}{2}\right)] \quad (27)$$

The influence of various parameters on the generated pressure head in the pump ends can be obtained from this expression.

The total current fringing into the end regions may be obtained now that $V(z)$ and $B(z)$ are known. The total fringing current including the wall current for <u>both</u> ends is then

$$I_E = 2 \int_0^\infty i_T dz = 2 \int_0^\infty (i + 2i_w) dz \quad (28)$$

Using Eq. (11) this may be written as

$$I_E = \frac{\sigma B_0 Q}{b} \int_0^\infty [V^*(1 + \bar{\sigma})\left(\frac{V}{V_0}\right) - \left(\frac{B}{B_0}\right)] dz \quad (29)$$

yielding

$$\frac{I_E^*}{2M(b/L)} = \frac{I_E}{(\sigma B_0 Q/2)} = \sqrt{2w} \ [V^*(1 + \bar{\sigma}) - \frac{1}{(1 + \xi b\sqrt{w/2})}] \quad (30)$$

The use of this expression, together with that for the generated pressure head, enables the overall pump efficiency to be evaluated.

Note that the above expression is the total current flowing in the end regions, that is, it includes currents both in the wall and in the fluid.

ANALYSIS OF AN ELECTROMAGNETIC PUMP 299

IV. Total Pump Design

Subject to the assumptions made in this derivation, the design of a dc EM pump, composed of a center region plus two end fringing regions in which the magnetic field decays exponentially, can be obtained from the equations given in Secs. II and III.

The total pressure rise can be obtained from Eqs. (1) and (27)

$$\frac{\Delta P_T^*}{M^2} = V^* [1 + \frac{2}{(\lambda/\nu + \tau)}] + \frac{(\lambda/\nu)^2}{\tau(1 + \bar{\sigma})(\lambda/\nu + \tau)^2}$$

$$- [1 + \frac{1}{\tau} + \frac{1}{M} + \frac{2}{M\tau}] \tag{31}$$

and, from Eqs. (2) and (30), the total current is

$$\frac{I^*}{2M} = V^*(1 + \bar{\sigma}) [1 + \frac{2}{\lambda/\nu}] - [1 + \frac{2}{(\lambda/\nu + \tau)}] \tag{32}$$

The total pump efficiency obtained from these equations and Eq. (4) is

$$\eta_T = \frac{V^* [1 + \frac{2}{(\lambda/\nu+\tau)}] + \frac{(\lambda/\nu)^2}{\tau(1+\bar{\sigma})(\lambda/\nu+\tau)^2} - [1 + \frac{1}{\tau} + \frac{1}{M} + \frac{2}{M\tau}]}{V^{*2}(1+\bar{\sigma})[1 + \frac{2}{(\lambda/\nu)}] - V^*[1 + \frac{2}{(\lambda/\nu+\tau)}]} \tag{33}$$

With the number of variables in this case, it is apparent that it is not possible to obtain a simple expression for the optimum pump efficiency, as was the case in Sec. II; in general, numerical techniques have to be used. For an insulated wall pump, $\sigma = 0$ and $\nu = 1/\sqrt{2}$, the efficiency takes the slightly simpler form,

$$\eta_T \big|_{ins} = \frac{V^*[1 + \frac{2}{(\sqrt{2}\lambda+\tau)}] + \frac{(\sqrt{2}\lambda)^2}{\tau(\sqrt{2}\lambda+\tau)^2} - [1 + \frac{1}{\tau} + \frac{1}{M} + \frac{2}{M\tau}]}{V^{*2}[1 + \frac{2}{\sqrt{2}\lambda}] - V^*[1 + \frac{2}{\sqrt{2}\lambda+\tau}]} \tag{34}$$

If the magnetic field could be cut off sharply at the ends of the electrodes, $\xi = \infty$ and hence $\tau = \infty$. For this limiting case, Eq. (34) becomes relatively simple

$$\eta_T \big|_{ins} = \frac{V^* - 1 - 1/M}{V^*[V^* - 1 + \sqrt{2}\, V^*/\lambda]} \tag{35}$$

Using Eq. (6), this may also be written in the same form as Eqs. (7) and (8),

$$\eta_T\big|_{ins} = \left[\frac{\overline{QM}(1 - \overline{Q})}{1 + \overline{QM}}\right]\left[\frac{1}{1 + (\sqrt{2}/\lambda)(1 + \overline{QM})}\right] \tag{36}$$

Comparison of this with Eq. (7) shows the general similarity of form, with the influence of end leakage for the insulated wall case being to reduce efficiency, as in the case of no end leakage but finite wall conductance.

The efficiency expression above shows how V^* is critical in determining the pump performance. In Eq. (35), if V^* is less than $(1 + 1/M)$, the efficiency becomes negative. That is, insufficient power is supplied to pump against the losses. When $V^* = (1 + 1/M)$, the pressure head is zero. When fringing of the magnetic field is included, [Eq. (33)] the situation becomes more complicated. Thus, for example, when $V^* = 1$, the efficiency may be positive or negative, depending on the value of the fringing parameter τ. In this case, no pumping occurs in the center region, but pumping can take place in the end regions to a greater or lesser extent depending on τ.

The general characterization of the MFD device is that it is a pump, flowmeter, or generator, depending on whether I^* is greater than, equal to, or less than zero. The corresponding values of V^* and η may be obtained from Eqs. (32) and (33).

V. Numerical Evaluation

Except in special cases where a relatively simple analytic expression can be obtained, it is generally necessary to resort to numerical techniques for pump design. The approach followed assumes that the pump flow rate, pressure head, and operating temperature (Q, ΔP, and T) are specified. The required fluid and wall parameters (σ, μ, ρ, and σ_w) are evaluated from the temperature and known fluid properties.

A multiple duct geometry with N ducts operating in parallel is assumed. In addition, the mean flow velocity $\bar{v}$, the duct aspect ratios b/a and $\lambda = L/b$, the magnetic field strength B_0, the magnetic field decay exponent ξ, and the conducting wall thickness t_w are chosen.

The dimensionless voltage V* is then found by inverting Eq. (31) to give

$$V^* = \frac{\Delta P_T^*/M^2 + [1+1/\tau+1/M+2/M\tau] - (\lambda/\nu)^2[\tau(1+\bar{\sigma})(\lambda/\nu+\tau)^2]^{-1}}{[1 + 2/(\lambda/\nu + \tau)]} \quad (37)$$

With this, the dimensionless current may be found from Eq. (32),

$$\frac{I^*}{2M} = V^*(1 + \bar{\sigma}) [1 + \frac{2}{(\lambda/\nu)}] - [1 + \frac{2}{(\lambda/\nu + \tau)}] \quad (38)$$

The total pump efficiency can then be found from the derived parameters and Eqs. (3) and (4) or directly from Eq. (33).

As an example, the above equations have been used to evaluate the performance of a large dc EM pump of the type which could be used in the primary coolant circuit of a large liquid-metal-cooled fast breeder reactor of the type now being designed for commercial operation.

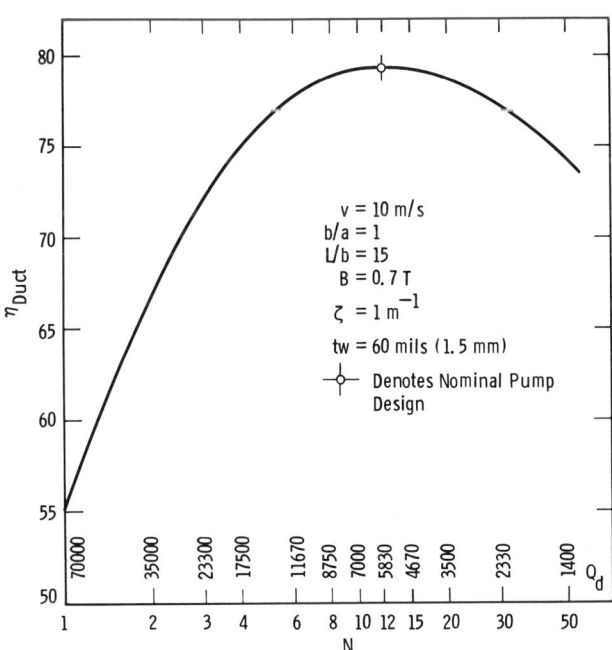

Fig. 2 Duct efficiency as a function of the number of ducts (N) and the volume flow per duct (Q_d in gallons per minute).

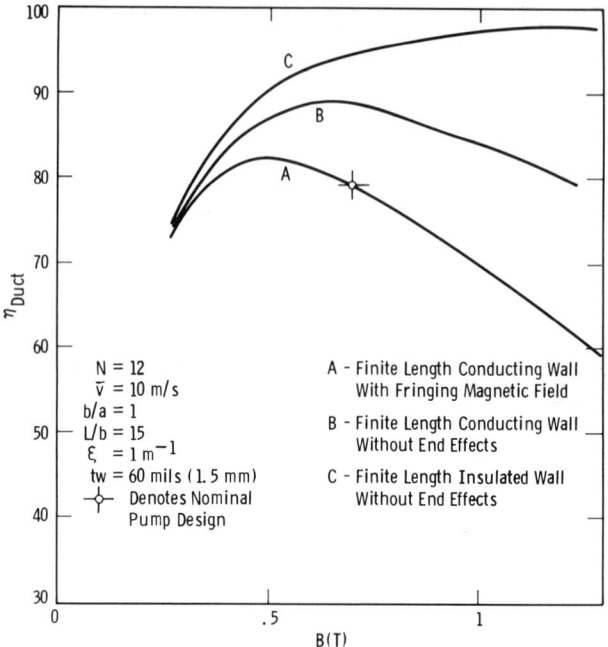

Fig. 3 Duct efficiency as a function of magnetic field strength, B (in T), for three analytic pump models.

The required output is 4.416 m^3/s (70,000 gal/min) with a developed pressure rise of 1.28 Mpa (500 ft head of liquid sodium). The operating temperature of the pump when located in the hot leg of the primary circuit is 537.8°C (1000°F). The required sodium properties at this temperature are obtained from the equations given by Foust (1972) and are σ = 3.421 x 10^6 mho/m, μ = 2.292 x 10^{-4} kg/ms, and ρ = 823.2 kg/m^3. It is assumed that the pump wall is made from Inconel 706, a typical high-temperature and sodium-compatible constructional material. At the operating temperature, its conductivity is σ_w = 8.511 x 10^5 S/m. The use of multiple-pump ducts is assumed, but only the performance of the single-duct unit is given here.

The performance evaluation centers around what is termed the "nominal" pump design. For this N = 12, $\bar{v}$ = 10 m/s, b/a = 1, L/b = 15, B = 0.7 T, ξ = 1, and t_w = 1.524 mm (60 mils). For this, the pump duct efficiency is predicted as 79.3 %, using the equations developed here.

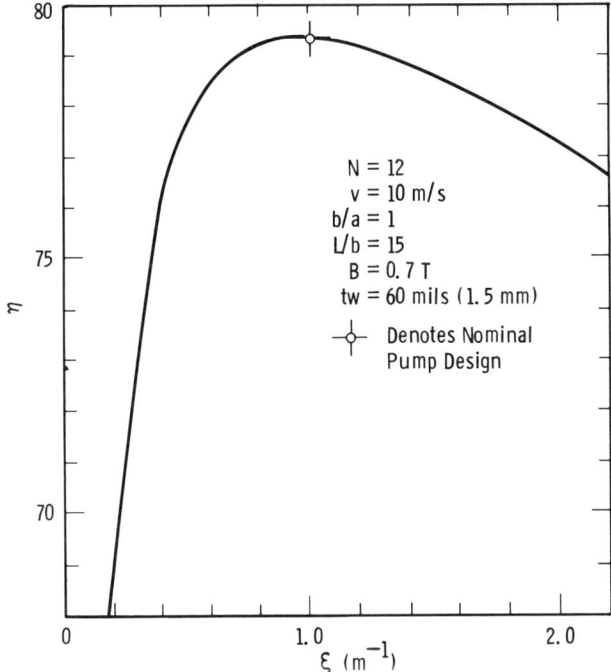

Fig. 4 Duct efficiency as a function of the fringing parameter (ξ).

The influence of varying each of the above parameters (separately) is shown in Figs. 2-8. Note that the quoted duct efficiencies are only for the quoted parameter values. Thus, for example, in Fig. 2, no attempt has been made to optimize the duct efficiency at each value of N. Note also that system considerations may necessitate operation of the pump duct at a point off its optimum efficiency value. The operating point for the "nominal" pump is denoted on each curve.

Figure 3 is of particular interest since additional data are used to allow comparison of three cases: 1) the present conducting wall duct with a fringing magnetic field; 2) a finite-length conducting wall duct, but with end losses omitted; and 3) a finite-length insulated wall duct, with end losses omitted. The influence of end losses in reducing efficiency can be seen. For the insulated wall case, the optimized efficiency [Eq. (10)] is not shown here; rather the dimensions for the present study are employed.

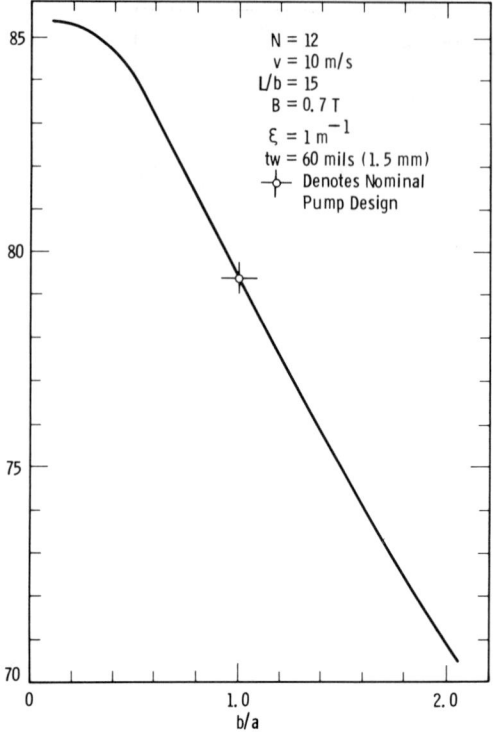

Fig. 5 Duct efficiency as a function of the duct cross-section aspect ratio (b/a).

Nevertheless, it is apparent that, for this case, theory predicts the use of as high a magnetic field as possible, consistent with practical considerations. It is the modification of this conclusion, as represented by curve C in Fig. 3, that it is the major influence of end effects on pump design.

The results presented herein are intended to be an approximate, yet adequate, analysis for engineering design. The exponential magnetic field decay was assumed because it is easy to handle analytically and corresponds closely to the natural decay from a sharply terminated pole piece. Certainly other shapes might give higher overall efficiency and should be investigated. Based on experimental results obtained at the Westinghouse Research and Development Center, we feel that the design procedure indicated here is indeed valid and other fringing field profiles may be treated by this lumped circuit method of analysis.

ANALYSIS OF AN ELECTROMAGNETIC PUMP

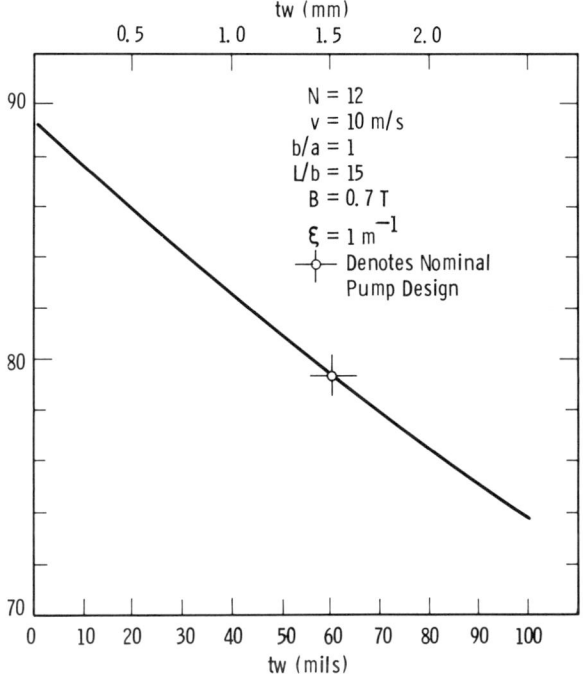

Fig. 6 Duct efficiency as a function of the thickness (t_w) of the conducting side wall.

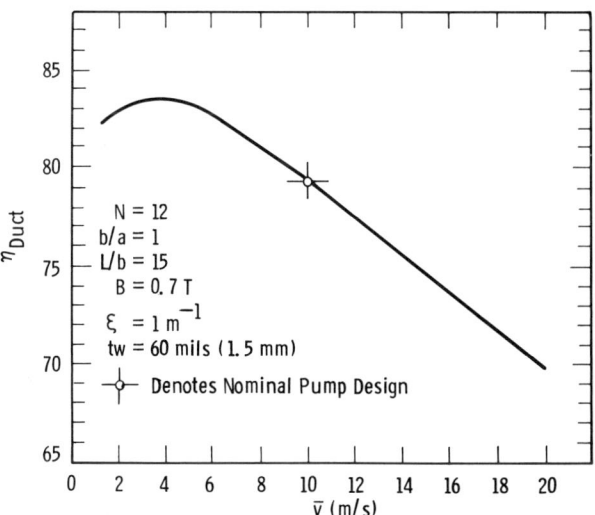

Fig. 7 Duct efficiency as a function of the mean fluid velocity ($\bar{v}$).

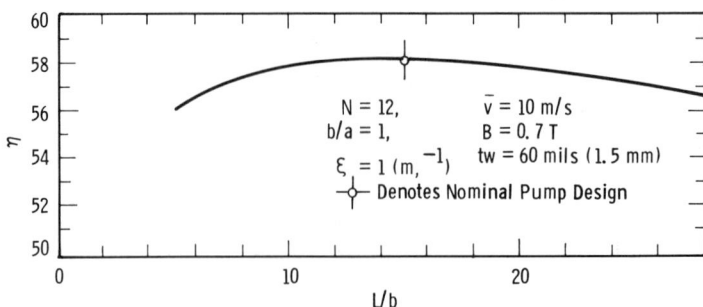

Fig. 8 Duct efficiency as a function of the length aspect ratio (L/b).

Acknowledgment

The work underlying this report was supported by the Westinghouse Electric Corporation and in part by the National Science Foundation.

Appendix: Two-Dimensional Approximations

A simple two-dimensional analysis can be effected if the velocity is assumed to be slug-like. This solution is the analytical equivalent of the numerical analysis of Gherson et al. (1980) with one relaxed boundary condition.

Assumptions

The electrical potential is assumed to vary two-dimensionally in the x-y plane. The applied magnetic field (in the y direction) is uniform in the central region between the electrodes but in the end regions the applied magnetic field is allowed to be an arbitrary function of z, the distance along the channel.

The central pump region consists of electrode walls (each at a uniform potential) at $x = \pm b$, connected by insulating side walls. The end regions (upstream and downstream) consist of walls which are extensions of the electrodes and have a conductivity σ_w. The side walls at $y = \pm a$ are assumed to be nonconducting. If they are conducting, the problem becomes three-dimensional and the present analysis is invalid.

The potential between the electrodes is V_0. For convenience, the potential $x = -b$ is taken as $V_0/2$ and the potential $x = +b$ is $-V_0/2$. Hence, the potential distribution in the duct may be referred to the potential in the y-z plane at x = 0, which is arbitrarily taken as zero.

At the inlet to the end region, z = 0, the potential between the electrodes is assumed to be linear. In the main central channel, this assumption is certainly valid, although near the ends of the electrodes, the potential will deviate somewhat from linearity but numerical studies show it to be a good approximation. The more exact two-dimensional field analysis of Sutton et al. (1962) could be used but the additional complexity seems unwarranted in view of the assumption about velocity.

The fluid velocity profile is assumed to be uniform, as slug flow with a quasifully developed boundary layer as discussed by Hunt and Stewartson (1965). The problem of the potential distribution, therefore, become uncoupled from the fluid dynamics. This assumption may not be valid because there is reason to believe that the velocity profile may undergo drastic changes in the end regions. The slug flow in the central region probably develops into an M-shaped profile with the possibility of flow reversal in the core.

However, the purpose of this analysis is to investigate the effect of a two-dimensional potential calculation on the efficiency of the pump, while assuming a uniform velocity profile and realizing that the results cannot be exact. Nevertheless, the results may be compared to the previous one-dimensional analysis in order to assess the effect of a two-dimensional potential and current calculation. This is an important consideration in the determination of what future direction to proceed for analytical refinements.

Under the slug flow assumption that v is uniform throughout the flow,

$$\frac{\partial^2 \phi}{\partial x^2} + \frac{\partial^2 \phi}{\partial y^2} = \nabla^2 \phi = 0 \tag{A1}$$

in the fluid.

The boundary conditions may be formulated as follows:

1) At z = 0, the potential ϕ is assumed to be of the form

$$\phi = - (V_0/2)(x/b) \tag{A2}$$

2) In order to simplify the problem, it becomes expedient to pose the problem over a finite domain. Far from the end regions the fringing currents all die out and the potential throughout the fluid is uniformly zero. Hence, we arbitrarily assume a distance $z = \ell$ at which the potential is taken as zero. The larger ℓ, the more accurate the solution. A convenient parameter is the ratio ℓ/L. For numerical calculations, a value of $\ell/L = 20$ has proved adequate, as will be discussed later.

3) Along the centerline $x = 0$, the potential ϕ is zero (for all z).

4) On the wall,

$$t\sigma_w \partial^2\phi/\partial x^2 \big|_{x=b} = -\sigma [\partial\phi/\partial x \big|_{x=b} + vB] \tag{A3}$$

The solution is

$$\phi = \sum_{n=1}^{\infty} C_n \sin \frac{n\pi z}{\ell} \sinh \frac{n\pi x}{\ell} - \frac{V_0}{2}(x/b)(1 - z/\ell) \tag{A4}$$

where C_n is given by

$$C_n = \frac{\bar{v}B_0[V^*\ell/n\pi - B_n^*]}{(n\pi/2)[\cosh n\pi b/\ell + a\bar{\sigma}(n\pi/\ell) \sinh n\pi b/\ell]} \tag{A5}$$

where

$$B_n^* = \int_0^{\infty} \frac{B(z)}{B_0} \cdot \sin \frac{n\pi z}{\ell} dz \tag{A6}$$

The result is valid for an arbitrarily graded magnetic field $B(z)$.

The total fringing current for one end $I_E/2$ can be found by integrating J_x on a plane through the center of the channel,

$$I_E^*/2 = -\frac{4\ell M}{L} \sum_{\substack{n=1 \\ n\,\text{odd}}}^{\infty} C_n^* + (V^*M\ell/L) - \frac{2M}{L} \int_0^{\infty} \frac{B(z)}{B_0} dz \tag{A7}$$

ANALYSIS OF AN ELECTROMAGNETIC PUMP 309

where C_n^* (the normalized Fourier coefficients) are defined as

$$C_n^* = C_n/\bar{v}B_0\ell$$

For a fully insulated channel, $\bar{\sigma} = 0$, which affects only the evaluation of C_n^*.

The Pressure Gradient in the End Regions

The assumption is made that flow velocity profiles are the same in the end regions as the central region so that

$$dp/dx = \bar{J}_x \cdot B(z) - [B(z) \cdot \sqrt{\sigma\mu}\ \bar{v}]/a \qquad (A8)$$

$\bar{J}_x$ is the mean value of J_x at any value of z. $\bar{J}_x$ may be evaluated from Ohm's law and the expression for ϕ

$$\bar{J}_x = \frac{1}{b}\int_0^b J_x dx = \frac{1}{b}\int_0^b \sigma[-\frac{\partial \phi}{\partial x} - \bar{v}\,B(z)] \cdot dx$$

$$= -\frac{\sigma}{b}\sum_{n=1}^{\infty} C_n \sin\frac{n\pi z}{\ell} \cdot \sinh\frac{n\pi b}{\ell} + \frac{\sigma V_0}{2b}(1 - \frac{z}{\ell}) - \sigma\bar{v}B(z) \quad (A9)$$

The total pressure rise in one fringing end region is then

$$\Delta P_E\Big|_{\substack{\text{one}\\\text{end}}} = -\left(\frac{\sigma}{b}\right)\int_0^\ell C_n \frac{\sin n\pi z}{\ell}\sinh\frac{n\pi b}{\ell}\cdot B(z)\cdot dz$$

$$+ \int_0^\ell \frac{\sigma V_0}{2b}(1-\frac{z}{\ell})B(z)dz - \sigma v\int_0^\ell [B(z)]^2 dz - \int_0^\ell \frac{B(z)\cdot\sqrt{\sigma\mu}\ \bar{v}}{a}dz$$

$$(A10)$$

The Insulated Pump with an Exponential Magnetic Field Decay

The results thus far have been derived for an arbitrarily shaped magnetic field $B(z)$ in the fringing region. We now continue the calculation for an explicitly specified $B(z)$. As discussed in an earlier report (Hughes and McNab

1977), an exponential decay seems appropriate. We take $B(z)$ as

$$B(z) = B_0 e^{-\xi z} \tag{A11}$$

and we further assume that $B(\ell)$ is negligibly small, such that

$$\int_0^\ell \frac{B(z)}{B_0} = \int_0^\ell e^{-\xi z}\, dz \sim \frac{1}{\xi} \tag{A12}$$

The fringing current equation (A7) becomes

$$I_E^* \Big|_{\text{one end}} = -\frac{4\ell M}{L} \sum_{\substack{n=1 \\ n\,\text{odd}}}^{\infty} C_n^* + \frac{\ell}{L} V^* M - \frac{2M}{\xi L} \tag{A13}$$

C_n^* may be evaluated for the insulated pump ($\sigma = 0$)

$$C_n^* = \frac{V^* - \{(n\pi/\ell)^2/[\xi^2 + (n\pi/\ell)^2]\}}{[(n\pi)^2/2]\cosh n\pi b/\ell} \tag{A14}$$

The pressure in the end region follows with $B = B_0 e^{-\xi z}$,

$$\Delta P_E^* \Big|_{\text{one end}} = -\frac{m^2 \ell^2}{(bL)^2} \sum_{n=1}^{\infty} C_n^* [\sinh \frac{n\pi b}{\ell}][\frac{n\pi/\ell^2}{\xi^2 + (n\pi/\ell)^2}]$$

$$+ \frac{V^* M^2}{\xi L} - \frac{V^* M^2}{L\ell\xi^2} - \frac{M^2}{2L\xi} - \frac{M}{L\xi} \tag{A15}$$

The region between the electrodes is assumed to be described by the model developed in the main section of the paper,

$$I^*\big|_c = 2M(V^* - 1) \tag{A16}$$

$$\Delta P^*\big|_c = M^2[V^* - 1 - 1/M] \tag{A17}$$

The final expression for η may be written in the following form;

$$V^* - 1 - \frac{1}{M} - 2\lambda \left(\frac{\ell}{L}\right)^2 \sum_{n=1}^{\infty} C_n^* [\sinh \frac{n\pi L}{\lambda \ell}] \cdot [\frac{n\pi}{(\ell/L\cdot\tau)^2 + (n\pi)^2}]$$

$$\eta = \frac{+ (2V^*/\tau)[(\ell\tau - L)/(\ell\tau)] - (2+M)/(M\tau)}{V^*[V^* - 1 + V^*(L/\ell) + 4(\ell/L) \cdot \sum_{\substack{n=1 \\ n_{odd}}}^{\infty} C_n^* - 2/\tau]}$$

where

$$C_n^* = \frac{V^* - [1 + (\tau\ell/n\pi L)^2]^{-1}}{[(n\pi)^2/2] \cosh(n\pi L/\lambda\ell)}$$

Figure A1 shows a plot of y vs V* for a typical pump design calculated from both the quasi-one- and two-dimensional theories. As can be seen the one-dimensional theory overpredicts by a few percent. Over a wide range of parameters studied the difference is generally less than about 5%. For V* 1 the efficiency may be 0 due to pumping in the end regions. For V* sufficiently small, the efficiency becomes negative corresponding to generator operation.

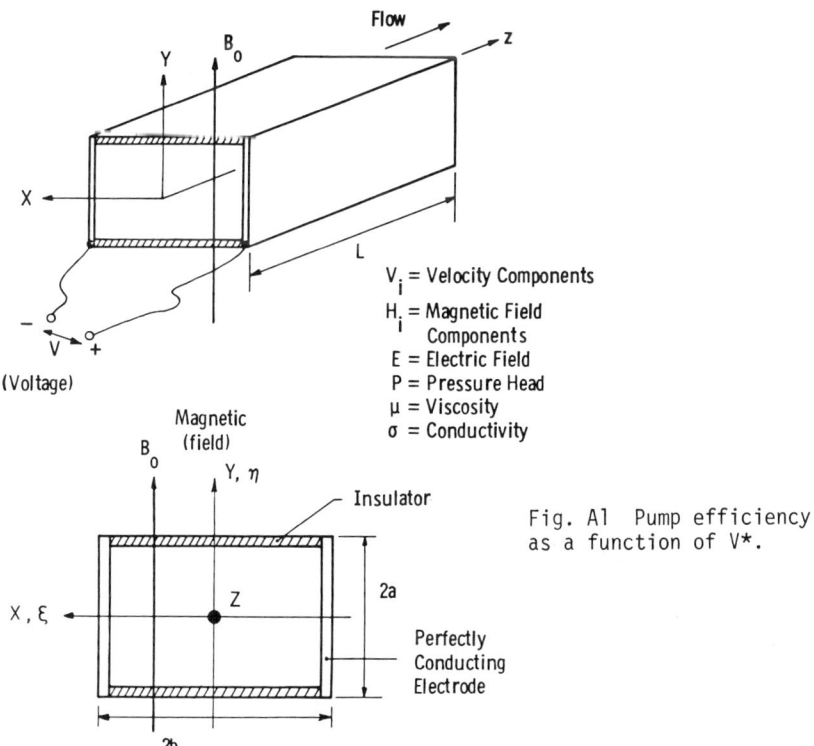

Fig. A1 Pump efficiency as a function of V*.

References

Barnes A.H. (1953) "Direct current electromagnetic pumps," Nucleonics, Vol. 11, p. 16.

Bewley L.V. (1948) Two-Dimensional Fields in Electrical Engineering, Macmillan, New York, p. 133.

Blake L.R. (1956) "Conduction and induction pumps for liquid metals," J. IEEE, Vol. 2, p. 429.

Childs B.M. (1973) "Electromagnetic pumps and flow meters," GEC J. Sci. Technol., Vol. 40, p. 10.

Foust O.J. (Ed.) (1972) Sodium-NaK Engineering Handbook, Vol. 1, Gordon and Beach, New York.

Gherson P., Lykoudis, P.S., and Lynch R.E. (1980) "Analytical study of end effects in liquid metal MHD generators," Proc. Eng. Aspects of MHD, Boston, Vol. II, p. 590.

Hoffman M.A. and Carlson G.A. (1971) "Calculation techniques for estimating the pressure losses for conducting fluid flows in magnetic fields," Univ. of California, Livermore, Rept. UCRL-51010, Feb. 4.

Hughes W.F. (1976) "A theoretical analysis of a dc electromagnetic pump for the LMFBR," Westinghouse Res. Rept. 76-1B6-ELMAG-R2, Dec. 14.

Hughes W.F. and McNab I.R. (1977) "A theoretical analysis of the dc electromagnetic pump end effects," Westinghouse Res. Rept. 77-1B6-ELMAG-R1, March 1.

Hunt J.C.R. and Stewartson K. (1965) "Magnetohydrodynamic flow in rectangular ducts," J. Fluid Mech., Vol. 23, p. 563.

Hunt J.C.R. and Stewartson K. (1965) "Magnetohydrodynamic flow in rectangular ducts, II," J. Fluid Mech., Vol. 23, p. 563.

Moszynski J.R. (1967) "Reduction of electrical end losses in MHD generator channels by insulating vanes," Argonne National Laboratory, Argonne, Ill., Rept. ANL 7188, Sept.

Shercliff J.A. (1962) Electromagnetic Flow Measurements, Cambridge University Press, Cambridge, England.

Sutton G.W., Hurwitz H., and Proitsky H. Jr. (1962) "Electrical and pressure losses in a MHD channel due to end current loops," Trans. Am. Inst. Electr. Eng. Part 1, Vol. 801, p. 687.

A Finite-Element Analysis of Two-Dimensional MHD Flow

N.S. Winowich* and W.F. Hughes†
Carnegie-Mellon University, Pittsburgh, Pa.

Abstract

A finite-element program has been developed for two-dimensional steady MFD viscous flows in channels with electrically conducting walls. The walls may contain electrodes with external current sources on sinks with direct application to dc electromagnetic pumps and generators. Low magnetic Reynolds number flow is assumed that the externally applied magnetic field is undisturbed by the flow. Sample results are presented in this paper.

Nomenclature

2b	=	distance between electrodes (see Fig. 1)
B	=	normalized magnetic flux density
B_o	=	characteristic magnetic field
N	=	interaction parameter
P	=	normalized pressure
Re	=	Reynolds number
U	=	average velocity
u,v	=	normalized velocity components
μ	=	viscosity
μ_m	=	magnetic permeability
ξ	=	magnetic field fringing parameter
ρ	=	mass density

Paper presented at Third Beer-Sheva International Seminar on Magnetohydrodynamic Flows and Turbulence, Ben-Gurion University of the Negev, Beer-Sheva, Israel, March 23-27, 1981. Copyright © American Institute of Aeronautics and Astronautics, Inc., 1982. All rights reserved.
  *Postdoctoral Research Fellow, Department of Mechanical Engineering.
  †Professor, Department of Mechanical Engineering.

ϕ = normalized electric potential
ϕ_0 = normalized electrode potential
Φ_n = finite-element interpolation functions
$\{\ \},\{\ \}^T$ = a column vector and its transpose
$[\ ],[\ ]^T$ = a matrix and its transpose

I. Introduction

The subject of the current research✢ is the numerical computer simulation of internal MHD flows occurring in engineering applications. Representative applications include: electromagnetic pumps, MHD power generation, lithium flows for fusion reactor cooling blankets, and molten metal stirring. The practical design of such devices requires the understanding of the associated three-dimensional electromechanical field interactions. Characteristically, these flows occur at high interaction parameters and exhibit unusual and unexpected flow patterns, particularly in regions of varying channel geometry and nonuniform electromagnetic fields.

The topic has been the object of considerable theoretical analysis by Hunt (1965, 1968, 1971), Hughes (1966), Turner (1973), Walker (1974), and Holroyd (1976) and their collaborators. Experimental efforts, of special mention the work of Branover et al. (1974, 1976), have examined many aspects of high Hartmann number and high interaction parameter flows. In general, theoretical analysis to date has, by necessity, involved assumptions which simplify the full governing equations for limiting ranges of operation. These have been useful in providing bounds on actual operation. However, intermediate ranges of operation require a numerical analysis of the system of MHD equations.

The finite-element method was chosen as the numerical technique because of its versatility and adaptability to arbitrary boundaries. In particular, mesh refinement in regions of anticipated high gradients are readily accommodated by elements of appropriate size, shape, and complexity. Illustrating the point is the effect of the electrical conductivity of the channel walls. Typically, most theoretical analysis considers the walls to be either perfectly conducting or insulating. However, the finite-element method

---

✢A more comprehensive formulation and review of the results are available in a doctoral dissertation by N.S. Winowich, Carnegie-Mellon University, Pittsburgh, Pa.

can easily allow for the joining of elements representing a wall of finite conductivity to those representing the fluid. The ultimate goal is a general-purpose computer code, with an appropriate "library" of elements, capable of modeling various channel geometries (rectangular, circular, converging-diverging, etc.) and electromagnetic field configurations.

The application of the finite-element method to fluid mechanics is relatively recent. However, analysis has already touched on virtually every major area of the subject. In particular the pioneering work of Oden (1969), Taylor and Hood (1973), and Gartling and Becker (1976) have influenced the current effort. Utilization of the method in relation to MHD includes an analysis of unsteady Hartmann flow by Wu (1973) and MHD instabilities associated with plasma confinement in nuclear fusion by Boyd et al. (1975).

In essence the finite-element method, as it pertains to nonlinear boundary value problems in fluid mechanics, is a weighted residual numerical technique such as the Rayleigh-Ritz or Galerkin methods. Historically, finite-element formulations were derived from variational principles, since the method originated in structural mechanics, a field governed by variational principles. However, derivations based on weighted residual techniques extend applications to problems for which no classical variational principle exists (e.g., Navier-Stokes equations) (Finlayson 1972). The basic difference of weighted residual methods, as used in finite-element formulations, and its conventional counterpart is that the approximating functions are constructed first in the subdomain (local elements) which then are assembled to form the global domain. This avoids the difficult task of satisfying complicated boundary conditions when the functions are applied directly to the global domain.

There are three basic formulations of the governing equations widely used in computational fluid mechanics. They are classified according to the physical significance of the dependent variables as: stream function, stream function-vorticity, and velocity-pressure (or primitive variable). The primitive variable approach was selected for this analysis. The main advantage of the primitive variable approach is the ease in which boundary conditions may be applied. In addition, the formulation is readily extendable to three dimensions, while the alternate approaches become exceedingly cumbersome.

The analysis focuses on the rectangular MHD channel configuration shown in Fig. 1 and is two-dimensional in nature. The configuration in Fig. 1 duplicates the electromagnetic pump in the experimental liquid-metal test loop in operation at Westinghouse Electric Research and Development Laboratories. The subsequent completion of the general three-dimensional program, in conjunction with experiment, should provide valuable insight into the design optimization of MHD devices.

In the present work the finite-element analysis has been applied to two-dimensional steady viscous MFD flow in channels with finite electrically conducting walls. Electrodes may be arbitrarily located in the walls and external current provided so that pumps, generators, or simple duct flow may be modeled.

The magnetic Reynolds number is assumed low so that the externally applied magnetic field is undisturbed by the flow. However, the external magnetic field may be arbitrarily profiled.

II. Description of the Finite-Element Formulation

A brief review of the mathematical formulation follows. Under the assumptions of laminar, steady, two-dimensional flow in the x-y plane the normalized equations are

Continuity:
$$\frac{\partial u}{\partial x} + \frac{\partial v}{\partial y} = 0 \qquad (1)$$

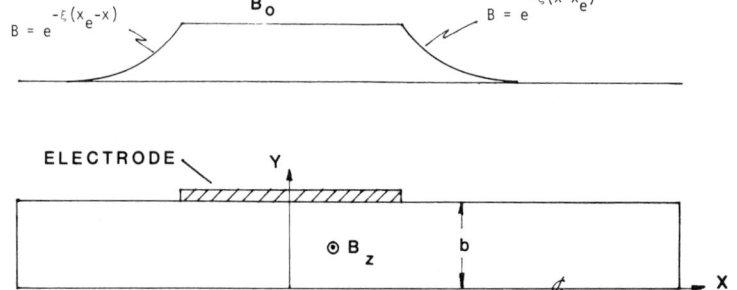

Fig. 1 The dc electromagnetic pump with the appropriate boundary conditions (the magnetic field is fringed exponentially in the end region).

Motion:

$$u\frac{\partial u}{\partial x} + v\frac{\partial u}{\partial y} = -\frac{\partial P}{\partial x} + \frac{1}{Re}\left(\frac{\partial^2 u}{\partial x^2} + \frac{\partial^2 u}{\partial y^2}\right) - N B_z^2 u - N B_z \frac{\partial \phi}{\partial y} \quad (2)$$

$$u\frac{\partial v}{\partial x} + v\frac{\partial v}{\partial y} = -\frac{\partial P}{\partial y} + \frac{1}{Re}\left(\frac{\partial^2 v}{\partial x^2} + \frac{\partial^2 v}{\partial y^2}\right) - N B_z^2 v + N B_z \frac{\partial \phi}{\partial x} \quad (3)$$

Potential:

$$\frac{\partial^2 \phi}{\partial x^2} + \frac{\partial^2 \phi}{\partial y^2} = B_z\left(\frac{\partial v}{\partial x} - \frac{\partial u}{\partial y}\right) + v\frac{\partial B_z}{\partial x} - u\frac{\partial B_z}{\partial y} \quad (4)$$

where the variables are normalized as

$$x = \frac{x^*}{b}, \quad y = \frac{y^*}{U}, \quad v = \frac{v^*}{U}, \quad P = \frac{P^*}{\rho U^2}, \quad \phi = \frac{\phi^*}{UbB_0}$$

The asterisks indicate dimensional variables. N is the interaction parameter defined as $N = (B_0^2 \sigma b)/\rho U$ and Re is the Reynolds number, $Re = \rho b U/\mu$, and $B_z = B_z^*/B_0$ is the normalized applied magnetic field which may be an arbitrary function of x and y. B_0 is a characteristic value of the applied magnetic field. The boundary conditions on velocity are that u and v are zero at the physical wall boundaries and the inlet velocity profile for upstream must be given along with the inlet pressure. Far downstream, out of the magnetic field, the velocity develops into a classical laminar profile and the pressure must be calculated there. The boundary conditions on potential are set on the electrodes. Note there are four equations with four unknowns u, v, P, and ϕ.

III. Finite-Element Formulation

The primitive variable finite-element technique used in this analysis is very similar to the well-known method of weighted residuals (or Galerkin method) of solving differential equations. In both cases the unknowns are expressed in terms of trial functions. In the classical method of weighted residuals these functions are specified over the entire global domain, while in the finite-element method the functions are specified over local elements. The total domain is then assembled from the various elements.

The variation of field variables over an element may be expressed in matrix form

$$\{u^e(s,t)\} = \Sigma \phi_n(s,t) u_n^e = \begin{Bmatrix} \phi_1 \\ \phi_2 \\ \vdots \\ \phi_n \end{Bmatrix} \{u_1^e u_2^e \cdots u_n^e\} = \{\phi_n\}^T \{u_n^e\}$$

where $\{\phi_n\}$ are the element shape or interpolation functions and $\{u_n^e\}$ are the unknown field values at the element nodes. Similarly

$$\{v^e\} = \{\phi_n\}^T \{v_n^e\} \tag{5}$$

$$\{P^e\} = \{\psi_n\}^T \{P_n^e\} \tag{6}$$

$$\{\phi^e\} = \{\phi_n\}^T \{\phi_n^e\} \tag{7}$$

It is well known that the interpolation for the velocity must be one order higher than the order of the pressure function. Here we use a quadratic form for the velocity and a linear form for pressure. A quadratic form is used for potential. An eight-node isoparametric element was used. The mesh was comprised of rectangular elements with refinement in regions of high gradients.

Analogous to the method of weighted residuals, the governing differential equations are multiplied by the weighting functions (chosen as the element interpolation functions) and integrated over an elemental volume

$$\int \{\psi\} \left(\frac{\partial u}{\partial x} + \frac{\partial v}{\partial y}\right) dV = 0 \tag{8}$$

$$\int \{\phi\} \left[u \frac{\partial u}{\partial x} + v \frac{\partial u}{\partial y} + \frac{\partial P}{\partial x} - \frac{1}{Re}\left(\frac{\partial^2 u}{\partial x^2} + \frac{\partial^2 u}{\partial y^2}\right) + NB_z^2 u + NB_z \frac{\partial \phi}{\partial y}\right] dV = 0 \tag{9}$$

$$\int \{\phi\} \left[u \frac{\partial v}{\partial x} + v \frac{\partial v}{\partial y} + \frac{\partial P}{\partial y} - \frac{1}{Re}\left(\frac{\partial^2 v}{\partial x^2} + \frac{\partial^2 v}{\partial y^2}\right) + NB_z^2 v - NB_z \frac{\partial \phi}{\partial x}\right] dV = 0 \tag{10}$$

$$\int \{\Phi\} \left[\frac{\partial^2 \phi}{\partial x^2} + \frac{\partial^2 \phi}{\partial y^2} - B_z \left(\frac{\partial v}{\partial x} - \frac{\partial u}{\partial y} \right) - v \frac{\partial B_z}{\partial x} + u \frac{\partial B_z}{\partial y} \right] dV = 0 \quad (11)$$

Once the aforementioned trial forms of the unknowns are substituted into the above equations, term-by-term integration results in a system of nonlinear algebraic equations of the following form:

$$[M(\{\chi\})]\{\chi\} = \{R\}$$

Here $\{\chi\}$ is a column vector of unknowns. The nonlinearities are associated with the convective inertia terms in the fluid equations of motion. The equations are solved by Newton's method. Let

$$\{E(\{\chi\})\} = \{[M(\{\chi\})]\{\chi\} - \{R\}\} \quad (12)$$

Then the algorithm for Newton's method is given by

$$\{\chi^{(i+1)}\} = \{\chi^{(i)}\} - [J^{(i)}]^{-1}\{E(\{\chi^{(i)}\})\} \quad (13)$$

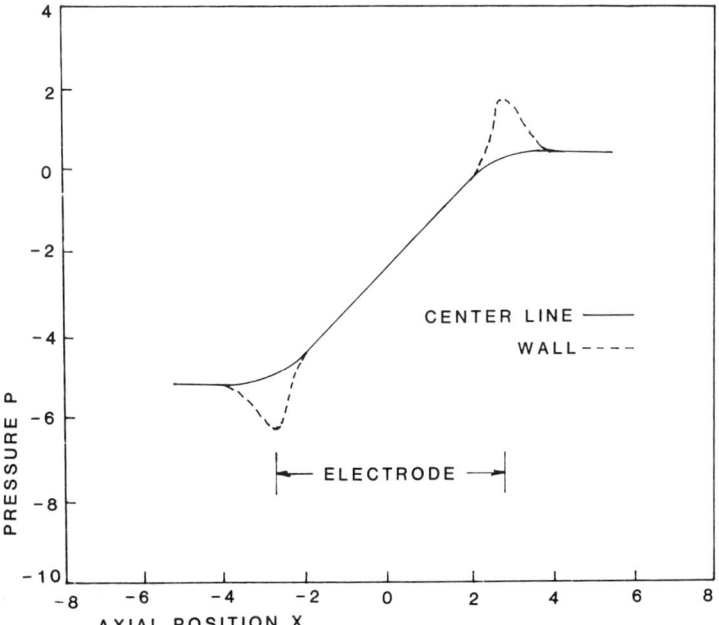

Fig. 2 Velocity profiles along the channel for Re = 2000, N = 3.5, and ϕ_0 = 1.34 (electrodes extend for x = -2.666 to +2.666, inlet velocity profile is assumed parabolic, fringing parameter ξ is 1· m^{-1}).

where
$$[J] = [\frac{\partial\{E(\{\chi\})\}}{\partial\{\chi\}}] \qquad (14)$$

It is impractical to invert the Jacobian in the above equation. Instead, a method used for linear systems is employed at each iteration. For our purpose the frontal solution method of Irons (1970) which is based on Gaussian elimination, was selected because it retains in a high-speed computer core only those degrees of freedom necessary as the individual elements are assembled in the mesh. Unnecessary unknowns are written onto a tape. This procedure aids computing economy and provides for large solution capabilities.

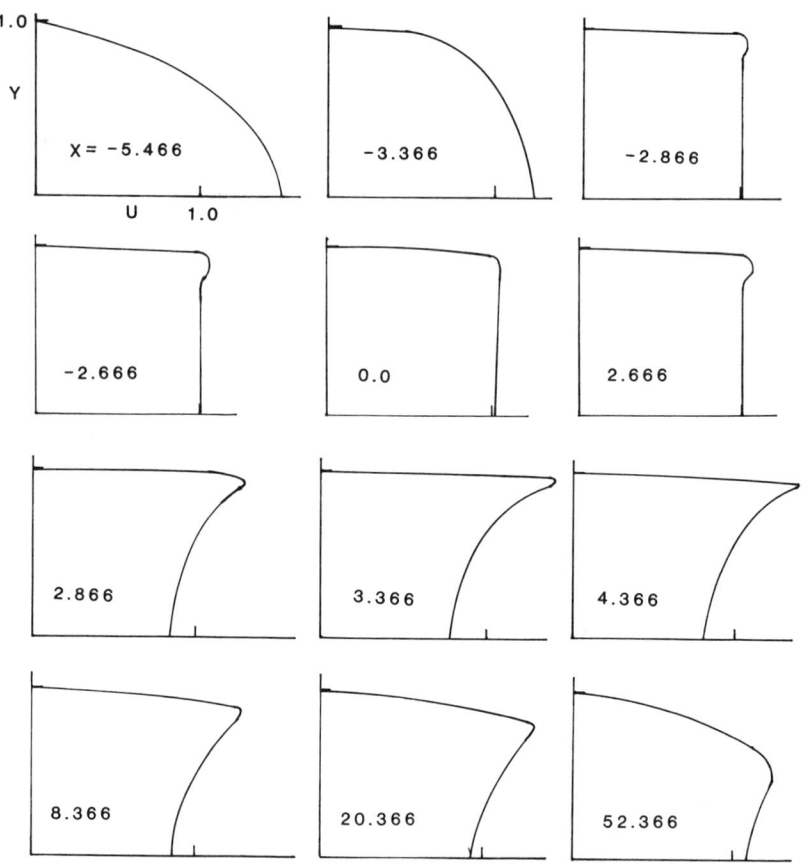

Fig. 3 Pressure distribution along the channel for Re = 2000, N = 3.5, ϕ_0 = 1.34 (fringing parameter ξ is $1 \cdot m^{-1}$).

IV. Sample Calculations and Results

Sample results are shown for a dc electromagnetic pump. The mathematical model of the pump is shown in Fig. 1 where the boundary conditions are indicated.

The velocity profile development for Re = 2000 and N = 3.5 are shown in Fig. 2. Here the electrode potential ϕ_0 is 1.34. The applied magnetic field is assumed to be uniform in the electrode region and decay exponentially in the fringing region with a fringing parameter ξ of unity (Fig. 1). The mesh field extended down to x = 70. The "M" character of the velocity near the electrode edges is quite evident. Although theoretical analyses based on high interaction parameters have predicted wall jets (Hunt and Shercliff 1971) and experimental work by Branover et al. (1974, 1976) has indicated that these "M" profiles do exist; this is to our knowledge the first analytical prediction of "M" profiles at low interaction parameters. The effects of the electromagnetic interaction persists far downstream and the "M" nature is evident at the last station plotted at x = 52.4. However, this evolution back into the classical parabola occurs outside the magnetic field and is occurring in essentially the classical laminar development length. Hence the finite-element mesh must extend far downstream to a station where the flow is close to being fully developed. Figure 3 shows the pressure profile along the channel centerline and along the wall. The sharply defined humps at the wall due to the transverse pressure gradient have been observed experimentally.

Acknowledgment

The work underlying this paper was supported by a grant from the National Science Foundation.

References

Boyd T.J.M., Gardner G.A., and Gardner L.R.T. (1975) "Hydromagnetic stability studies using the finite element method," Finite Elements in Fluids, Vol. 2, Gallgher, Oden, Taylor, and Zienkiewicz (Eds.), pp. 255-274.

Branover H. (1974) "On some important effects in laboratory MHD flows in rectangular ducts in transverse magnetic fields," Proc. 14th Conf. Eng. Aspects of MHD, Tullahoma, Tenn.

Branover H. and Gershon P. (1976) "MHD turbulence study," Ben-Gurion University, Rept. BGUN-RDA-100-76.

Finlayson B.A. (1972) The Method of Weighted Residuals and Variational Principles, Academic Press, New York.

Gartling D.K. and Becker E.B. (1976a) "Finite element analysis of viscous incompressible fluid flow, Part 1: Basic methodology," Comput. Meth. Appl. Mech. Eng., Vol. 8, No. 1, pp. 51-60.

Gartling D.K. and Becker E.B. (1976b) "Finite element analysis of viscous incompressible fluid flow, Part 2: Applications," Comput. Meth. Appl. Mech. Eng., Vol. 8, No. 2, pp. 127-138.

Holroyd E.T. and Hunt J.C.R. (1976) "A review of MHD flows in ducts with changing cross section areas and non-uniform magnetic fields," Euromech Colloquium 70, March 16-19.

Hughes W.F. and Young F.J. (1966) The Electromagnetodynamics of Fluids, John Wiley & Sons, New York.

Hunt J.C.R. and Ludford G.S.S. (1968) "Three dimensional MHD duct flows with strong transverse magnetic fields, Part 1: Obstacles in a constant area channel," J. Fluid Mech., Vol. 33, p. 693.

Hunt J.C.R. and Shercliff J.A. (1971) "MHD at high Hartmann number," Ann. Rev. Fluid Mech., Vol. 3, p. 37.

Hunt J.C.R. and Stewartson K. (1965) "MHD flow in rectangular ducts, II," J. Fluid Mech., Vol. 23, p. 563.

Irons B.M. (1970) "A frontal solution for finite element analysis," Int. J. Numer. Methods Eng., Vol. 2, pp. 5-32.

Oden J.T. (1969) "A general theory of finite elements, II: Applications," Int. J. Numer. Methods Eng., Vol. 1, No. 3, pp. 247-259.

Taylor C. and Hood P. (1973) "A numerical solution of the Navier-Stokes equations using the finite element technique," Comput. Fluids, Vol. 1, pp. 73-100.

Turner R.B. (1973) "Aspects of MHD duct flow at high magnetic Reynolds number," PhD Thesis, Univ. of Warwick, England.

Walker J.S. and Ludford G.S.S. (1974) "MHD flow in insulating circular expansions with strong transverse magnetic fields," Int. J. Eng. Sci., Vol. 12, pp. 1045-1061.

Wu S.T. (1973) "Unsteady MHD duct flow by the finite element method," Int. J. Numer. Methods Eng., Vol. 6, No. 1, pp. 3-10.

Fusion Application of an Imploding Shell Initially Formed by Falling Liquid Metal

Yasuyuki Itoh,* Takashi Kanagawa,† Nobuo Yamaoka,‡ and Keiji Miyazaki§
Osaka University, Osaka, Japan
and
Yoichi Fujii-e¶
Nagoya University, Nagoya, Japan

Abstract

Applications of a liquid-metal shell are considered in this paper for wall protection in ICF reactors and for plasma compression in a liner fusion reactor. A liquid-metal envelope to protect the first structural wall of the ICF reactor against released energy from an exploding pellet can be formed instantaneously in free space by the implosion of a falling cylindrical shell of liquid metal with a cusp-driver field. The formation of a waterfall-type liquid-metal shell for liner fusion can also be achieved by introducing a blade lattice inclined to the radial direction, which reduces the terminal velocity of the fall and makes it possible to rotate the imploding liner for stabilization of the inner surface. An experimental device was constructed to study the magnetohydrodynamics of a falling and imploding liquid-metal shell with free surfaces.

Paper presented at Third Beer-Sheva International Seminar on Magnetohydrodynamic Flows and Turbulence, Ben-Gurion University of the Negev, Beer-Sheva, Israel, March 23-27, 1981. Copyright © American Institute of Aeronautics and Astronautics, Inc., 1982. All rights reserved.
*Graduate Student, Department of Nuclear Engineering (presently with Toshiba Corp.).
†Graduate Student, Department of Nuclear Engineering (presently with Mitubushi Atomic Power Industry Co).
‡Technical Research, Department of Nuclear Engineering.
§Associate Professor, Department of Nuclear Engineering.
¶Professor, Institute of Plasma Physics.

I. Introduction

Recently, a liquid-metal cylindrical shell has been considered to be used in fusion reactors. It can be applied not only to wall protection in an inertial confinement fusion (ICF) reactor [Maniscalco et al. (1978)] but also to the compression of high-β plasma in a liner fusion reactor [Book et al. (1978a)], where the shell is imploded in the radial direction. In each case, a shell of liquid lithium or its alloy protects the first structural wall of the reactor from direct exposure to released energy from the exploding pellets or plasma and has the functions of a heat-transfer medium and blanket to absorb the major fraction of thermonuclear energy and to breed tritium.

The present paper covers the applications of an imploding shell initially formed by a falling liquid metal in ICF [(Itoh and Fujii-e (1981)] and liner fusion reactors [Itoh and Fujii-e (1980)].

II. Application to ICF Reactors

In the design study of ICF reactors, discussions have been made on wall protection against energy released from an exploding pellet, which consists of energetic neutrons, charged particles, pellet debris, and x-rays. Kulcinski (1979) has summarized many earlier proposed concepts for wall protection, namely Blascon, wetted wall, fluidized wall, rotating drums, dry sacrificial wall, magnetic protection, and gas protection.

In the fluidized wall concept [Maniscalco et al. (1978), Monsler et al. (1978), Yamanaka et al. (1979)], wall protection must be achieved against spherical energy release and the resultant shock wave by forming a closed shell of liquid metal isolated from the reactor components, i.e., by forming a liquid-metal envelope in free space.

Now let us consider the method of its formation. Initially, a falling liquid-metal cylindrical shell is introduced into a reactor cavity as shown in Fig. 1. The single-turn cusp-field driver coils with a circular cross section are placed outside the liquid shell, which are connected to the capacitor banks. When the electrical circuit is closed, the liquid shell near the coil is locally imploded by the magnetic pressure of the cusp field. In this manner, the liquid-metal envelope is formed in free space instantaneous-

FUSION APPLICATION OF AN IMPLODING SHELL

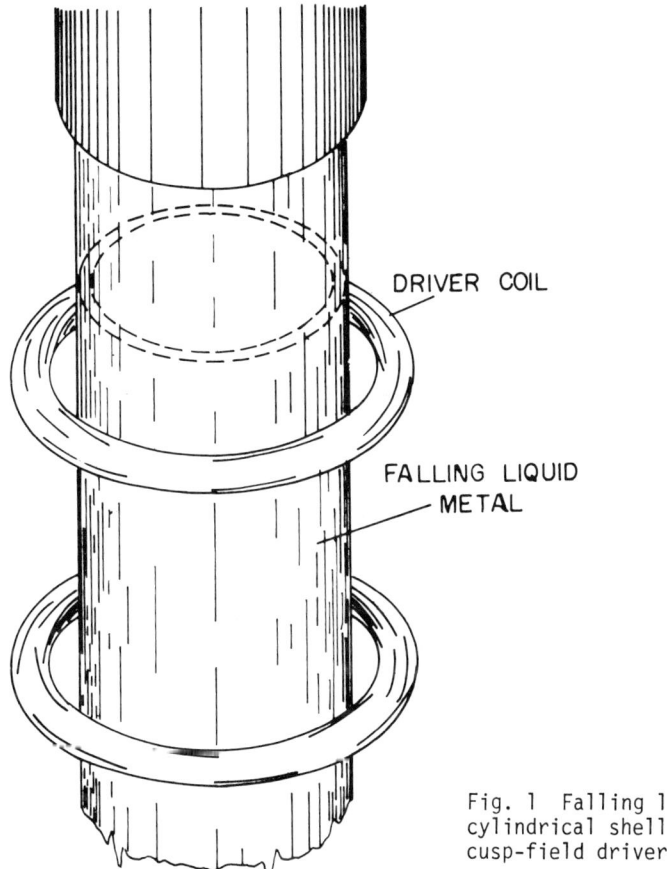

Fig. 1 Falling liquid-metal cylindrical shell through cusp-field driver coils.

ly before each shot and shields the first structural wall from the released energy (especially of short-range deposition, e.g., energies of charged particles and pellet debris) of an exploding pellet in it.

In numerical simulation of the motion of the imploding liquid shell, it is assumed for simplicity that the liquid shell is infinitely long, infinitesimally thin, has infinite electrical conductivity, and its velocity of fall is negligible in the time scale of implosion. The cusp-field driver coils are placed at $z = \pm b$ in the cylindrical coordinates, where its major and minor radii are a and ρ, respectively. The coil current is determined from the conservation law of magnetic flux trapped between the coil and the shell.

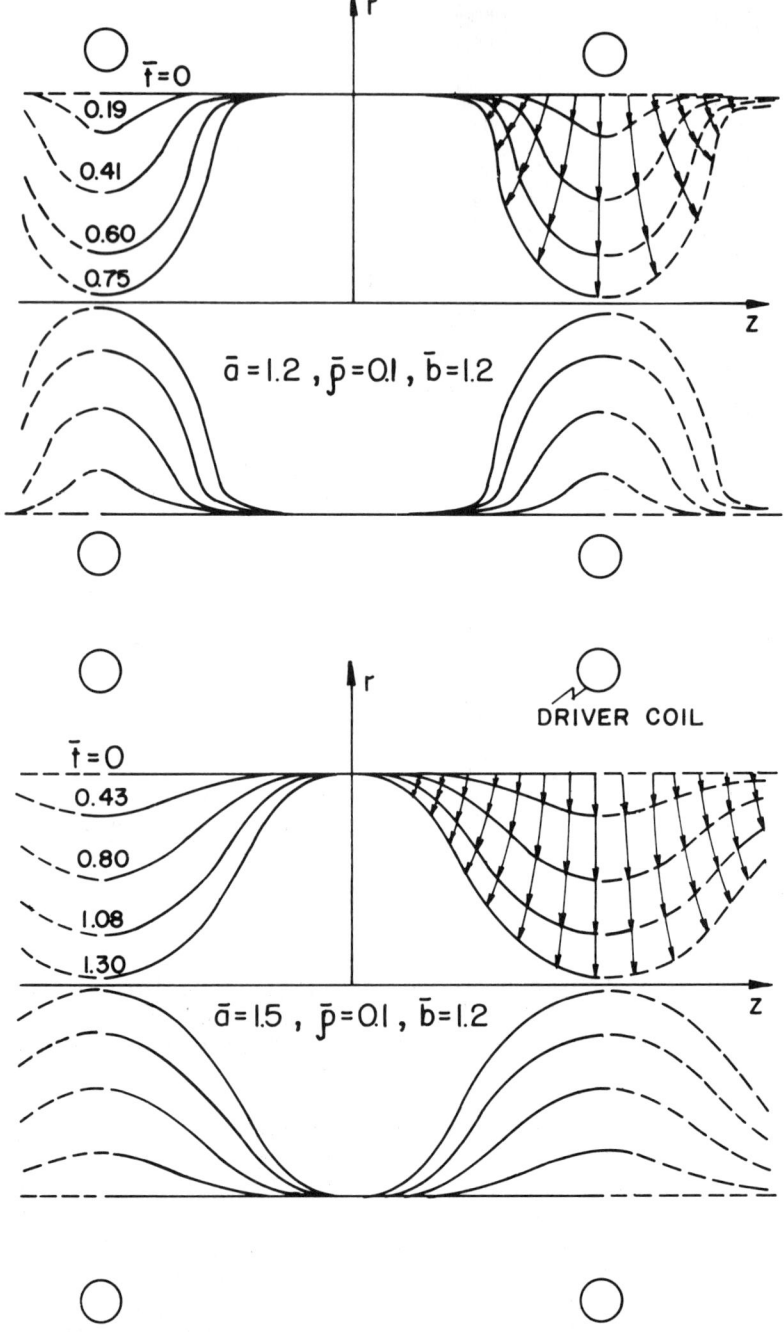

Fig. 2 Profiles of imploding liquid shell in r,z plane.

FUSION APPLICATION OF AN IMPLODING SHELL

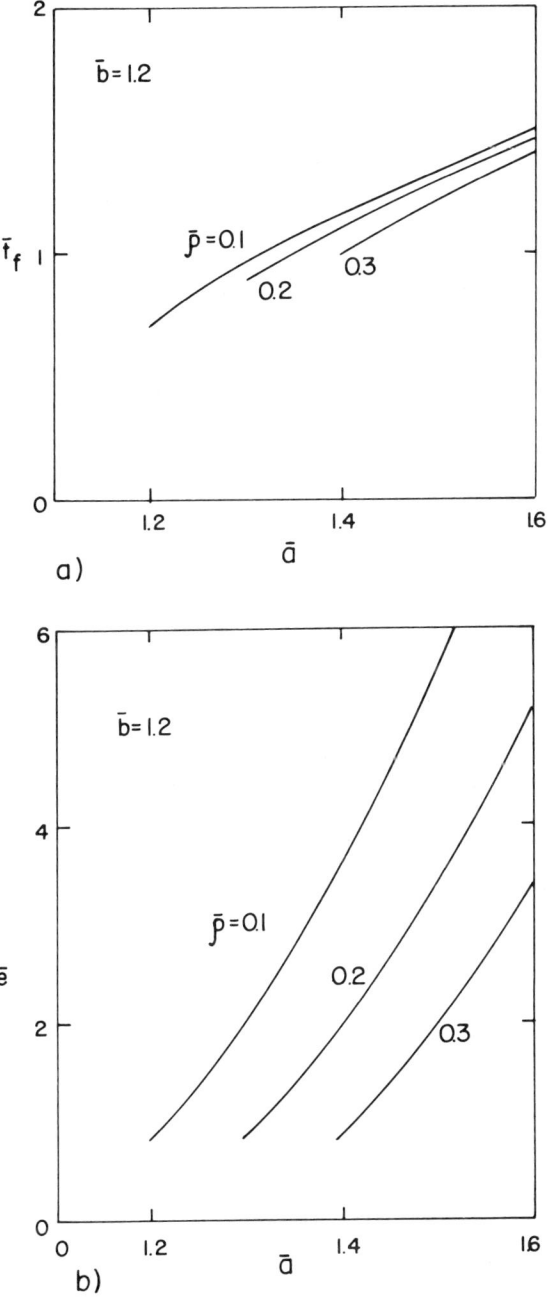

Fig. 3 Quantities $\bar{t}_f$ and $\bar{e}$.

Figure 2 shows profiles of the imploding liquid shell in the r,z plane at time t = t/τ,

$$\tau = (8\pi^2 \sigma_i R^3 / \mu_0 I_i^2)^{1/2}$$

where σ_i is the initial shell density, R the initial shell radius, μ_0 the magnetic permeability of free space, and the I_i the initial coil current. The required coil current I_i and the driving field energy e for a given implosion time T (time to close the liquid shell) are found from the numerical calculations as follows

$$I_i = (8\pi^2 \sigma_i R^3 / \mu_0)^{1/2} \bar{t}_f(\bar{a},\bar{b},\bar{\rho})/T \qquad (1)$$

$$e = (8^2 \sigma_i R^4 / T^2) \bar{e}(\bar{a},\bar{b},\bar{\rho}) \qquad (2)$$

where $\bar{a}$ = a/R, $\bar{b}$ = b/R, $\bar{\rho}$ = ρ/R, and the values of $\bar{t}_f$ and $\bar{e}$ are shown in Fig. 3.

Since the imploding liquid shell is subjected to the Rayleigh-Taylor instability, the disruption of its geometrical configuration tends to occur when the perturbation amplitude grows up to a value comparable to the shell thickness during the period of motion. Figure 4 shows the distribution of normalized shell thickness [σ(Z)/σ$_i$], where Z is the initial value of the axial position and $\bar{Z}$ = Z/R. When the

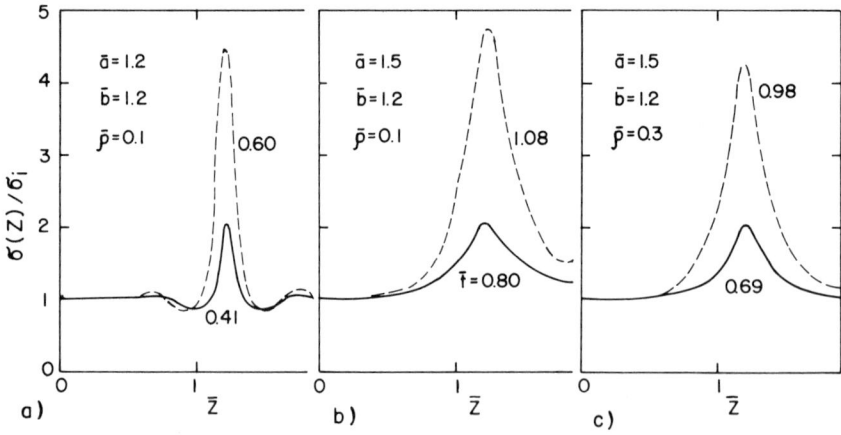

Fig. 4 Distributions of surface mass density σ(Z)/σ$_i$ of imploding liquid shell.

Table 1 Design parameters of present system

Shell material	Liquid lithium
(mass density)	($\sim$500 kg/m^3)
(heat capacity)	($\sim$4 kJ/kg°C)
(resistivity)	($\sim$3x10^{-7} Ωm for 300°C)
Initial shell thickness	5 cm
(surface mass density)	(25 kg/m^2)
Initial shell radius	1 m
Coil major radius	1.5 m
Coil minor radius	0.3 m
Coil distance	2.4 m
Implosion time	30 ms
Driving field energy	4.4 MJ
Initial coil current	1.9 MA

shell is imploded toward the center axis, its averaged thickness is increased due to the mass conservation law. However, we find from Figs. 2 and 4 that for a small major radius of driver coil, the liquid shell is imploded so locally that the thickness of a certain region becomes small compared with the initial value because of rapid axial motion. This reduction in thickness is considered to enhance the disruption of the imploding shell.

Therefore, the decrease in the major radius of the coil is restricted from the viewpoint of shell integrity, while it is desirable for the reduction in the required driving energy as seen from Fig. 3. Figures 3 and 4 suggest that a decrease in the driving energy should be achieved by increasing the minor radius ρ.

Table 1 shows an example of the design parameters of the present system. The lithium shell with a 5 cm thickness absorbs $\sim$40% of the energy released from the exploding pellet including all energy of short-range deposition [Maniscalco et al. (1978)]. When the length of the initial shell ($\simeq$ the height of the reactor cavity) is greater than 5 m and the thermonuclear energy E_f per pulse is $\sim$1000 MJ [Maniscalco et al. (1978)], the initial shell radius should be less than 1 m to obtain the mean temperature rising of the shell (i.e., the difference ΔT between inlet and outlet temperatures of liquid lithium) being greater than 100°C [cf. $\Delta T \simeq 100\sim200$°C for the primary coolant in fast breeder reactor (FBR) designs]. For a pulse repetition rate of $\sim$1 Hz [Maniscalco et al. (1978)], the implosion time T of the liquid shell must be sufficiently smaller than 1 s because most of the time interval between each shot will be consumed to re-establish the liquid shell. For T = 30 ms, the required driving energy and the initial coil current are found from Eqs. (1) and (2) to be e = $2.2\bar{e}$ MJ and I_i = $1.6\bar{t}_f$ MA, respectively. If the coil parameters are chosen to be $\bar{a}$ = 1.5, $\bar{b}$ = 1.2, and $\bar{\rho}$ = 0.3, we obtain e = 4.4 MJ and I_i = 1.9 MA, where the energy of 4.4 MJ is considerably smaller compared to E_f (~1000 MJ). The initial hoop stress P_θ acting on each driver coil is calculated by $P_\theta \cong 8\pi\sigma_i R(\bar{\ell}/t_f)/(\bar{\rho}T)^2$ in the present range of coil parameters and we obtain $P_\theta \cong$ 12 MPa for $\bar{\rho}$ = 0.3. This value of hoop stress will be acceptable if stainless steel is used for the driver coils. Furthermore, the repulsive force between the coils is found to be small compared with the hoop force.

Consequently, we can conclude that it is possible to use the present system for wall protection in ICF reactors.

III. Application to Liner Fusion

The system concept for the liner fusion reactor has been presented in Book et al. (1978a), Turchi et al. (1980), Robson (1978), where a rigidly rotating liner is imploded by high-pressure gas through free pistons to eliminate the outer surface instability (captive liner concept). The inner surface instability is then suppressed by centrifugal force, in which that stability criterion is $\dot{U} \leq V^2/r$, where U and V are the radial and azimuthal velocities. In this system, there are a few complicated problems: it is necessary not only to rotate the reactor structure but also to give the liquid metal a substantial amount of rotational energy as it enters a reactor and also to recover it upon leaving.

In contrast to the captive liner, the liner formation with a falling liquid metal [Itoh and Fujii-e (1980)] (see Fig. 5a) is desirable from the viewpoint of liner recycle as a heat-transfer medium. To obtain a liquid liner with a uniform thickness, it is necessary to satisfy the condition $Dw^2/2d$ = g, where w is the terminal velocity of fall, d the hydraulic diameter, D = 0.3164 $(wd/\nu)^{-1/4}$ the friction factor [Blasius formula, Bird et al. (1960)], ν the kinematic viscosity, and g the gravitational acceleration. If a blade lattice is introduced in the fall as shown in Fig. 5, the terminal velocity w becomes

$$w = 10.6[\{2\chi(2+\chi)r_{1i}/(1+\chi)\}^5/\nu]^{1/7}[1+m_B\chi/\pi/(1+\chi)]^{-5/7}$$

FUSION APPLICATION OF AN IMPLODING SHELL

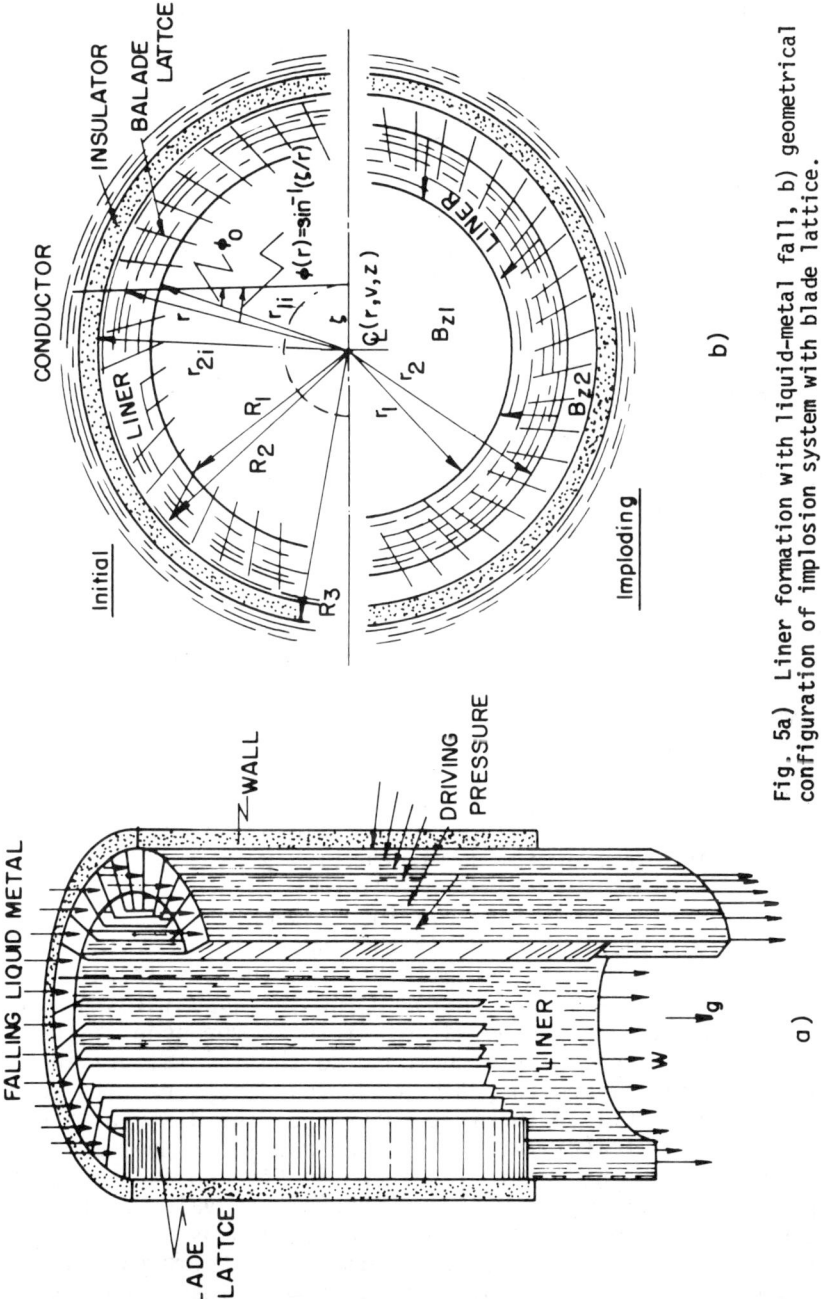

Fig. 5a) Liner formation with liquid-metal fall, b) geometrical configuration of implosion system with blade lattice.

where $\chi = r_{2i}/r_{1i} - 1$, r_{1i} and r_{2i} are the initial inner and outer radii of the liquid shell (liner), m_B the number of blades, and all quantities are in mks units. Thus owing to the presence of a blade lattice, the terminal velocity of fall is reduced by the factor $[1+m_B\chi/\pi/(1+\chi)]^{-5/7}$ because of the decrease in the hydraulic diameter.

As a measure to mitigate the Rayleigh-Taylor instability of the irrotational liner in the present system, it is considered to introduce the MHD dynamic stabilization method [Itoh and Fujii-e (1979)], where oscillating azimuthal fields are superposed on the driving axial field for the outer surface and on the compressed axial field for the inner surface. These azimuthal fields are produced by axial currents on the outer liner and plasma surfaces. For example, when this dynamic stabilization is applied to the outer surface perturbation, the azimuthal fraction γ_θ of the driving field pressure is necessary to satisfy the criterion $\Delta_F/\Delta_L \lesssim \gamma_\theta \lesssim 1$, where Δ_L and Δ_F are the thicknesses of the liner and of the driving field layer.

The rotational stabilization of the inner surface can also be achieved by giving angular momentum to the imploding liner by using the aforementioned blade lattice inclined in the radial direction as shown in Fig. 5b, where the angular momentum A per unit mass of the liner in the lattice is given by $A = rU\tan\phi$. The motion of the liner magnetically imploded on the axial magnetic flux was numerically analyzed, using the conservation laws of mass, magnetic flux, angular momentum, and energy on the assumption of an inviscid liner with an infinite electrical conductivity.

Figure 6a shows the relation between the blade angle ϕ_0 $[\phi(r_{1i})]$ and $1 - \bar{R}_1$ (R_1 is the inner radius of the blade lattice, $\bar{R}_1 = R_1/r_{1i}$) to satisfy the stability criterion $(\dot{U} - V^2/r)_{r=r_1} = 0$ at turnaround, and the quantities η and δ which are defined as the ratios of the compressed field and rotational energies at turnaround to the initial driving field energy, where the dots-on-curves denote results for the liner initially in rigid rotation. The distribution functions h $[=A(r_1)/A(r_{1i})]$ of the angular momentum of liner leaving the lattice is shown in Fig. 6b as a function $r_i = (1+\bar{r}^2-\bar{r}_1^2)^{1/2}$, where r_i is the initial value of $r(t)$ and the symbol $-$ denotes the radius normalized by r_{1i}. For the approach run of liner to rotate its inner surface, the inner radius of the lattice, of course, must be less than that of the initial liner that is at rest. This leads to a reduction in the initial compression volume to be filled with a

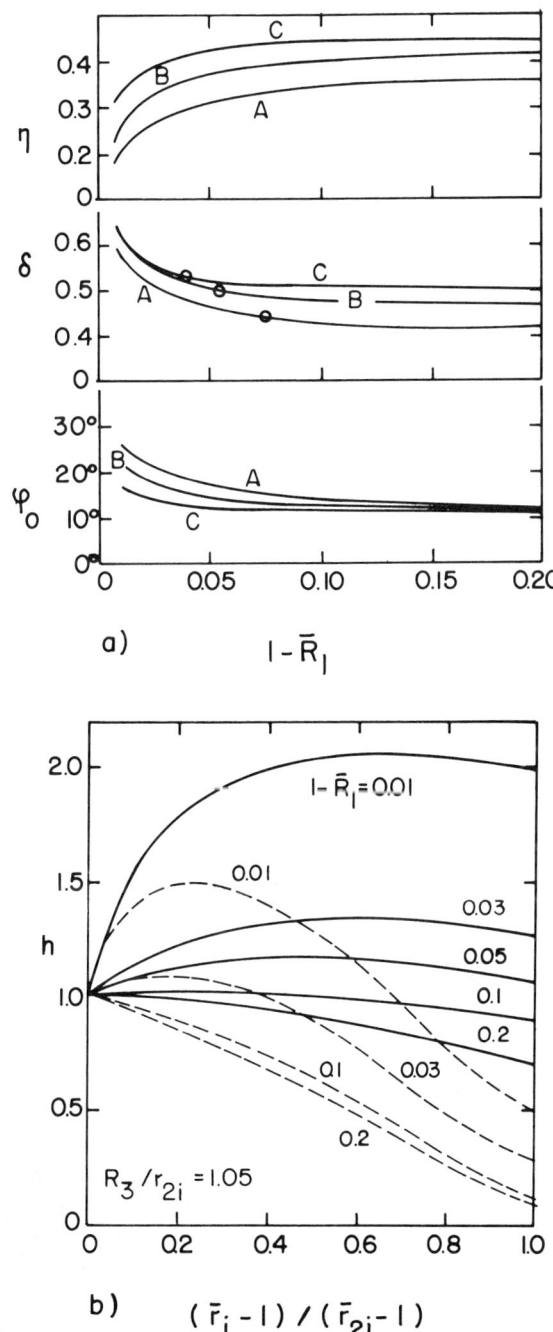

Fig. 6a) Relation between inner radius and angle of blade lattice required for rotational stabilization and quantities η and δ (A is R_3/r_{2i} = 1.1, B is 1.05, C is 1.02, R_3 is the conductor radius), b) distribution function h of angular momentum, where r_{2i}/r_{1i} = 1.2, r_{1i}^2/r_{1f}^2 = 1000 (f is the turnaround value).

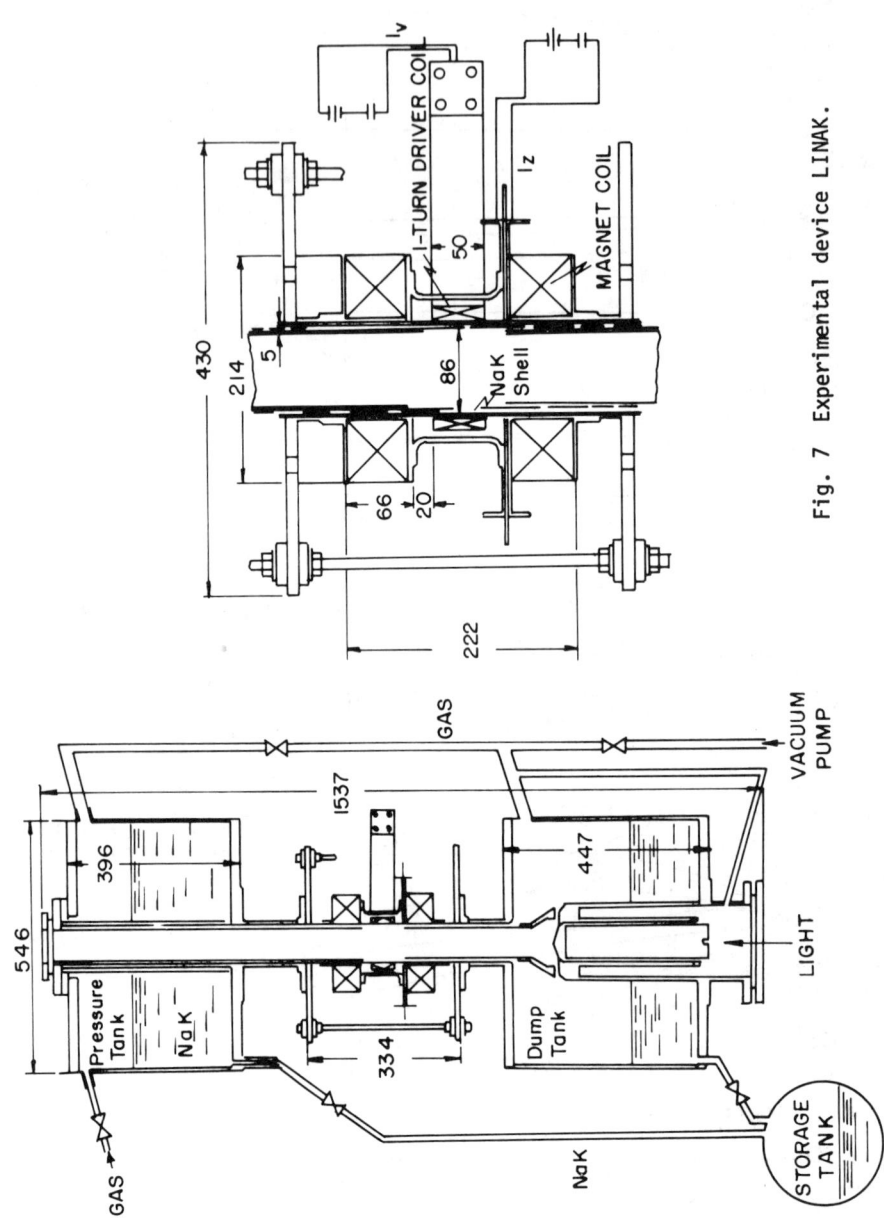

Fig. 7 Experimental device LINAK.

plasma because the lattice acts as a limiter. However, we see from Fig. 6 that a value of $\sim 0.9 r_{li}$ for the inner radius of the lattice is enough to rotate the imploding liner and then the blades do not encroach on the initial plasma whose radius is considered to be $0.8-0.9 r_{li}$ [Book et al. (1978b), Boris et al. (1972)].

IV. Experiment

For the purpose of studying the magnetohydrodynamics of a falling and imploding liquid-metal shell with free surfaces, the experimental device shown in Fig. 7 was constructed as a daughter loop of the NaK Blowdown MHD Experimental Facility of Osaka University. The device is called LINAK and a working fluid of sodium-potassium alloy (NaK) is supplied from its mother loop. In this device, liquid NaK in the upper vessel is driven by pressurized argon gas into an annular channel of 86 mm outer and 76 mm inner diameters with a velocity up to 3 m/s. The inner guide tube of the channel is removed at the central part of the test section to form a cylindrical liquid shell with a free inner surface. The outer tube of the annular channel is made of fiber reinforced plastic (FRP) to attain electrical insulation and is reinforced by an epoxy resin mold. To apply the θ-pinch drive field, a single-turn driving coil is embedded in the mold on the same elevation as the free surface flow. A couple of coils are also embedded in the resin above and below the single-turn coil and serve to produce the axial magnetic flux inside the cylinder of liquid metal. A contrived circuit of low impedance can provide the liquid-metal shell with an axial electric current to apply the Z-pinch.

Thus, the liquid-metal shell can be imploded by means of θ and Z pinches and can compress the magnetic flux in-

Table 2 Parameters of LINAK device

Shell material	NaK (50 ℓ)
Outer shell diameter	86 mm
Shell thickness	5 mm
Velocity of fall	2-3 m/s
Driver coil length	50 mm
Driver coil diameter	92 mm
Condenser	13 kV, 7 kJ, 1 μH
Implosion velocity	~ 10 m/s
Implosion time	~ 1 m/s
Initial magnetic field	~ 500 G
Compressed magnetic field	$\sim 10^4$ G

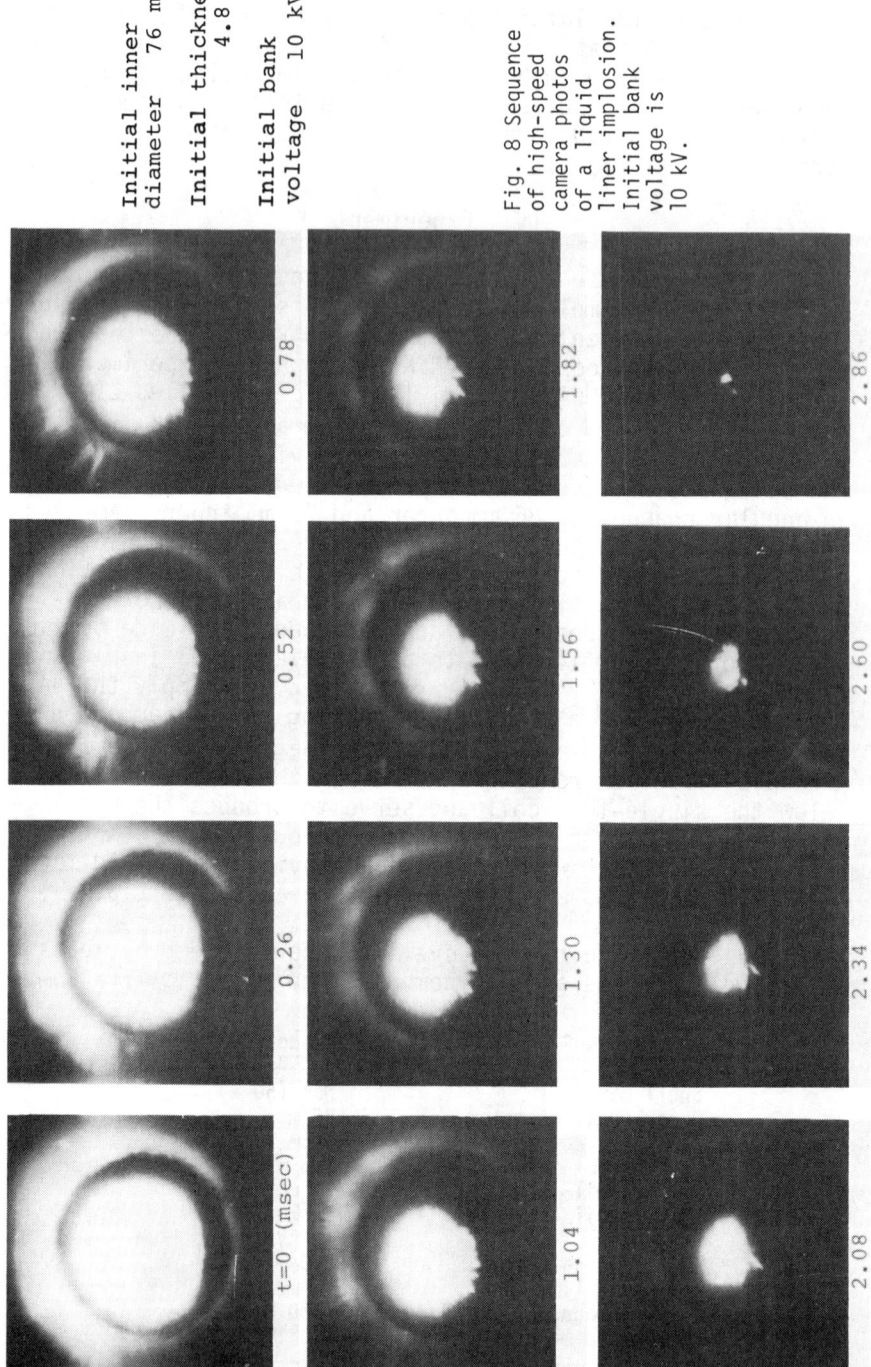

Fig. 8 Sequence of high-speed camera photos of a liquid liner implosion. Initial bank voltage is 10 kV.

Initial inner diameter 76 mm
Initial thickness 4.8 mm
Initial bank voltage 10 kV

side the cylinder. The motion of a liquid shell can be observed through the upper glass window. A high-speed camera is used with the aid of back-lighting of halogen bulbs through the lower windows.

The inner guide tube has a 4 mm smaller diameter and the width of the liquid annulus is 7 mm, while that of the upper is 5 mm. The lower observation port is a double window type to prevent the direct contact between air and NaK. The outline of LINAK is given in Table 2.

Figure 8 shows a photographic sequence of the moving shell when it is imploded by the θ-pinch effect with a capacitor bank voltage of 10 kV and no magnetic flux inside the shell. The boundary of the blank circles in the photographs indicate the inner surface of the liquid shell. Further experiments and analysis are under way.

V. Summary

For wall protection with a fluidized wall in ICF reactors, it is necessary to form a closed liquid shell isolated from the reactor components. This envelope can be obtained instantaneously in free space by the implosion of a falling cylindrical shell of liquid metal with a cusp-driver field. Numerical results show that the required driving field energy is considerably smaller than the thermonuclear energy per pulse, and the electromagnetic forces acting on the coils are acceptable in the case where a cylindrical shell of liquid lithium with a 1 m radius is formed into the envelope in a time of 30 ms.

The application of liquid-metal fall to the liner fusion forming a cylindrical shell (liner) enables us to exclude the rotary mechanism from the implosion system. In obtaining a liner with uniform thickness, the terminal velocity of fall can be decreased by the introduction of a blade lattice that is inclined to the radial direction and makes it possible to rotate the imploding liner for the stabilization of its inner surface.

An experiment is now underway to investigate the dynamics of a falling liquid-metal shell magnetically imploded on an axial magnetic flux.

Acknowledgement

The authors wish to express their thanks to S. Inoue and S. Umezawa for their invaluable discussions and technical assistance.

References

Bird, R.B., Stewart W.E., and Lightfoot E.N. (1960) <u>Transport Phenomena</u>, John Wiley & Sons, New York, p. 187.

Book D.L. et al. (1978a) <u>Plasma Physics and Controlled Nuclear Fusion Research</u>, Vol. 2, IAEA, p. 93

Book D.L., Hammer D.A., and Turchi, P.J. (1978b) "Theoretical studies of the formation and adiabatic compression of reversed-field," <u>Nucl. Fusion</u>, Vol. 18, p. 159.

Boris J.P. and Shanny R.A. (1972) "Parametric studies of LINUS, an ultra-high magnetic field theta-pinch configuration in imploding liners," <u>Proc. 5th Eur. Conf. Controlled Fusion Plasma Physics</u>, Grenoble, p. 20.

Itoh Y. and Fujii-E Y. (1979) "Dynamic stabilization of imploding liquid metal liner," J. Nucl. Sci. Technol., Vol. 16, No. 3, p. 175.

Itoh Y. and Fujii-E Y. (1980) "Liquid metal liner implosion systems with blade lattice for fusion," <u>J. Nucl. Sci. Technol.</u>, Vol. 17 No. 3, p. 167.

Itoh Y. and Fujii-E Y. (1981) "Constricted liquid metal curtain for inertial confinement fusion reactors," <u>J. Nucl. Sci. Technol.</u>, Vol. 18, No. 4, p. 261.

Kulcinski G.L. (1979) "First wall protection schemes for inertial confinement fusion reactors," Univ. of Wisconsin, Madison, Rept. UWFDM-281.

Maniscalco M. and Walker P. (1978) "Civilian applications of laser fusion," Lawrence Livermore Laboratory, Livermore, Calif., Rept. UCRL-52349, Rev. 1.

Monsler M. et al. (1978) "Electric power from inertial confinement fusion: the hylite concept," Lawrence Livermore Laboratory, Livermore, Calif., Rept. UCRL-81866.

Robson A.E. (1978) "A conceptual design for an imploding liner fusion reactor (LINUS)," Naval Research Laboratory Memorandum, Rept. 3861.

Turchi P.J. et al. (1980) <u>Megagauss Physics and Technology</u>, Plenum, New York, p. 375.

Yamanaka C. et al. (1979) "Report of system design of ICF reactor," (in Japanese), Institute of Laser Engineering, Osaka Univ., p. 74.

Geometrical Integrity of a Metallic Cylindrical Shell Magnetically Imploded on an Axial Magnetic Field

Yasuyuki Itoh,* Takashi Kanagawa,† Sigemitsu Umezawa,‡ and Keiji Miyazaki§
Osaka University, Osaka, Japan
and
Yoichi Fujii-e¶
Nagoya University, Nagoya, Japan

Abstract

A cylindrical metal shell (liner) imploded on a payload (compound of magnetic flux and plasma) tends to disrupt because of the Rayleigh-Taylor instability and the cavitation. Generally, a magnetic field has no effect on the Rayleigh-Taylor instability where the deformation wave vectors are perpendicular to the lines of force. This flute type instability, which is also observed in an experiment conducted on the θ pinch of a solid-potassium liner, is theoretically shown to be reduced by superposing oscillating azimuthal fields on all of the driving and compressed axial fields except the eccentric displacement (m=1) of the inner surface. It is also found from the analytical and numerical

---
Paper presented at Third Beer-Sheva International Seminar on Magnetohydrodynamic Flows and Turbulence, Ben-Gurion University of the Negev, Beer-Sheva, Israel, March 23-27, 1981. Copyright © American Institute of Aeronautics and Astronautics, Inc., 1982. All rights reserved.
 *Graduate Student, Department of Nuclear Engineering, Faculty of Engineering (Presently, Ph.D., Toshiba Corp.).
 †Graduate Student, Department of Nuclear Engineering, Faculty of Engineering (Presently, Mitubushi Atomic Power Industry Co).
 ‡Department of Nuclear Engineering, Faculty of Engineering.
 §Associate Professor, Department of Nuclear Engineering, Faculty of Engineering.
 ¶Professor, Institute of Plasma Physics.

treatments of compressible fluid dynamics that the cavitation of a liner is caused by the pressure wave and by the tension of the diffused magnetic field, but that the cavitation can be prevented by making use of a rotating double-layer liner with an inner layer highly compressible in comparison with the outer layer.

I. Introduction

In the liner fusion concept (Book et al. 1978), which is schematically illustrated in Fig. 1, the geometrical integrity of an imploding cylindrical metal shell (liner) is one of the most important problems. Two mechanisms play a part in the loss of liner integrity, namely, the Rayleigh-Taylor instability (Harris 1962, Somon 1969, Book 1974, Barcilon 1974) and the cavitation.

Since the liner is accelerated inward and outward, its free surfaces cause the Rayleigh-Taylor instability: the outer surface is perturbed during the inward acceleration phase and the inner surface during the outward acceleration. On the other hand, the pressure transient at the inner surface produces the compression wave in the liner and the resultant rarefaction wave tends to generate the negative pressure with consequent liability of cavitation.

These phenomena lead not only to the distortion but also to the disruption of an imploding liner that cause the contamination of compression volume with the liner material and prevent the recovery of liner kinetic energy during the

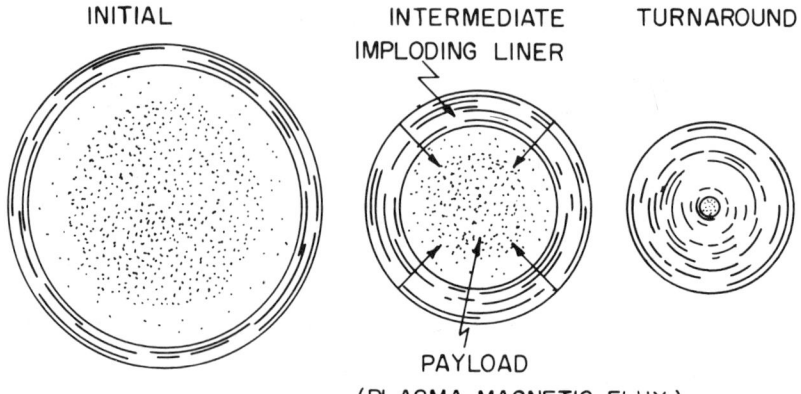

Fig. 1 Schematic illustration of liner fusion concept.

expansion. In this paper, investigations are made of the suppression of the Rayleigh-Taylor instability and of the cavitation of the liner to sustain its cylindrical geometry.

II. Suppression of Rayleigh-Taylor Instability

First to investigate the Rayleigh-Taylor instability (Itoh et al. 1980, Fujii-e 1979) of an imploding liner, a cylindrical shell of solid potassium with a 66 mm o.d. (47 mm long, 3-4 mm thick) was collapsed by θ pinch in the experimental apparatus shown in Fig. 2.

The experimental results reveal that the deformation wave appearing in the collapsed liner due to the instability has its vector in the azimuthal direction. Hence, this instability is considered to be of the flute type where the wave is free from any bending actions exerted by the magnetic lines of force. The displacement ξ of the position of the liner from the unperturbed state is then expressed by

$$\xi \propto \exp(\omega t + im\theta)$$

where m is the azimuthal mode number and ω the growth rate of the deformation wave. Figure 3 presents the results of mode analysis (for $M \gtrsim 2$) based on the examination of photographs of a liner (Fig. 4). In the case of an inviscid liner, the growth rate of instability is known from linear theory to be approximately proportional to the square root of the mode number (Harris 1962), which contrasts with the behavior of the amplitude ξ_m in Fig. 3, where it is seen to decrease with increasing mode number m, while beyond $m \sim 10$ the perturbation is negligible.

This behavior of ξ_m may be attributed to the effect of viscosity. On the assumption of an infinitesimally thin liner (Harris 1962) and in consideration of the effect of viscosity, the equation of the perturbed motion in the (r,θ) plane is expressed (Itoh, Umezawa, Yamaoka, and Fujii-e 1980) as

$$d\underline{u}/dt = -\underline{n} \cdot P/\sigma + \nu [\nabla^2 \underline{u}]_{\Delta \to 0}$$

where $\underline{u}$ ($\propto \exp(\omega t + im\theta)$) and $\underline{n}$ are the deviations of the liner velocity and the normal vector to the outer surface, respectively, from those of the unperturbed state, P the driving pressure, ν the kinematic viscosity, and Δ the liner thickness. The maximum growth rate ω of the perturbation

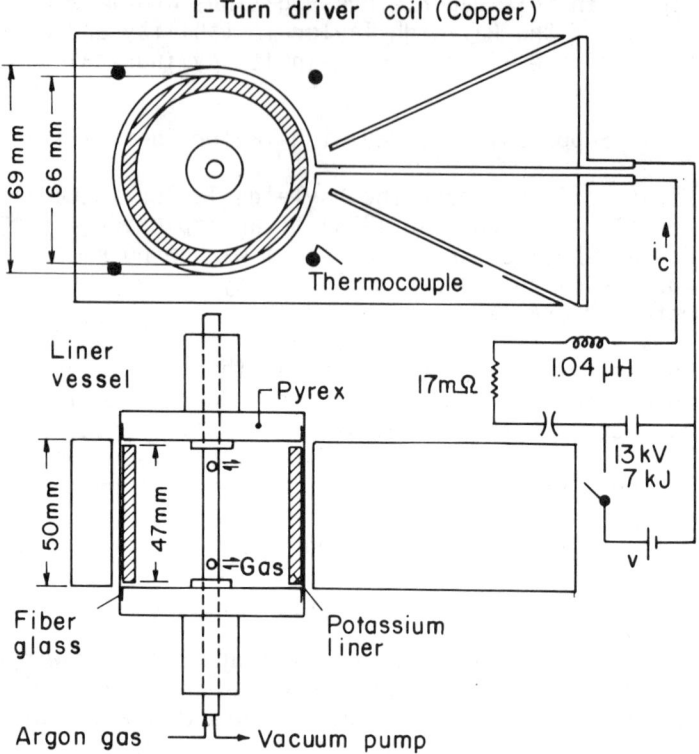

Fig. 2 Experimental apparatus for θ pinch of solid potassium liner.

for $|m| = 1$ is obtained from the above equation and the result is given by (Itoh, Umezawa, Yamaoka, and Fujii-e 1980).

$$2\hat{\omega} = - R_\mu^{-1}(|m|-1)^2 + [R_\mu^{-2}(|m|-1)^4 + 4(|m|-1)]^{1/2} \quad (1)$$

where $\hat{\omega} = \omega(r/U^*)$, $R_\mu = rU^*/\nu$, $U^* = (rP/\sigma)^{1/2}$, and r is the liner radius. We find from Eq. (1) that the effect of viscosity becomes appreciable for mode number $m > m_c = (2R_\mu)^{3/2} + 1$, and that for m satisfying the condition $(m/m_c)^3 \gg 1$ the growth rate is decreased with increasing viscosity in inverse proportion to m ($\omega \sim R_\mu m^{-1}$), while $\omega \sim \sqrt{m}$ ($m \gg 1$) for $R_\mu \to \infty$, i.e., and inviscid liner. Assuming that the solid potassium is Bingham fluid, its kinematic viscosity can be estimated from experimental data of the collapsed liner to be on the order of 1 m²/s (Itoh, Umezawa, Yamaoka, and

GEOMETRICAL INTEGRITY OF AN IMPLODING SHELL

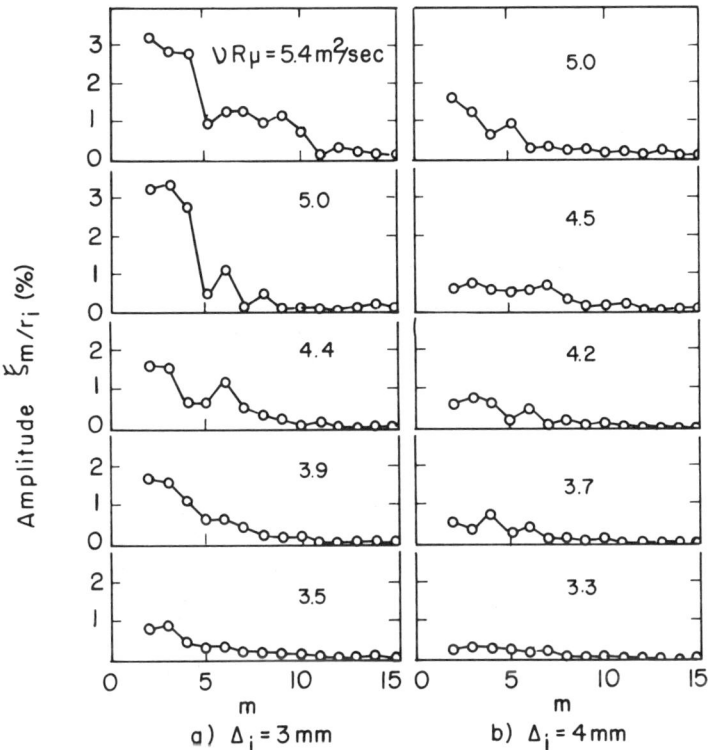

Fig. 3 Results of mode analysis applied to deformation wave (I = 10°C, Δ_i = initial liner thickness).

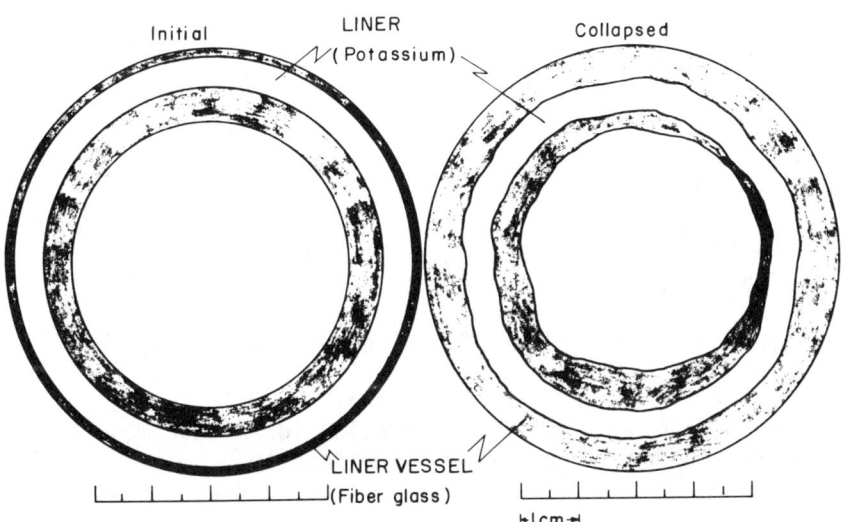

Fig. 4 Solid potassium liner collapsed at 50°C with condenser charged to 11 kV.

Fujii-e 1980) in the temperature range 10-57°C (melting point 63°C). This value of viscosity is large enough to reduce the perturbations of higher mode (m $\geq$ 10) under the present experimental conditions ($R_\mu \simeq$ 4-5).

Although the effect of viscosity reduces the perturbations, it irreversibly dissipates the kinetic energy of a liner. A solid liner is therefore inferior to a liquid liner, the viscosity of which is negligible under the condition of liner implosion. The rotation of a liquid liner has been proposed for the suppression of the perturbation of inner surface during the outward acceleration phase (Book et al. 1974, Barcilon et al. 1974). This rotational stabilization, however, requires a large amount of energy to satisfy the stability criterion $\dot{U} \leq V^2/r$ at the inner surface, where U and V are the radial and azimuthal velocities of a liner, respectively. On the other hand, the effect of the magnetic field on the instability appears only when the deforming fluid surface bends the magnetic lines of force. It is considered that the above-mentioned flute-type instability is suppressed by changing the direction of the lines of force on the fluid surface with respect to time. Likewise, the stabilization of liner surfaces with the magnetic shear (Suydan 1958) cannot be acceptable because the production of magnetic shear requires the deep diffusion of the field into the liner, i.e., the resultant Joule heating not only causes the energy loss but also vaporizes the liner material to release high-Z impurities into the plasma.

Now let us consider the MHD dynamic stabilization of liner surfaces, where oscillating fields are superposed on the driving axial field during the inward acceleration phase and on the compressed axial field during the outward acceleration as shown in Fig. 5. These azimuthal fields are produced by the axial currents on the liner outer and plasma surfaces.

The liner stability analysis for this problem was made on the basis of the following assumptions: 1) the liner and the plasma are incompressible, inviscid, infinitely long, and have infinite electrical conductivity; 2) the plasma has a sharp boundary and a uniform pressure; and 3) the angular frequency of the azimuthal field is so high that the possible parametric excitation for higher mode can be suppressed due to the effect of viscosity.

GEOMETRICAL INTEGRITY OF AN IMPLODING SHELL   345

The stability criterion can be analytically found when the liner velocity vanishes, i.e., at the start of motion and the turnaround point. Letting the axial field strength be constant and the azimuthal field strength be $\bar{B}_\theta \cos(\Omega_\theta t)$ where Ω_θ is the angular frequency, a set of Hill's equations are obtained by the combination of linearized MHD equations and boundary conditions which describe the perturbed motion of liner and plasma. If $\Omega_\theta \gg T^{-1}$, where T is the characteristic time of the fluid motion, the following results (Itoh and Fujii-e 1979) are obtained.

For stabilization of the outer surface at the start of motion, the required azimuthal fraction $\gamma_{\theta i}$ of the driving magnetic pressure is then found for $kr_{1i} \ll 1$ (k = axial wave number) to be

$$\gamma_{\theta i} = [\frac{B_\theta^2/2}{B_\theta^2/2 + B_z^2}] > [\ln(r_{2i}/r_{1i})^2 \{m \frac{1+(r^4/r^{2i})^{2m}}{1-(r^4/r^{2i})^{2m}} -1\}]^{-1} \quad (2)$$

where B_θ is the amplitude of oscillating azimuthal field strength and B_z the axial field strength, r_1, r_2, r_4 are the radii of the inner, outer, and conductor surfaces, respectively, and the subscript i denotes the initial value. This criterion can be simplified by $2\gamma_{\theta i} \gtrsim \Delta_F/\Delta_L$ for lower mode numbers, where Δ_F and Δ_L are the thicknesses of the driving field layer and the liner.

This result is also predicted from the following estimation. The pressure gradient in the liner is approximated by $p(r_{2i})/\Delta_L$. The increment of the azimuthal field due to the radial displacement ξ_2 of the outer surface is estimated from the conservation of magnetic flux, $B_\theta(r_{2i})\xi_2/\Delta_F$. Then, the continuity of pressure at the outer surface is written by

$$p(r_{2i}) + P(r_{2i})\xi_2/\Delta_L \simeq 2\gamma_{\theta i}(\Delta_F^{-1} + r_{2i}^{-1}) \cdot P(r_{2i})\xi_2$$

where P is the pressure. Since the outer surface is stable when the deviation $p(r_{2i})$ of the pressure is positive for $\xi_2 > 0$, we obtain the criterion $2\gamma_{\theta i} \geq \Delta_F/\Delta_L$ by neglecting r_{2i}^{-1} in comparison with Δ_F^{-1}.

The increase in the azimuthal fraction $\gamma_{\theta i}$ thus appreciably reduces the growth rate $\hat{\omega}$ for $kr_{1i} \ll 1$ as shown in

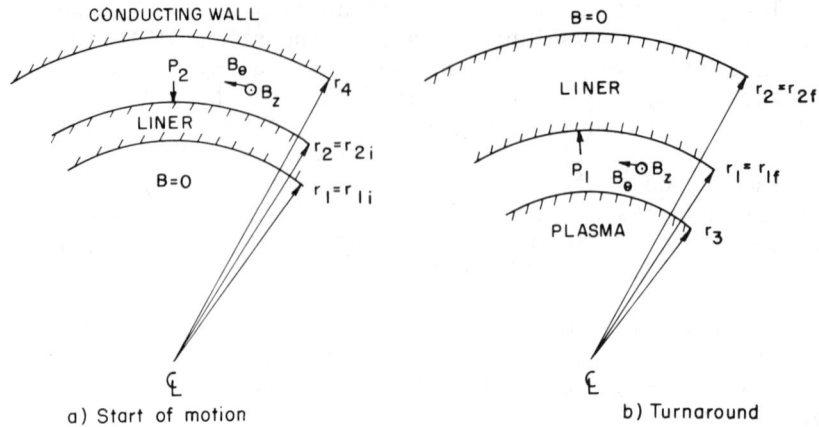

Fig. 5 Schematic diagrams of model for MHD dynamic stabilization.

Fig. 6a, where $\hat{\omega} = \omega r_{1i}/[P(r_{2i})/\rho]^{1/2}$, where P is the pressure and ρ the density. However, we see from Fig. 6b that the stability for $kr_{1i} \gtrsim 1$ becomes worse if $\gamma_{\theta i}$ exceeds 1/2. Consequently, the considerable reduction in the instability at the start of motion can be achieved only when

$$\Delta_F/\Delta_L \lesssim 2\gamma_{\theta i} \lesssim 1$$

The required fraction $\gamma_{\theta f}$ for the inner surface at turnaround is also found for $kr_{1f} \ll 1$ to be

$$\gamma_\theta > \frac{[1+\Omega^2(r_{1f})/2]/\ln(r_{2f}/r_{1f})^2 - \Omega^2(r_{1f})}{1+m\dfrac{m-1-(r_3/r_{1f})^{2m}(m+1)}{m-1+(r_3/r_{1f})^{2m}(m+1)}} \quad (3)$$

where the rotation of the liner is taken into account. r_3 is the plasma radius, Ω the rotation angular frequency normalized by $[\rho/P(r_{1f})]^{1/2} \cdot r_{1f}$, and subscript f the turnaround value. Since the imploding liner becomes thick near turnaround due to the mass conservation law, i.e., $r_{2f}/r_{1f} \gg 1$, the growth rate $\hat{\omega} = \omega r_{1f}/[P(r_{1f})/\rho]^{1/2}$ for $kr_{1f} \ll 1$ is considerably reduced with a small $\gamma_{\theta f}(<1/2)$ for the parameters used in the calculations (Fig. 7). Thus, the liner is shown to be stable. The plasma surface in the present model serves only as a boundary of the free space magnetic field because its density is small in comparison

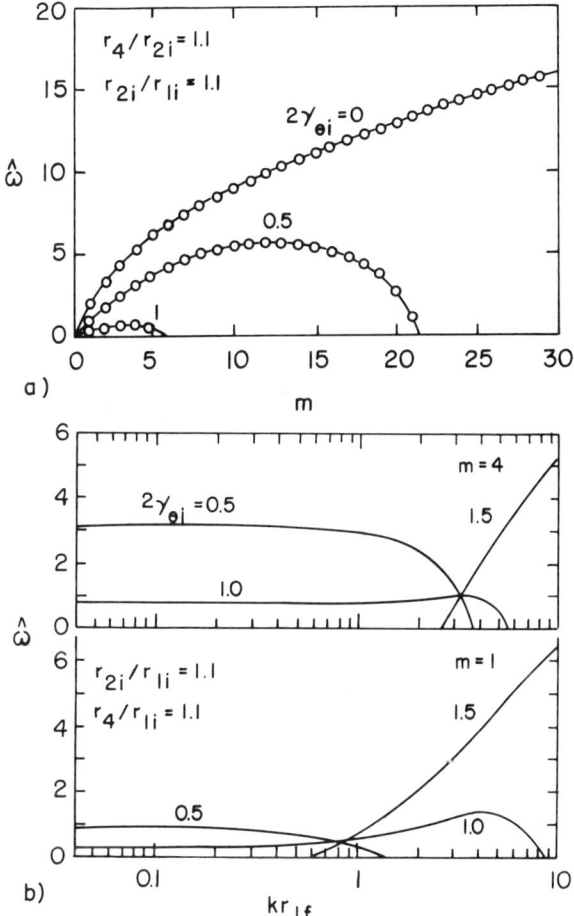

Fig. 6 Growth rate of perturbation at start of motion.

with that of the liner. The surface is then deformed quasi-statically in accordance with the motion of inner surface with reducing the restoring force produced by the magnetic field against the distortion of the liner. Due to this effect, as found from Eq. (3) and also from Fig. 7, the growth rate of perturbation of m = 1 cannot be reduced and the stabilization effect of the field becomes weak with the increasing ratio of the plasma radius to that of the inner surface.

Compared with the rotational stabilization, the perturbation of higher mode can be suppressed more effectively in MHD dynamic stabilization. Furthermore, this dynamic stabilization is desirable to suppress the instability because the required energy is also used to drive the liner and to confine the plasma.

III. Cavitation of the Liner and its Suppression

The compression and expansion undergone by the payload causes the compression wave to propagate through the liner from the inner to the outer surface. At the outer free surface, the compression wave is reflected in the form of a rarefaction wave that tends to generate the negative pressure with consequent liability of cavitation.

The trajectory of the inner surface of an incompressible liner near turnaround can be approximated by $r_1^2(t) = r_{1f}^2(1+t^2/\tau^2)$, where the instant of turnaround is taken as the initial point in time and

$$\tau^2 \simeq (P_{1f}/\rho/r_{1f}^2)^{-1} \ln(r_{2f}/r_{1f})$$

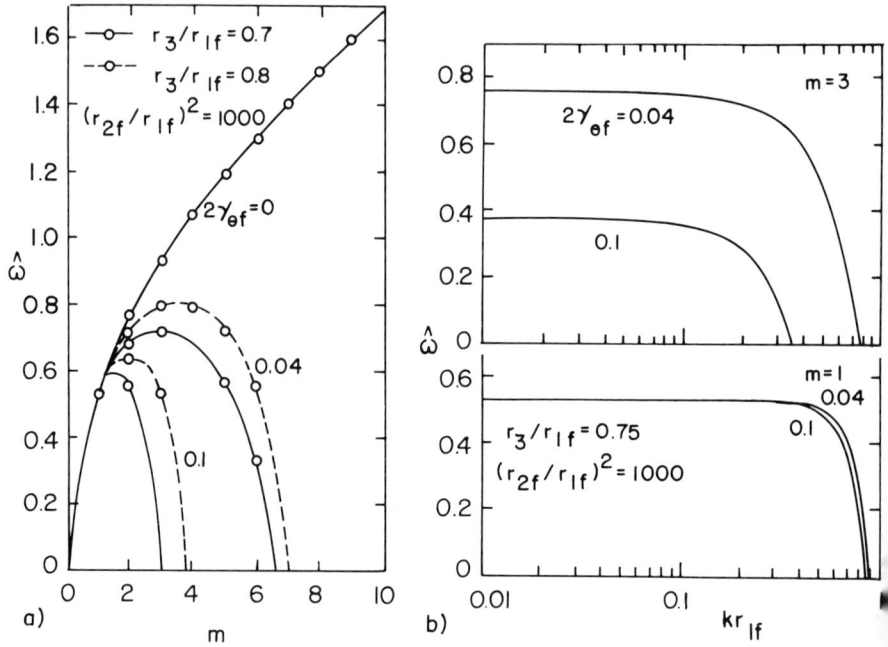

Fig. 7 Growth rate of perturbation at turnaround.

Under the conditions $r_{1f}/c\tau < 1$ and $(r_{2f}-r_{1f})/c\tau < 1$ (c = acoustic velocity), the pressure $P(r,t)$ in the liner near the turnaround can be approximated by (Itoh and Fujii-e 1980, Itoh et al. 1982)

$$P = P_M + \tilde{P} \qquad (4)$$

where P_M is the pressure due to the radial motion with the velocity U in the cylindrical geometry and is given by (Yamagishi and Schaffer 1978)

$$P_M(r,t) = \frac{1}{2} U^2(r_1)\{1-(r/r_1)^2 - [1-(r_1/r_2)^2]\ln(r/r_1)/\ln(r_2/r_1)\}$$

The second term $\tilde{P}$ in Eq. (4) is the pressure determined by the history of the boundary pressure $P(r_1)$ that influences the pressure of a fluid element r with a time delay

$$\tau_n^{\pm} = [\pm(r-r_2) + (2n+1)(r_2-r_1)]/c$$

Thus,

$$\tilde{P}(r,t) = \kappa^{-1}(r) \sum_{n=0}^{\infty} [\Pi(t-\tau_n^+) - \Pi(t-\tau_n^-)]$$

where $\Pi - \kappa(r_1)P(r_1,t)$ and $\kappa(r) = (r_2-r)/\ln(r_2/r)$ (also see Itoh and Fujii-e 1980).

In the case of the adiabatic compression of payload of which the ratio of specific heat is $5/3 \sim 2$, the criterion for the generation of negative pressure is found from Eq. (4) to be

$$\alpha = M_\infty(\zeta-1)[(\zeta^2-1)/\ln\zeta^2]^{1/2} \geq 0.6$$

where $M_\infty = U_\infty/c$, $\zeta = r_{2f}/r_{1f}$, and U_∞ is the initial radial velocity.

Since a liquid has no tensile strength, the geometrical configuration of the liner is not sustained against the negative pressure. To clarify the magnitude of liner disruption, the dynamics of a compressible liner imploded on a trapped magnetic flux (Fig. 8) was numerically analyzed with using a finite electrical conductivity and the follow-

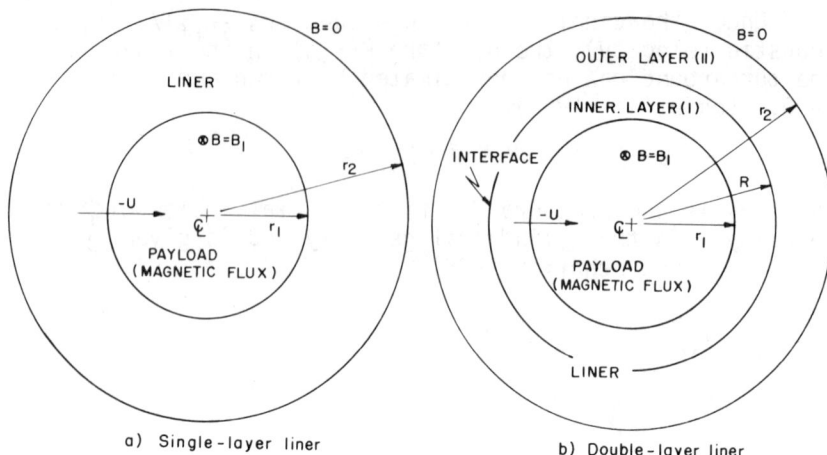

Fig. 8 Schematic diagrams of model for numerical analysis of liner motion.

ing equation of state: $P = K\hat{\rho}^2(\hat{\rho}-1)$ (Book and Turchi 1979, Yamagishi and Schaffer 1978) for $\hat{\rho} \geq 1$ and $P = 0$ for $\hat{\rho} < 1$, where $\hat{\rho} = \rho/\bar{\rho}$, K is the bulk modulus, and $\bar{\rho}$ the density at $P = 0$. Figure 9 shows the density profiles of the liner, where $\hat{t} = tc/r^*_{1f}$, $\hat{B}^2 = B^2(r_1)/(2\mu_0 K)$, $\hat{r} = r/r^*_{1f}$, B is the magnetic flux density, and the asterisk * denotes the quantity for the incompressible liner with infinite conductivity. We see from Fig. 9 that cavitation of the liner ($\rho < 1$) accompanies the propagation of a rarefaction wave for a large value of α defined by Eq. (5).

In feasibility considerations of a liner fusion reactor, the energy gain is often defined as the ratio of the fusion energy per cycle to the initial kinetic energy of the liner. Since the energy gain is proportional to $\rho U_\infty \sqrt{S}$ (Yamagishi and Schaffer 1978), where S is the cross-sectional area of the liner in the (r,θ) plane, the condition of liner implosion in the reactor is liable to satisfy the criterion of Eq. (5).

We find from Eq. (5) that the generation of negative pressure is avoided by reducing the liner thickness. In the reactor, the liner thickness, however, will be determined to be ∿1 m at turnaround from the viewpoint of the absorption of fusion neutrons (Robson 1978). The reduction in the liner thickness, then, gives a larger inner radius at turnaround and hence it gives a large reactor size for a fixed volume compression ratio of payload. It is con-

GEOMETRICAL INTEGRITY OF AN IMPLODING SHELL 351

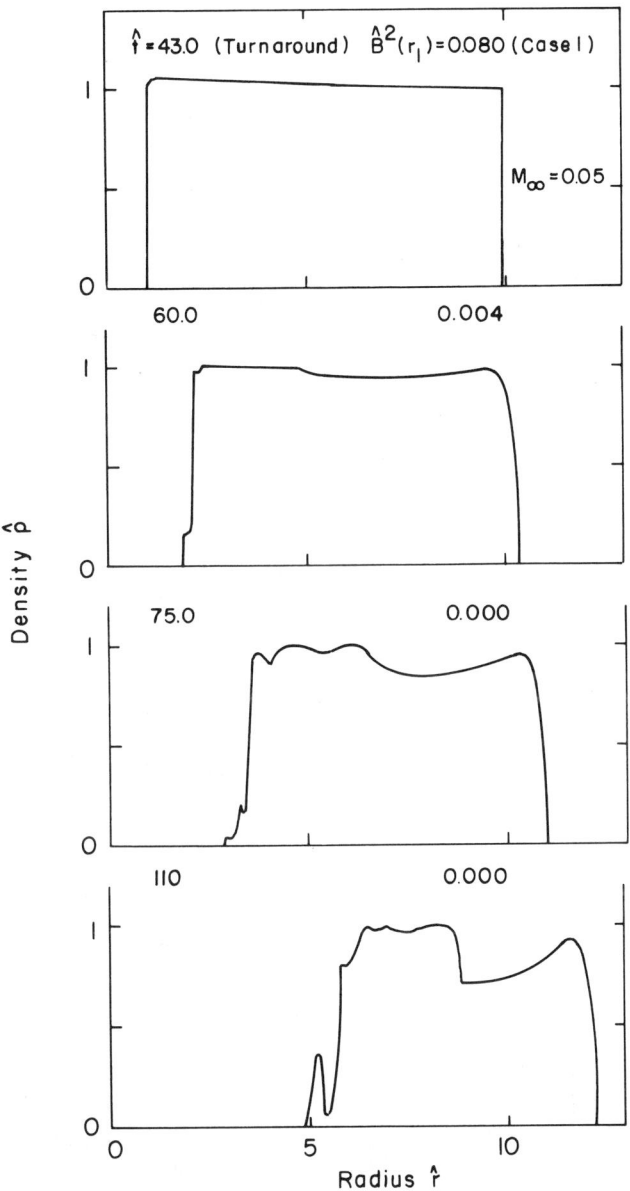

Fig. 9 Density profiles of single-layer liner for $r^*_{2f}/r^*_{1f} = 10$, and $r^*_{1f} c/\nu_m = 1000$ (ν_m = magnetic diffusivity).

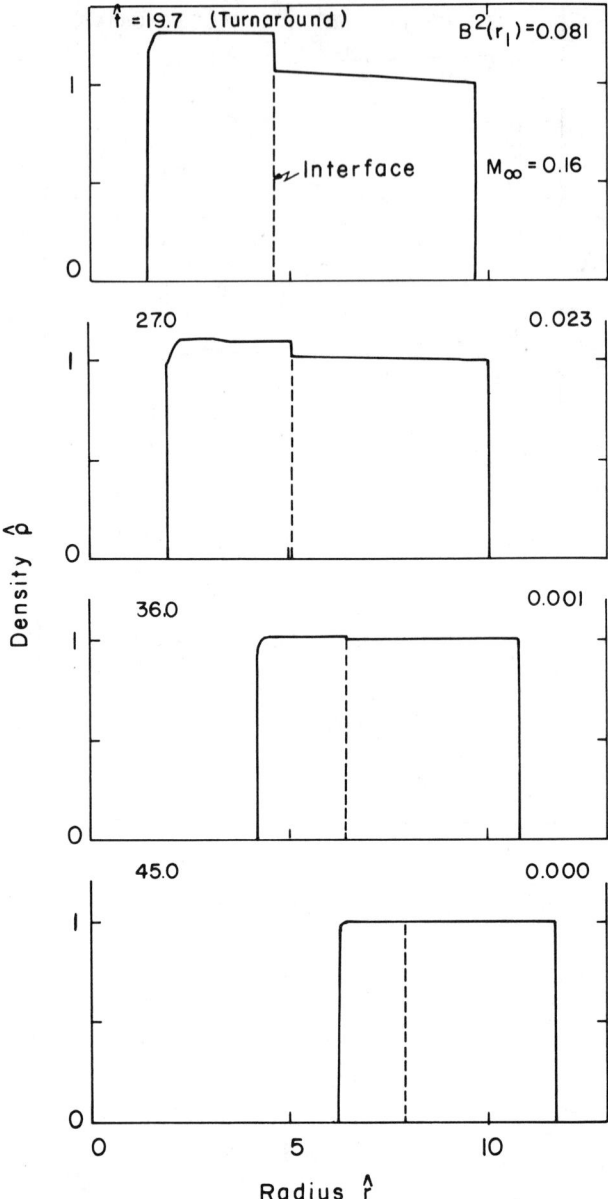

Fig. 10 Density profiles of rotating double-layer liner for $r_{2f}^*/r_{1f}^* = 10$, $r_{1f}^* c/\nu_m = 1000$, $\bar{\rho}_I/\bar{\rho}_{II} = 0.05$ $K_I/K_{II} = 0.2$, and $R_f^*/r_{1f}^* = 5$ (R = interface radius, ν_m = magnetic diffusivity).

sidered that this difficulty can be overcome by adding a
more compressible layer to the inner region of the liner,
i.e., by forming a double-layer liner such as is shown in
Fig. 8b, to compensate for the reduction in the original
layer for the absorption of fusion neutrons. When the compressibility of the inner layer is sufficiently larger than
that of the outer layer, the interface can be regarded as a
free boundary for the outer layer. Then, the criterion of
Eq. (5) says that the generation of negative pressure in
the outer layer can be prevented by reducing its thickness
for a given total liner thickness. It should be noted that
the concept of double-layer liner has already been considered to satisfy both of the requirements in the physical properties of the liner, i.e., a low compressibility and a
high electrical conductivity (Robson 1978) and to enhance
the dwell time of the liner near turnaround to extend the
fusion reaction time (Yamagishi and Schaffer 1978).

The numerical results (Fig. 9) also show that the layer
near the inner surface ablates due to the tension of the
lines of force when the field strength on the expanding inner surface becomes smaller than that of the diffused field
in the liner. If the liner is rotated, the centrifugal
force not only increases the pressure in the liner to compensate for the tension of the lines of force but also
pushes the ablating layer toward the inner surface under
the conservation of angular momentum.

Consequently, the disruption and the ablation can be
avoided by making use of the rotating double-layer liner as
shown in Fig. 10, where half of the initial kinetic energy
is consumed in the rotation at turnaround. To suppress the
cavitation in this manner, a certain amount of energy is
consumed in rotating the liner and in compressing the inner
layer with a small bulk modulus. If the liner geometry is
thus sustained to succeed in the recovery of its radial
kinetic energy, the additional energy for the implosion will
not give a significant problem on the reactor energy balance
of liner fusion.

IV. Summary

A solid potassium liner collapsed by θ pinch reveals
that the deformation wave generated due to the Rayleigh-Taylor instability is of the flute type, i.e., its vector
is parallel to the azimuthal direction. Although the effect of viscosity reduces the perturbation as observed in

the experiment, it irreversibly dissipates the kinetic energy of the liner. Therefore, a solid liner is inferior to that of a liquid in which viscosity is negligible. To reduce the surface perturbations of a liquid liner, the MHD dynamic stabilization against the flute type was considered on the basis of the linear perturbation theory, where oscillating azimuthal fields are superposed on driving and compressed axial fields. The results are as follows: 1) the considerable reduction in the growth rate ω of the perturbation of the outer surface can be achieved when the thickness of the driving field layer is less than that of the liner and 2) the reduction in ω of the inner surface is also possible except for the eccentric displacement ($m = 1$).

As for the cavitation, the numerical results show that it appears in the liner with the propagation of rarefaction wave if the initial radial velocity and the turnaround thickness is large enough to satisfy the criterion derived analytically for the generation of negative pressure. The ablation of the layer near the inner surface also takes place due to the tension of the diffused field when the field strength at the expanding inner surface becomes smaller than that in the liner. However, it is possible to prevent the cavitation by making use of the rotating double-layer liner of which inner layer is highly compressible in comparison with the outer layer.

Acknowledgment

The authors wish to sincerely thank N. Yamaoka for his invaluable discussions and technical assistance.

References

Barcilon A. et al. (1974) Phys. Fluids, Vol. 17, p. 1707.

Book D.L. et al. (1974) Phys. Fluids, Vol. 17, p. 662.

Book D.L. et al. (1978) Plasma Physics and Controlled Nuclear Fusion Research, Vol. 2, IAEA, p. 93.

Book D.L. and Turchi P.J. (1979) Phys. Fluids, Vol. 22, p. 68.

Harris J.G. (1962) Phys. Fluids, Vol. 5, p. 1057.

Itoh Y. and Fujii-e Y. (1979) J. Nucl. Sci. Technol., Vol. 16, p. 175.

Itoh Y. and Fujii-e Y. (1980) J. Nucl. Sci. Technol., Vol. 17, p. 801.

Itoh Y., Umezawa S., Yamaoka N., and Fujii-e Y. (1980) J. Nucl. Sci. Technol., Vol. 17, p. 573.

Itoh Y., Kanagawa T., Miyazaki K., and Fujii-e Y. (1982) J. Nucl. Sci. Technol., Vol. 19, p. 1

Robson A.E. (1978) Naval Research Laboratory Memorandum Rept. 3861.

Somon J.P. (1969) J. Fluid Mech., Vol. 38, p. 769.

Suydam B.R. (1958) Proc. 2nd Geneva Conf., Vol. 31, p. 157.

Yamagishi T. and Schaffer M.J. (1978) GA-A Rept. 14792.

Chapter IV   Metallurgical Magnetohydrodynamics

Electromagnetic Stirring in the Coreless Induction Furnace

D.J. Moore* and J.C.R. Hunt†
University of Cambridge, Cambridge, England

Abstract

Recirculating flows in coreless induction furnaces require investigation in order to understand and predict both their advantages with regard to the heating and mixing capacity of the furnace and their disadvantages, i.e., lining wear. A laboratory model was constructed to study the main features of the flow and to compare the results with approximate analysis and computation. The flow was measured using a novel velocity probe which measures the drag force of a perforated tantalum spherical shell. The spatial variation of the mean flow velocity and turbulent intensity have been recorded and the linearity between the coil current and the mean velocity, as predicted theoretically, confirmed. It has also been observed that, as in real furnaces, the flow exhibits intense, low-frequency fluctuations containing large swirl components. These fluctuations are found to occur on a time scale which is much larger than the time scale of the shear flow turbulence in the shear layers at the wall.

I. Introduction

The coreless induction furnace is an important metal melting tool for ferrous and nonferrous metals. It is manu-

---

Paper presented at Third Beer-Sheva International Seminar on Magnetohydrodynamic Flows and Turbulence, Ben-Gurion University of the Negev, Beer-Sheva, Israel, March 23-27, 1981. Copyright © American Institute of Aeronautics and Astronautics, Inc., 1982. All rights reserved.
*Department of Engineering.
†Professor, Department of Engineering, and Department of Applied Mathematics and Theoretical Physics.

factured in a wide range of charge capacities and power ratings and is usually classified by the frequency of operation. A typical single-phase configuration is shown in Fig. 1.

An alternating current in the coil causes a concave electromagnetic wavefront to diffuse into the crucible with the result that large azimuthal eddy currents $\bar{J}$ are induced in the charge, which eventually melts through Joule dissipation (J^2/σ).

The resultant melt in such furnaces is subject to electromagnetic body forces arising from the interaction of the induced currents with the magnetic flux density $\bar{B}$ produced by the surrounding coil. This ($\bar{J} \times \bar{B}$) force field has an irrotational component which is primarily radial and results in the formation of a convex surface meniscus and a much smaller (and largely vertical) rotational component which gives rise to a strong recirculating flow. The pattern of the flow is influenced by the shape of the crucible, coil length, the position of the melt relative to the coil, and the frequency of operation.

The stirring action is important because:

1) It leads to homogeneity of melt and uniform temperature distribution.

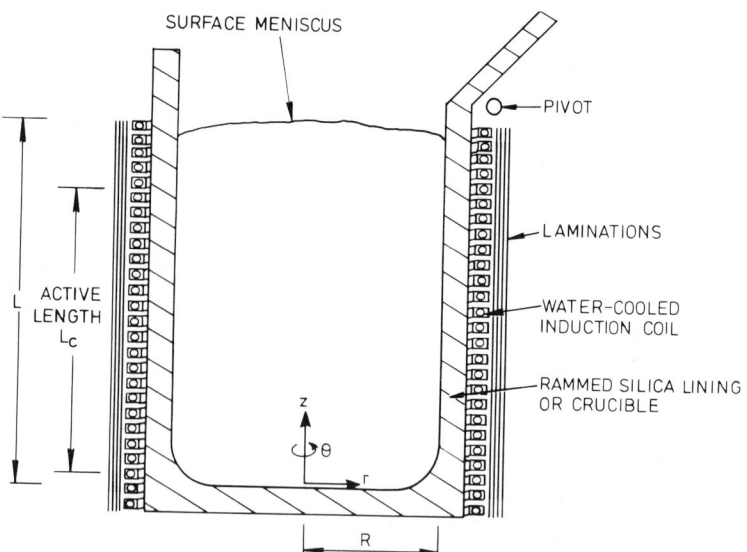

Fig. 1 Schematic outline of coreless induction furnace.

2) It helps rapidly entrain and mix chemicals added to the melt at the surface.

3) It affects heat-transfer coefficient, which is particularly significant at mains frequency operation where continued melting by dissolution is current practice.

4) It encourages undesirable oxidation and entrainment of hydrogen.

5) It promotes lining wear which leads to a decrease in electrical efficiency and ultimately costly replacement (Edgerley and Langman 1979).

A better understanding of the nature of the stirring action and how it can be related to furnace parameters that are under the control of the furnace designer and operator is essential if more effective and economical furnaces are to be produced. This is particularly important today since cheaper solid-state frequency conversion has made the medium frequency furnace, with its reduced stirring action, an attractive alternative to the 50 Hz mains frequency furnace.

To investigate experimentally the flow in a typical furnace, a laboratory simulation has been designed and constructed. This paper describes the simulation and presents some initial experimental results.

II. Experimental Rig

The salient features of the apparatus are shown in Fig. 2. Essentially it consists of a low-permeability ($\mu_r < 1.05$) stainless steel tank which, when fully charged, contains about 300 kg of mercury. The tank is surrounded by an 11 turn, water-cooled induction coil and is supplied from a 50 Hz, single-phase power transformer which is fed from a variable voltage supply. The coil current is measured with a calibrated current transformer and the height of the coil relative to the mercury can be varied, which allows a range of coil/melt geometries to be examined.

To control the temperature of the mercury a curtain of cold water is maintained over the outside of the tank, supplied from a water gallery situated around the top. After leaving the tank sidewall the water, before it leaves the apparatus, has to find its way through a series of suitably perforated annular stainless steel baffles. The actual wa-

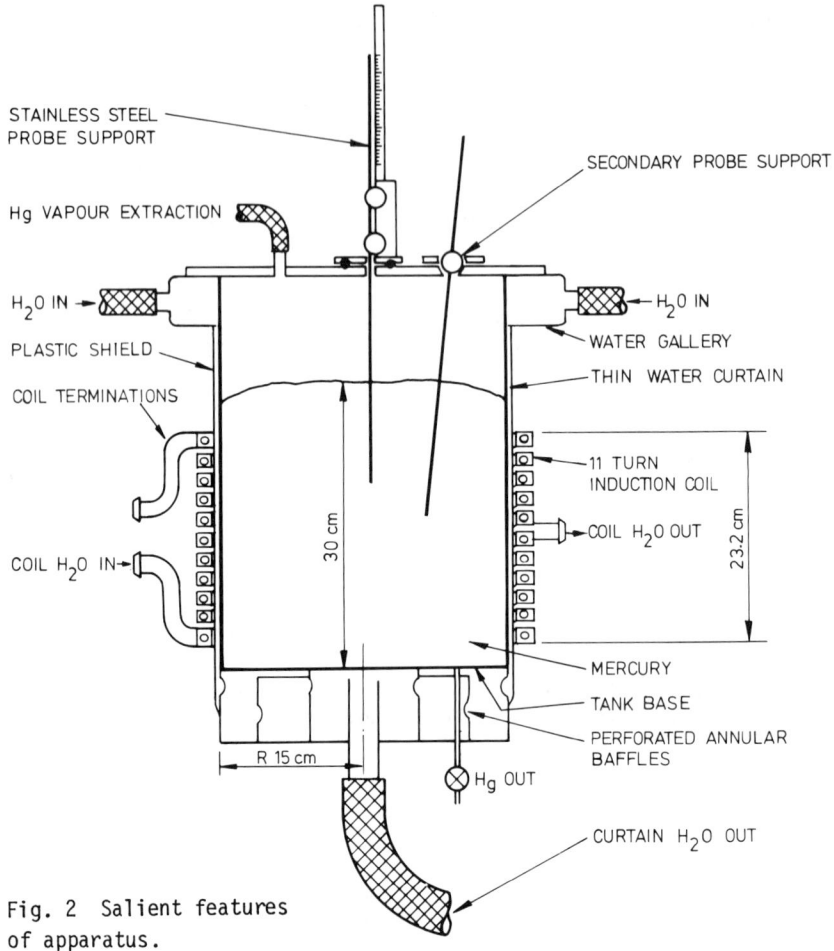

Fig. 2 Salient features of apparatus.

ter exit is provided by a central weir arrangement located in the second base of the tank (see Fig. 2). During its passage to the central exit the water is in continuous contact with the actual tank base, thus providing the essential cooling over this region. This technique of cooling is effective and allows the apparatus to run continuously. Also because of the high thermal conductivity of the mercury and the turbulent nature of the flow (Re $\sim 10^5$), a negligible temperature gradient (<0.01°C cm^{-1}) exists across the tank radius.

The traverse on the Perspex lid of the tank is designed to locate accurately (±1mm) the end of a stainless steel tube anywhere within the recirculating flow. This tube rigidly supports the variety of probes used to monitor the variables of the flow.

III. Instrumentation

For the measurement of the velocity in liquid-metal MHD flows, the hot-film anemometer has a number of limitations, including poor directional sensitivity (especially significant to highly turbulent flows) and acute sensitivity to temperature variations and deposits of dirt on the probe. Because of these limitations and the intense turbulence of the recirculating flow (in some regions >100% of the mean flow), a new type of probe was designed and constructed.

It consists of a 0.5 cm diam perforated tantalum spherical shell having an open area ratio of 0.6. The perforations effectively eliminate vortex shedding and given an increased drag (see Fig. 3). The technique was first developed by Rosenbrock and Tagg (1951) to measure wind speeds.

High-fatigue-life strain gages are bonded to either side of a 19.9×10^{-3} cm thick beam to provide increased sensitivity and good temperature compensation, typically <0.5 $\mu\varepsilon$/°C. The gages are adequately protected from the Hg with a thin film of a silicon rubber compound and the signal-to-noise ratio of the arrangement is not appreciably affected by the presence of the 0.07 T, 50 Hz electromagnetic field within the tank.

The signals from such a probe arrangement will obviously be a measure of the drag force experienced by the spherical shell and the connecting rod. Since the drag on the rod is <2.3% that of the total, the arrangement has a spatial resolution of ∿0.25 cm. Moreover, within the Re range

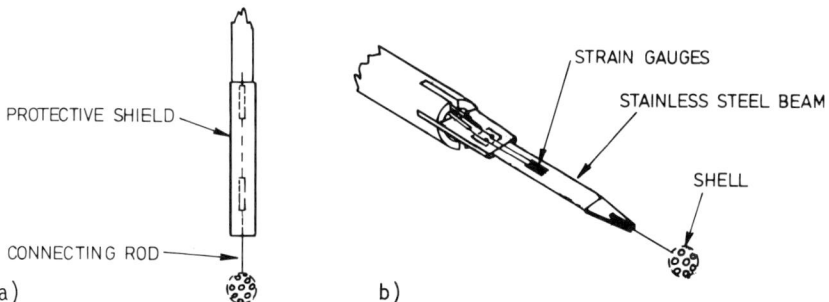

Fig. 3 The probe: a) with protective shield; b) with shield removed.

(2 - 21) x 10^3, drag force measurements reveal that the relation between the mean velocity $\bar{U}$ and the mean drag force $\bar{D}$ has the simple form

$$\bar{D} = k'\bar{U}^2 \qquad (1)$$

In order to deal directly with signals that are proportional to flow velocity, the output from the strain gage amplifier was fed to a circuit designed to give an output S directly proportional to the square root of the input and having the same sign. The output signals were processed by a Solatron microprocessor voltmeter to obtain $\bar{S}$ and σ_S which are direct measures of the mean velocity and corresponding standard deviation.

The probe was calibrated by lowering it into an annular channel of mercury which rotated at a series of known speeds. Due allowance was made for initial transients and the error due to slip was compensated. A typical calibration curve is shown in Fig. 4.

IV. Experimental Results

With the tank containing 288 kg of mercury and a coil current of 1930 A, local values of U_r and U_z were measured over a 1 cm mesh. The results shown in Fig. 5 reveal that, at low frequencies and for symmetrical coil/melt geometries,

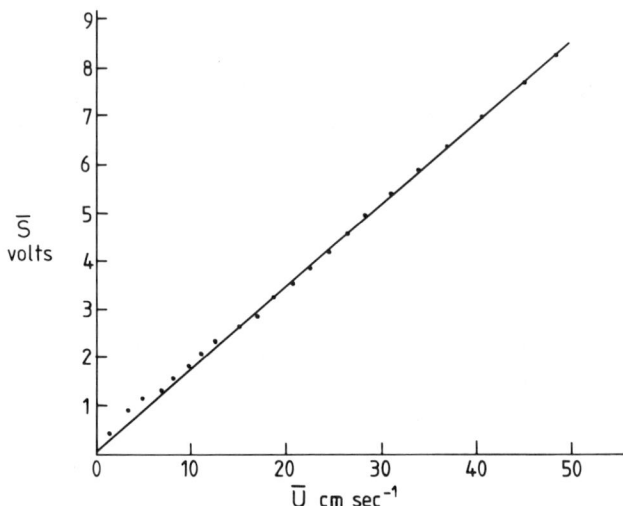

Fig. 4 Typical calibration for spherical shell probe.

ELECTROMAGNETIC STIRRING IN INDUCTION FURNACE 365

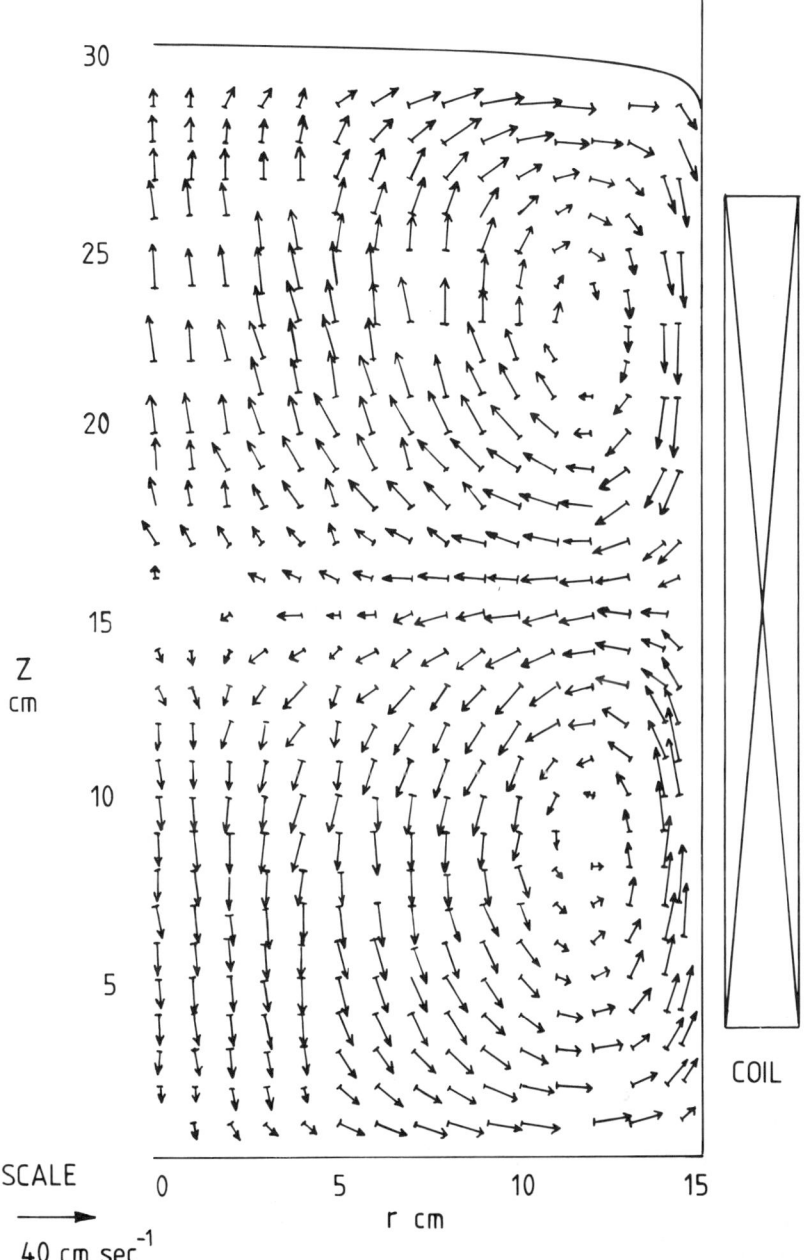

Fig. 5 Experimentally determined mean velocity profile.

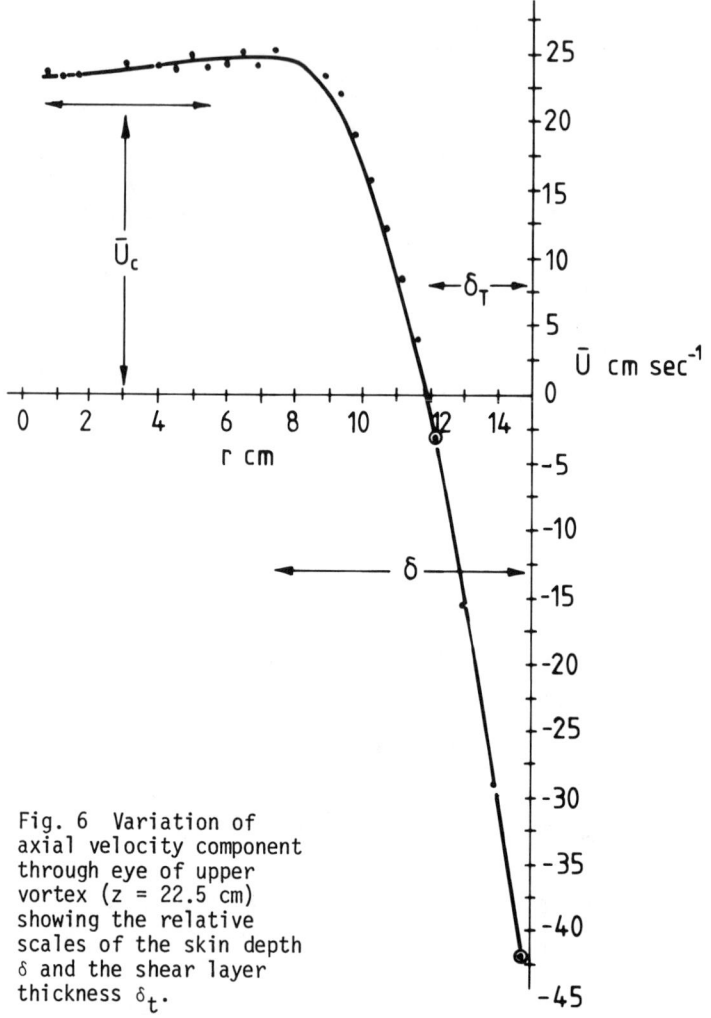

Fig. 6 Variation of axial velocity component through eye of upper vortex ($z = 22.5$ cm) showing the relative scales of the skin depth δ and the shear layer thickness δ_t.

the flow divides into <u>two</u> distinct cells of recirculating flow.

Figure 6 shows the variation with r of U_z through the eye of the upper vortex (results ⊖ were obtained with the small "wall probe" described by Moore 1982) and Fig. 7 shows the corresponding variation in the turbulent intensity $(\sigma_u/\overline{U})$ and (σ_u/U_c) where U_c is the average core velocity $(0 < r < 5)$.

From the results of Fig. 7, the flow may be divided into two principal regions, viz. 1) a central core region over

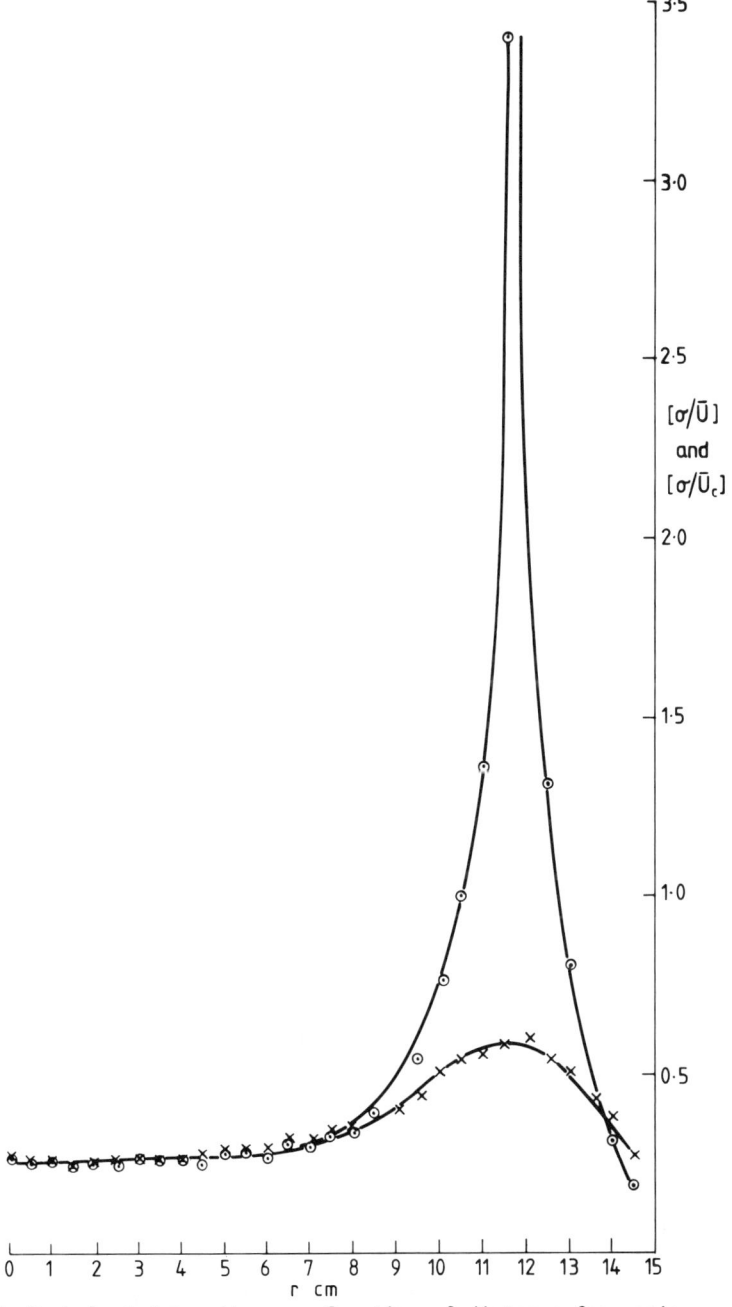

Fig. 7 Turbulent intensity as a function of distance from axis: a) (X) = $|\sigma_u/\bar{U}|$, based on local velocity; b) (·) = $|\sigma_u/\bar{U}_c|$, core velocity.

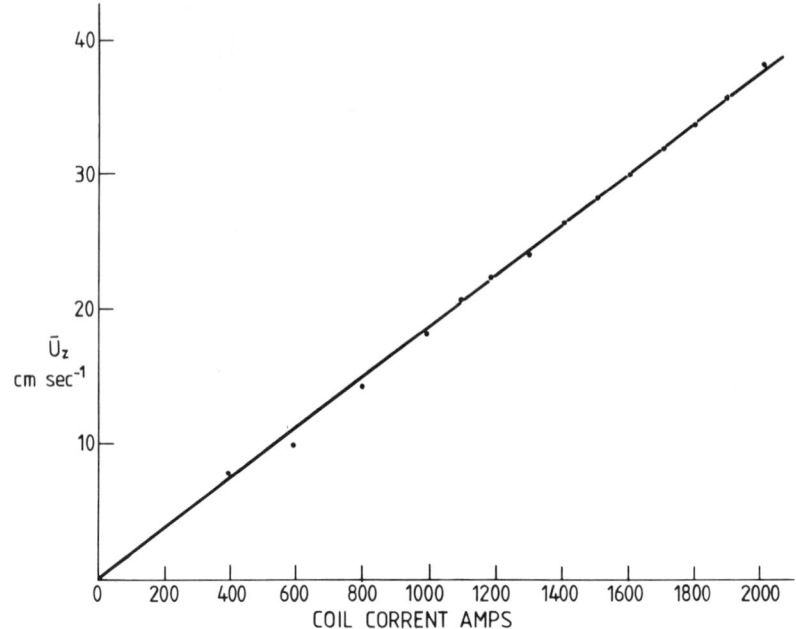

Fig. 8 Variation of time mean axial component of velocity $\bar{U}_z$ with coil current (at r = 13, z = 22.5).

which the axial velocity is effectively constant and 2) a turbulent shear layer of width δ_T close to the wall.

The variation of the mean axial velocity near the wall with coil current is presented in Fig. 8.

The linearity between these two variables is quite marked, especially at the higher values of coil current. It was also observed that the flow possessed intense, low-frequency $\sim$(1-3) Hz swirl components. These components occur on a time scale $L/\bar{U}$ which is $>> \delta/\bar{U}$, the time scale of the shear eddies. It is felt that the origin of this component of flow lies with the unstable nature of the toroidal vortex. Swirl components have not been reported in other experimental studies of low-frequency coreless furnaces, although they have been reported by engineers operating such devices. However, there are many flow arrangements where vortex rings occur and where they are found to be unstable. The general theory of this instability has been discussed by Escudier and Merkli (1979) and Ekchian and Hoult (1979).

V. Order-of-Magnitude Analysis

Having obtained some physical insight into the nature of the flow within the coreless furnace it is possible to carry out an order-of-magnitude analysis which reveals how variables of the flow such as mean velocity and Reynolds stress are related to the furnace parameters.

The intense $(\bar{J} \times \bar{B})$ body force which drives the melt has a time mean and oscillatory component, but it has been shown by Hunt and Maxey (1978) that for the frequency range and length scales normally encountered in induction furnaces, the latter component is of no hydrodynamical significance. Hence, in terms of mean values, the balance of forces within the furnace will be described by the time-smoothed turbulent Navier-Stokes equation

$$\rho(\bar{U}\cdot\nabla)\bar{U} = -\nabla p + \nabla\cdot\bar{\tau} + \overline{(\bar{J} \times \bar{B})} \tag{2}$$

and when expressed in terms of vorticity

$$\rho\nabla \times (\bar{U}\cdot\nabla)\bar{U} = \nabla \times \nabla\cdot\bar{\tau} + \nabla \times \overline{(\bar{J} \times \bar{B})} \tag{3}$$

But for a furnace $\bar{B} = (B_r, 0, B_z)$ and $\partial/\partial\theta = 0$, so $\nabla \times (\overline{\bar{J} \times \bar{B}}) = [\nabla \times (\overline{\bar{J} \times \bar{B}})]_\theta$ which means $U_\theta = 0$. Assuming $1/r \ll \partial/\partial r$, then the equation for the mean vorticity reduces to

$$\rho[\bar{U}\cdot\nabla\Omega_\theta] = \frac{-\partial^2}{\partial z \partial r}(\tau_{zz} - \tau_{rr}) - \left(\frac{\partial^2}{\partial r^2} - \frac{\partial^2}{\partial z^2}\right)\tau_{rz} + [\nabla \times (\overline{\bar{J} \times \bar{B}})]_\theta \tag{4}$$

Taking U_0 to be a typical side layer mean velocity, the order of magnitude of the terms in the above equation can be found as outlined below.

Inertial Terms

$$\rho\bar{U}\cdot\nabla\Omega_\theta = \rho U_z \frac{\partial\Omega_\theta}{\partial z} + \rho U_r \frac{\partial\Omega_\theta}{\partial r} \tag{5}$$

In the shear layer $U_r \ll U_z$, so

$$\rho\bar{U}\cdot\nabla\Omega_\theta \sim \rho U_0 \frac{\partial\Omega_\theta}{\partial z}$$

But

$$\Delta\Omega_\theta = \frac{\partial\Omega_\theta}{\partial z}\frac{L}{2}$$

Hence

$$\rho \overline{U} \cdot \nabla \Omega_\theta \sim \rho \frac{U_o}{L} \Delta\Omega_\theta \qquad (6)$$

where $\Delta\Omega_\theta$ is the change in Ω_θ along a streamline during its passage through the side layer.

Electromagnetic Terms

$$[\nabla \times (\overline{\overline{J \times B}})]_\theta = \frac{\partial}{\partial r}(\overline{\overline{J \times B}})_z - \frac{\partial}{\partial z}(\overline{\overline{J \times B}})_r \qquad (7)$$

But within the shear layer, as confirmed by experiment with an orthogonal search coil arrangement (as reported by Moore 1982) and supported by the computations of Hodgkins (1972), the second term is dominant, so

$$[\nabla \times (\overline{\overline{J \times B}})]_\theta \simeq -\frac{\partial}{\partial z}(\overline{\overline{J \times B}})_r \qquad (8)$$

Now

$$(\overline{\overline{J \times B}})_r = \frac{B_z}{\mu}\left(\frac{\partial B_r}{\partial z} - \frac{\partial B_z}{\partial r}\right) \qquad (9)$$

and since in a typical furnace configuration $\partial B_r/\partial z \ll \partial B_z/\partial r$, then

$$(\overline{\overline{J \times B}})_r \simeq -\frac{1}{2\mu}\frac{\partial B_z^2}{\partial r} \qquad (10)$$

Hence

$$[\nabla \times (\overline{\overline{J \times B}})]_\theta \simeq -\frac{B_o^2}{\mu L \delta} \qquad (11)$$

where δ is the skin depth $[(2/\mu\sigma\omega)^{1/2}]$ and B_o a typical value in the shear layer.

Reynolds Stress Terms

Townsend (1976) has shown that in turbulent jets and shear layers

$$(\tau_{zz} - \tau_{rr}) \sim 0\,(\tau_{rz}) \qquad (12)$$

and since in the shear layer $(\partial/\partial r)\tau_{rz} \gg (\partial/\partial z)\tau_{rz}$

$$-\left(\frac{\partial^2}{\partial r \partial z} + \frac{\partial^2}{\partial z^2} + \frac{\partial^2}{\partial r^2}\right)\tau_{rz} \sim -\frac{\partial^2}{\partial r^2}\tau_{rz} \qquad (13)$$

An order of magnitude for τ_{rz} may be obtained by the following argument. Within the furnace turbulent kinetic energy is generated and dissipated in the turbulent shear layer at the wall, but the latter process also continues as the melt is carried around the furnace. At equilibrium where $(\partial/\partial t)(1/2\overline{q^2}) = 0$ the turbulent kinetic energy equation reduces to

$$\frac{1}{\rho} \tau_{rz} \frac{\partial \overline{U}}{\partial r} + \varepsilon = 0 \qquad (14)$$

The kinetic energy generated within a fluid element in the shear layer is

$$\int \frac{1}{\rho} \tau_{rz} \frac{\partial \overline{U}}{\partial r} dz \sim \left(\frac{\tau_{rz}}{\rho}\right) \frac{U_0}{\delta_t} \frac{L}{2} \qquad (15)$$

Assuming the dissipation to be the same as that for homogeneous grid turbulence

$$\varepsilon \simeq \left(\frac{\tau_{rz}}{\rho}\right)^{3/2} \frac{1}{KL_x} \qquad (16)$$

where K is the von Kármán constant (0.4) and L_x is the integral length scale of the component of the turbulence normal to the wall.

Taking $L_x \sim \delta_t$, then

$$\varepsilon \simeq \oint \left(\frac{\tau_{rz}}{\rho}\right)^{3/2} \frac{1}{K\delta_t} dS \simeq \left(\frac{\tau_{rz}}{\rho}\right)^{3/2} \frac{1}{K\delta_t} 2(R + \frac{L}{2}) \qquad (17)$$

From Eqs. (14), (15), and (17) we obtain

$$\tau_{rz} \simeq \frac{\rho K^2 U_0^2}{4[(1+2R/L)]^2} \qquad (18)$$

Taking $\partial^2/\partial r^2 \sim 1/\delta_t^2$ since δ_t is effectively the scale over which τ_{rz} changes its slope, we have

$$\frac{\partial^2 \tau_{rz}}{\partial r^2} = \frac{\rho K^2 U_0^2}{r\delta_t^2[(1+2R/L)]^2} \qquad (19)$$

Thus the equation for the mean vorticity reduces to

$$2\rho \frac{U_0}{L} \Delta\Omega_\theta = \frac{-\rho K^2 U_0^2}{4\delta_t^2 [(1+2R/L)]^2} + \frac{B_0^2}{L\mu\delta} \quad (20)$$

But since the streamlines are closed, then around any given streamline $\Delta\Omega_\theta = 0$, in which case a fundamental dynamical balance must exist between the shear stress and $(\bar{J} \times \bar{B})$ terms. Hence in the shear layer

$$\frac{\rho K U_0^2}{4\delta_t^2 [(1+2R/L)]^2} \simeq \frac{B_0^2}{\mu L \delta}$$

So

$$U_0 \simeq \frac{B_0}{\sqrt{\mu\rho}} \frac{2\delta_t [(1+2R/L)]}{K\sqrt{L\delta}} \quad (21)$$

At a coil current of 1930 A where $B_0 \simeq 0.06$ T (hence $B_0/\sqrt{\mu\rho} = 0.46$ ms^{-1}) and $\delta_t \simeq 0.03$ (see Fig. 6), this gives $U_0 \simeq 0.95$ ms^{-1}. However, in order for Eq. (21) to be completely predictive, i.e., not relying upon an experimental determination of δ_t, it may be possible to determine δ_t from detailed calculations in the shear layer.

VI. Conclusion

The linear relation between axial velocity and coil current predicted by Eq. (21) is confirmed by the experimental results presented in Fig. 8. Figure 6 shows that the peak velocity occurs very close to the wall (<2 mm) which suggests that when furnace designers consider the flow near the wall (e.g., when estimating lining wear), they should take account of wall roughness. It is instructive to compare this flow with that of a conventional wall jet without any body force, where the peak velocity occurs further from the wall than in our case.

The estimated mean velocity given by the order-of-magnitude analysis is on the high side but has served to highlight the significance of this region of the flow. It will be interesting to see how these results, especially the measurements near the wall, compare with numerical computations based on the "k-ε" turbulence models, such as those of Tarapore and Evans (1976) and Szekely and Chang (1977).

Acknowledgments

The authors would like to express their gratitude to the Electricity Council Research Centre, Capenhurst, for providing one of us (DJM) with full financial support during the course of this investigation, and to P.W. Turner of Cambridge University Engineering Department for help with the design of the apparatus.

References

Edgerley C.J. and Langman R.D. (1979) "Progress in the design of furnace control and power supply equipment for induction furnaces," Electricity Council Research Centre, Rept. M1305.

Ekchian A. and Hoult D.P. (1979) "Flow visualisation study of the intake process of an internal combustion engine," Trans. Soc. Auto. Eng., Vol. 88, p. 383.

Escudier M.P. and Merkli P. (1979) "Observations of the oscillatory behavior of a confined ring vortex," AIAA J., Vol. 17, p. 253.

Hodgkins W.R. (1972) "Mathematical calculations on electromagnetic stirring," Electricity Council Research Centre, Rept. ECRC/MM12.

Hunt J.C.R. and Maxey M. (1978) "Estimating velocities and shear stresses in turbulent flows of liquid metals driven by low frequency electromagnetic fields," Proc. 2nd Beer-Sheva Int. Seminar on MHD Flows and Turbulence, Israel Universities Press, Jerusalem, pp. 249-269.

Moore D.J. (1982) "Magnetohydrodynamics of the coreless induction furnace," PhD Dissertation, Cambridge University, Cambridge, England.

Rosenbrock H.H. and Tagg J.R. (1951) "Wind and gust-measuring instruments developed for a wind-power survey," Proc. Inst. Electr. Eng., Vol. 98, 438.

Szekely J. and Chang C.W. (1977) "Turbulent electromagnetically driven flow in metals processing, Pt. II," Iron Steelmaking, Vol. 3, p. 196.

Tarapore E.D. and Evans W.E. (1976) "Fluid velocities in induction melting furnaces," Metall. Trans., Vol. 7B, p. 345.

Townsend A.A. (1976) The Structure of Turbulent Shear Flow, Cambridge University Press, Cambridge, England.

Single-Phase Electromagnetic Stirring in Coreless Induction Furnaces

Y. Fautrelle*
Université de Grenoble, Grenoble, France

Abstract

Electromagnetic stirring in coreless induction furnaces is investigated both analytically and numerically. First, considering the simple case of a circular cylinder in a transverse alternating magnetic field, the motion equations may be solved in the high-frequency limit. It is shown that the electromagnetic forces, although localized in the electromagnetic layer, generate a flow in the bulk of the domain, and that there exit two primary flow regimes. The motion is driven either by viscous stresses in the weak field limit or via a pressure gradient in the strong field case. Then, dealing with the more realistic geometry of the truncated cylinder in a coil, the magnetic field and the turbulent motion are computed using simple classical eddy viscosity models. A comparison of the results with previous measurements shows that the agreement is only qualitative for both the flow pattern and the magnitude of the velocities.

I. Introduction

Induction-melting furnaces are designed to heat and melt a metal. Electric currents are induced by an external coil supplied with alternating electric currents. The eddy currents create ohmic losses but also electromagnetic forces

---

Paper presented at Third Beer-Sheva International Seminar on Magnetohydrodynamic Flows and Turbulence, Ben-Gurion University of the Negev, Beer-Sheva, Israel, March 23-27, 1981. Copyright © American Institute of Aeronautics and Astronautics, Inc., 1982. All rights reserved.
*Maître Assistant, Institut de Mécanique.

because of their interaction with the magnetic field. These forces lead to a vigorous stirring of the bath. Stirring may have both useful and adverse effects. It is a means of improving the homogeneity of the melt, but in turn it may shorten the lifetime of the refractory walls due to erosion.

Stirring is well known by engineers, but it has been investigated theoretically only quite recently by Sneyd (1971, 1979), Tir (1976), Tarapore and Evans (1976), Moffatt (1978), Moreau (1980), and Fautrelle (1981). From the various observations the flow is turbulent, but it is commonly admitted that there exists large-scale mean circulation which may be predicted. So far, the various approaches mainly consist in numerical or seminumerical investigations. This is due to the complexity of the problem in which we have to deal with complex geometries and turbulent motions.

A fruitful analysis may be carried out at large frequencies in the case of an infinite circular cylinder in a transverse magnetic field. This case has been studied first by Sneyd (1971) and Fautrelle (1981) referred to hereafter as I. Its simplicity allows us to solve both Maxwell and Navier-Stokes equations in the laminar approximation (Sec. II). The solutions highlight the physics of the phenomenon by showing the flow pattern organization and the relationship between the magnetic field and the velocity magnitude.

A second step must be made toward the prediction ability in realistic situations. The main source of complexity comes from the turbulence of the motion. It is interesting to compare the results of simple eddy viscosity models with measurements achieved by Cremer and Alemany (1981).

II. Heuristic Approach

Some analytical investigations may be carried out in the high-frequency limit, i.e., $\delta/a \ll 1$, where a denotes the radius of the pool and δ is the so-called skin depth,

$$\delta = [(\mu\sigma\omega)/2]^{-1/2} \qquad (1)$$

where μ is the magnetic permeability of the vacuum, σ the electrical conductivity of the liquid metal, and $f = \omega/2\pi$ the frequency of the applied currents.

In this asymptotic situation, the electric currents induced by the pulsation are confined in the skin depth along the walls of the pool [for further details the reader is referred to Sneyd (1979) or I]. By means of local coordinates (x,y) defined from the boundary (cf., Fig. 1) an analytical expression of the mean electromagnetic forces (averaged over one period) may be obtained, namely

$$\overline{F} = \frac{1}{2} R_\omega \frac{B_o^2}{\mu a^2} A^2(x) e^{-2y} \overline{i}_y \qquad (2)$$

where $A(x)$ is the nondimensional vector potential modulus (scaled by $B_o \delta$) along the boundary of the bath, B_o a typical value of the magnetic field, and R_ω a dimensionless parameter defined from δ by

$$R_\omega = \mu\sigma\omega a^2 = 2a^2/\delta^2 \qquad (3)$$

x and y are dimensionless coordinates respectively scaled by a and δ. Throughout this section, we shall restrict the analysis to the bidimensional case of the infinite circular cylinder in a transverse uniform magnetic field.

To simplify matters, we shall assume that the turbulent stresses may be represented by an effective constant viscosity. Furthermore, we shall focus our investigations on the weak viscosity limit (or strong magnetic field limit). The domain may be split into two regions (see I):

1) An interior invisicid region where the single vorticity component ζ is constant.

2) An electromagnetic and viscous wall layer.

In the latter region, the classical boundary-layer approximations may be used to simplify the motion equations (see, for example, I). In the local coordinates, the dimensionless motion equations reduce to

$$u \frac{\partial u}{\partial x} + v \frac{\partial u}{\partial y} = -\frac{dP}{dx} + A \frac{dA}{dx} e^{-2y} + \frac{1}{R_\delta} \frac{\partial^2 u}{\partial y^2} \qquad (4)$$

$$\frac{\partial u}{\partial x} + \frac{\partial v}{\partial y} = 0$$

$$p + \frac{A^2(x)}{2} e^{-2y} = P(x), \quad P(x) = \text{unknown function} \qquad (5)$$

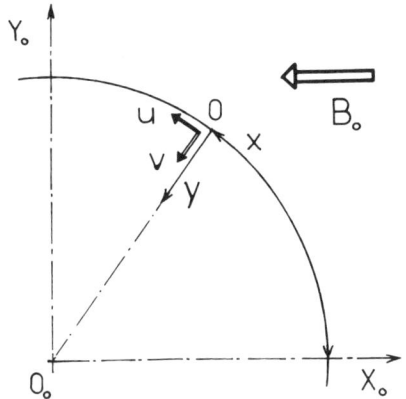

Fig. 1 Local frame defined from the boundary of the circular cylinder.

where p denotes the pressure, the velocities are normalized by u_0 such that

$$u_0 = B_0/(\mu\rho)^{1/2}, \quad \rho \text{ being the density} \tag{6}$$

and R_δ is a dimensionless parameter defined by $R_\delta = u_0\delta^2/\nu a$.

Because of the various symmetries, the domain is restricted to a quarter of the cross section of the cylinder. Therefore, the boundary conditions on the velocity field are

$$u(0,y) = u(\pi/2, y) = 0 \tag{7a}$$

$$u(x,0) = v(x,0) = 0 \tag{7b}$$

$$u(x,\infty) = U(x) \tag{7c}$$

where $U(x)$ is slip velocity of the interior flow at the wall. Applying Eq. (7c) to Eq. (4) shows that $P(x)$ is related to $U(x)$ by

$$\frac{dP}{dx} = -U\frac{dU}{dx} \tag{8}$$

As for the interior region, if we assume that the flow consists of one vortex in a quarter of the cross section in consistency with the transient problem (Sneyd 1971), the constant vorticity may be expanded in the following way (cf. I)

$$\zeta = \sum_{n=1}^{\infty} \zeta_n \sin 2nx \tag{9}$$

with $\zeta_n = (2U_1/n)(1 - \cos n\pi)$ up to a constant U_1. The slip velocity $U(x)$ is readily obtained from Eq. (9), namely

$$U(x) = U_1 \tilde{U}(x) \text{ with } \tilde{U}(x) = \sin 2x - \sum_{n=2}^{\infty} \frac{\zeta_n}{2(n+1)} \sin 2nx \qquad (10)$$

As for the wall layers two asymptotic cases may be distinguished according to the value of R_δ in Eq. (4).

Case $R_\delta \ll 1$

In the electromagnetic layer, Eq. (4) reduces simply to

$$A \frac{dA}{dx} e^{-2y} + \frac{1}{R_\delta} \frac{\partial^2 u}{\partial y^2} = 0 \qquad (11)$$

and the solution of Eq. (11) satisfying Eqs. (7a) and (7b) is

$$u(x,y) = \frac{R_\delta}{4} A \frac{dA}{dx} (1 - e^{-2y}) \qquad (12)$$

There exists a boundary layer of depth δ_v, such that

$$\delta_v = \frac{\nu a}{u_0 \delta} \gg \delta \qquad (13)$$

where Eq. (12) matches the interior solution given by Eq. (10).

It is checked that the weak viscosity limit is valid only if the central region remains inviscid, namely

$$\delta_v/a = \nu/u_0 \delta \ll 1 \qquad (14)$$

Note that if $u_0 \delta/\nu$ is much less than unity, the Reynolds number of the flow

$$Re = (u_0 \delta/\nu)^2$$

is much less than one and this corresponds to the viscous limit.

The equation governing the flow in the matching region is similar to Eq. (4) with zero forces. Approximate solution may be sought by truncating the expansion of Eq. (10). For example, retaining the first harmonic in Eq. (10) leads to a simple solution of Eq. (4) which matches Eq. (12), namely

$$U_1 = R_\delta/4 \tag{15}$$

Case $R_\delta \gg 1$

From Eq. (13), the viscous layer depth δ_v is less than the skin depth δ. Two methods of investigation may be used to study Eq. (4) with the boundary conditions of Eq. (7).

The first method developed in I consists of expanding the electromagnetic layer solutions in the Fourier series analogous to Eq. (10)

$$u(x,y) = \sum_{n=1}^{\infty} f_n'(y) \sin 2nx \tag{16}$$

Inserting Eq. (16) in Eq. (4) and retaining the first two harmonics yields a system of two ordinary differential equations for the functions f_1 and f_2, namely,

$$f_1'^2 - f_1 f_1'' = U_1^2 + \frac{1}{R_\delta} f_2'''$$

$$2f_2 f_1'' - f_1' f_2' - f_1 f_2'' = e^{-2y} + \frac{1}{R_\delta} f_1''' \tag{17}$$

with boundary conditions

$$f_1(0) = f_1'(0) = f_2(0) = f_2'(0), \quad f_2'(\infty) = 0, \quad f_1'(\infty) = U_1 \tag{18}$$

Numerical solutions of Eqs. (17) and (18) obtained by a Runge-Kutta method show that U_1 is an eigenvalue of the system and that there exist two kinds of flow regimes according to the value of R_δ. The latter statement is shown in Fig. 2 where U_1 has been plotted vs R_δ. The first branch corresponding to $R_\delta < 1$ is the continuation of Eq. (15), whereas the second one indicates approximately an asymptotic behavior in the form

$$U_1 \propto R_\delta^{1/3} \tag{19}$$

The harmonics f_1' and f_2' behave like wall jets, as shown in Fig. 3.

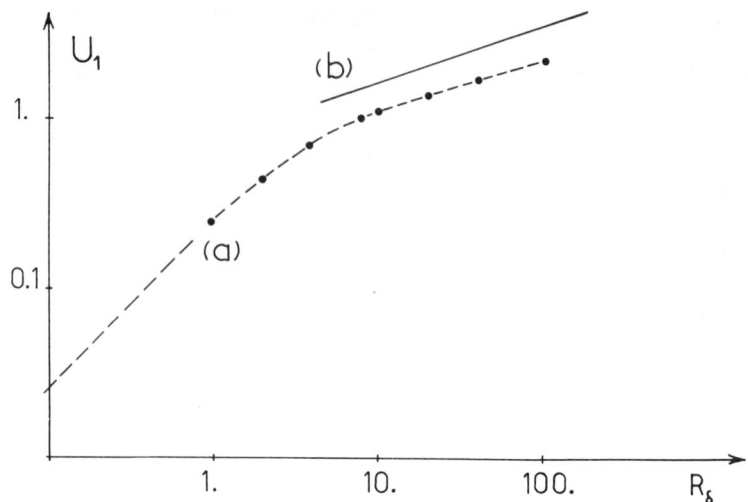

Fig. 2 First coefficient of Fourier expansion U_1 of the flow in the interior region vs R_δ: curve a corresponds to the truncated model and curve b corresponds to the global model in the asymptotic limit $R_\delta \gg 1$.

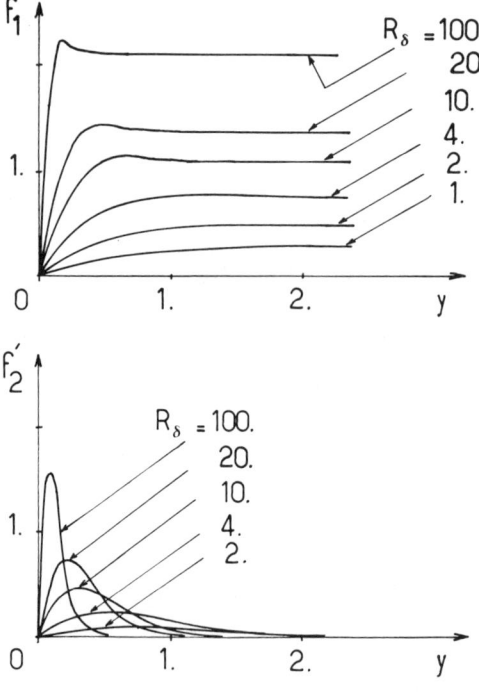

Fig. 3 Tangential velocity profiles given by the truncated model in the electromagnetic layer (f_1' and f_2' correspond to the first harmonics).

The above results are confirmed by the second method which consists in adapting the Karman-Polhausen method to the boundary-layer type of Eq. (4). Integrating Eq. (4) in the electromagnetic layer and using the boundary conditions of Eqs. (7) and (8) yields

$$\frac{d}{dx}\int_0^\infty U(u-U)dy + U'\int_0^\infty (u-U)dy = \sin 2x - \frac{1}{R_\delta}\left(\frac{\partial u}{\partial y}\right)_0 \tag{20}$$

Let us set

$$u = U_1[\tilde{U}(x)g(\eta) + (f(x) - \tilde{U}(x))h(\eta)] \tag{21}$$

where $\eta = \lambda y$, $\tilde{U}(x)$ is given by Eq. (10) and $g(\eta)$, $h(\eta)$ are shape functions. The parameter λ may be identified with the ratio δ/δ_v, while $f(x)$ is an unknown function which has to satisfy

$$f(0) = f(\pi/2) = 0 \tag{22}$$

The shape functions have been chosen as

$$g(\eta) = 1 - e^{-\eta}, \quad h(\eta) = \eta e^{-\eta} \tag{23}$$

the conditions of Eqs. (7b) and (7c) being satisfied. Inserting Eqs. (21) and (23) into Eq. (20) and setting $\lambda = U_1^2$ yields a differential equation for $f(x)$

$$\frac{1}{2}\frac{d}{dx}f^2 + 2(\tilde{U}+\mu)f = \sin 2x + \tilde{U}\tilde{U}' \tag{24}$$

where $\mu = U_1^3/R_\delta$. It is noticeable that Eq. (24) is a first-order differential equation containing a single parameter μ. Because of the boundary conditions [Eq. (22)], μ appears as an eigenvalue of the problem. Equation (24) with Eq. (22) may be solved numerically by using a shooting method. The smallest eigenvalue of Eqs. (22) and (24) is

$$\mu = 0.6261 \tag{25}$$

and the profile of $f(x)$ is shown in Fig. 4. Then the asymptotic behavior of U_1 is

$$U_1 = (\mu R_\delta)^{1/3} = 0.855\, R_\delta^{1/3} \tag{26a}$$

$$\frac{\delta_v}{\delta} = \frac{1}{\lambda} = (\mu R_\delta)^{-2/3} = 1.366\, R_\delta^{-2/3} \tag{26b}$$

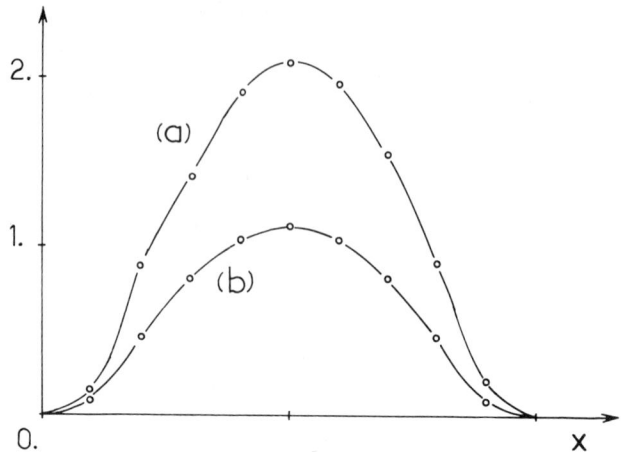

Fig. 4 Solution of the global boundary-layer Eq. (24); curves a and b correspond, respectively, to the eigenfunction $f(x)$ and the slip velocity $\tilde{U}(x)$.

The asymptotic curve $U_1(R_\delta)$ given by Eq. (26a) is plotted in Fig. 2, and the comparison with the curve given by the truncated model shows that the agreement is qualitatively good.

To interpret physically the meaning of an eigenvalue U_1, we must recall that U_1 is involved in the electromagnetic-layer force balance via the wall pressure gradient [cf., Eq. (8)]. In the limit of weak viscosity, viscous stresses are unable to equilibrate the electromagnetic forces in the electromagnetic region. The balance is achieved thanks to the pressure gradient which, in turn, drives a motion in the central inviscid region.

III. Modeling

From a practical point of view, it is important to give good predictions in situations close to reality. The magnetic field computation techniques provide fairly accurate results, even in complex geometries. Therefore, the main difficulty comes from the turbulence of the motion. Indeed, for example in an iron furnace of radius a = 0.5 m velocities of the order of 0.3 m/s are commonly observed, and the corresponding Reynolds numbers are of the order of 10^5. It is likely that turbulence is weakly influenced by the damping effects of the magnetic field. The parameter interaction N which quantifies those effects (see, for example, Alemany 1978), namely

$$N = (\sigma B_0^2/\rho)/(\ell'/u') \tag{27}$$

is at most of the order of unity even in large-scale furnaces. There, in a first approach one may use the "ordinary" turbulence models to predict the mean motion.

Among the so-called one-point closure models we have chosen for simplicity the eddy viscosity model, ν_t being given by a Prandtl mixing length theory. The nondimensional motion equations governing the mean velocity field $\overline{U}$ are

$$\overline{U} \cdot \nabla \overline{U} + \nabla p = \overline{F} + \nabla \cdot (\varepsilon \overrightarrow{\overline{e}}) \tag{28}$$

$$\nabla \cdot \overline{U} = 0$$

where $\overrightarrow{\overline{e}}$ denotes the deformation rate tensor whose components are e_{ij} and the function ε is the reduced eddy viscosity de-

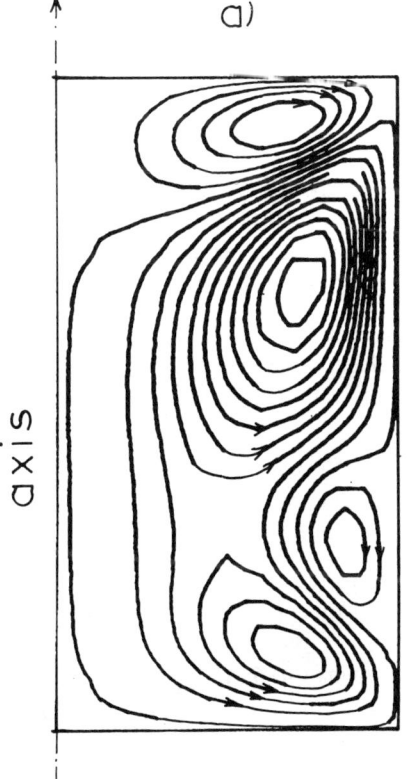

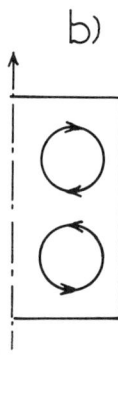

Fig. 5 Streamline pattern of the flow in the mercury pool: a) computed flow pattern, ε given by Prandtl mixing length theory; b) experimental configuration.

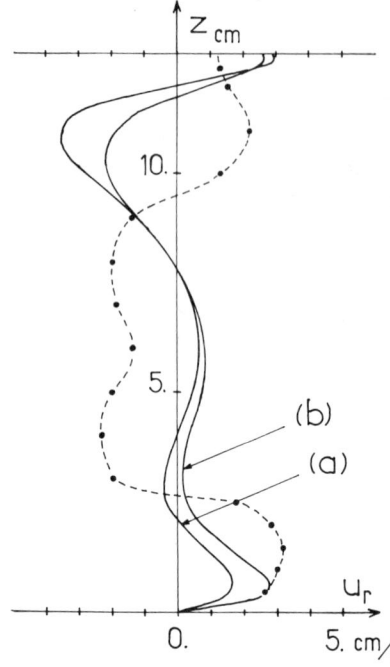

Fig. 6 Radial velocity profiles $U_r(a/2,z)$; ——computed profiles for a) ε given by Eq. (29) and b) $\varepsilon = 300$; ---- experimental profile measured by Cremer and Alemany (1981).

fined as

$$\varepsilon = \nu_t/\nu = \ell^2 (e_{ij}\, e_{ij})^{1/2} \qquad (29)$$

ℓ being the nondimensional typical eddy length. The velocities and the lengths in Eq. (28) are normalized by u_0 defined in Eq. (6) and the radius a, respectively. The model depends on only one parameter - the eddy length ℓ.

A classical finite-difference scheme as been used to solve Eqs. (28) and (29) [for further details see Gosman et al. (1969)]. The runs have been performed in the following conditions:

  Five-turn coil: height = 290 mm
         radius = 135 mm
         maximum intensity = 1840 A

  Mercury pool: height = 130 mm
         radius = 78 mm

The corresponding values of the nondimensional parameters introduced in Sec. II are $R_\omega = 88.6$ and $R_s = 10^4$. These numerical values correspond to the experiments carried out by Cremer and Alemany (1981) who achieved velocity measurements. The conjectured experimental flow pattern consists of two vortices in a half-meridian plane. It may be compared with the flow configuration obtained numerically in Fig. 5 with $\ell = 0.1$. We have also compared the numerical and experimental radial velocity profiles in Fig. 6. We observe that the numerical results are consistent with the force distribution along the wall (cf. I) in the case of a truncated circular cylinder in an axial magnetic field. However, the computed pattern is not in good agreement with the experimental one. Comparison of the velocity profiles also shows that one must expect only qualitative results from such a modeling. Incidentally, note that a run corresponding to $\varepsilon = 300$ (a constant) yields results which are not much different from those given by Eq. 29 (cf. Fig. 6).

IV. Conclusion

In summary, we have shown in Sec. II how the forces, although localized in a wall layer, may create a motion in the central regions via a pressure gradient. It is also noticeable that in the laminar case there appears a break in the flow regime according to the ratio of the viscous layer thickness to the skin depth.

As for the modeling, simple eddy viscosity models yield only qualitative results. However, it must be pointed out that the measurements in such a geometry are difficult and quite uncertain. Therefore, further improvements are necessary concerning both the numerical models and the measurements.

References

Alemany A. (1978) "MHD à l'echelle du laboratoire, quelques résultats, quelques applications," Thèse de Doctorat d'Etat, Université de Grenoble, France.

Cremer P. and Alemany A. (1981) "Aspects expérimentaux du brassage electromagnétique en creuset," J. Méc. Appl., Vol. 5, No. 1, pp. 37-50.

Fautrelle Y. (1981) "Analytical and numerical aspects of the electromagnetic stirring," J. Fluid Mech., Vol. 102, pp. 405-430.

Gosman A.D., Run W.M., Runchal A.K., Spalding D.B., and Wolfshtein M. (1969) Heat and Mass Transfer in Recirculating Flows, Academic Press, London, pp. 18-137.

Moffatt H.K. (1978) "Some problems in the magnetohydrodynamics of liquid metals," Z. Angew. Math. Mech., Vol. 58, pp. 65-70.

Moreau R. (1980) "MHD flows driven by alternating magnetic field," Proc. 2nd Bat-Sheva Seminar on MHD Flows and Turbulence, 1978, H. Branover (Ed.), pp. 65-82.

Sneyd A. (1971) "Generation of fluid motion in a circular cylinder by an unsteady applied magnetic field," J. Fluid Mech., Vol. 49, pp. 817-827.

Sneyd A. (1979) "Fluid flow induced by a rapidly alternating or rotating magnetic field," J. Fluid Mech., Vol. 92, pp. 35-51.

Tarapore E. and Evans J.W. (1976) "Fluid velocities in induction melting furnaces, Part I: theory and laboratory experiments," Metall. Trans. B, Vol. 7, pp. 343-351.

Tir L.L. (1976) "Features of mechanical energy transfer to a closed metal circuit in electromagnetic systems with azimuthal currents," Magn. Gidrodin., Vol. 2, pp. 100-109.

Liquid-Solid Separation in a Molten Metal by a Stationary Electromagnetic Field

Ch. Vives* and R. Ricou†
Centre Universitaire d'Avignon, Avignon, France

Abstract

The experimental determination of electromagnetic resulting forces applied to particles of various shapes, orientations, and electrical conductivities, suspended in a still or moving liquid metal, submitted to a vertical, stationary, and uniform to infinite electromagnetic force field are presented. The breaking of magnetohydrostatics equilibrium are also revealed. Provided that the Reynolds number of the particle is less or equal to 1500, experimental results obtained in magnetohydromagnetic situations are comparable to those obtained in the corresponding magnetohydrostatic cases. Experiments show, through the appearance of very strong instabilities, that separation may be effective only if both the direction of the electromagnetic resulting force and the velocity are opposite. The conclusions of this systematic study are confirmed by observations of ingots elaborated by means of molten metal (Sn, Pb, Al) and artificial cylindrical and spherical solid suspensions (Cu, brass, silica).

I. Introduction

By usual solid-liquid separation processes, it is frequently impossible to salvage particles suspended in a mol-

---

Paper presented at Third Beer-Sheva International Seminar on Magnetohydrodynamic Flows and Turbulence, Ben-Gurion University of the Negev, Beer-Sheva, Israel, March 23-27, 1981. Copyright © 1982 by Ch. Vives. Published by the American Institute of Aeronautics and Astronautics with permission.
*Professor, Faculté des Sciences, Laboratoire de magnetohydrodynamique.
+Assistant, Faculté des Sciences, Laboratoire de magnetohydrodynamique.

ten metal, particularly if the density of the inclusion is nearly the same as the bath. Separation tests have been done (Pavlov et al. 1976; Andres 1976; Menchikov et al. 1978) utilizing the electric conductivity difference between the suspensions and the liquid metal. The central idea behind these tests consists of imposing an electromagnetic force field able to increase or decrease the apparent weight of the particles. This method offers manifold advantages: purification of the liquid metal before solidification, salvage of inclusions which may be rare materials, and decrease of the decantation time (which permits an economy of energy).

In this systematic study, horizontal, uniform, and stationary magnetic and electric fields are imposed on a molten metal of density ρ. This molten metal is initially at rest and includes particles of volume v and of density ρ' of oblong shape (schematized by cylinders or ellipsoids) or without particular dimensions (represented by spheres).

The first study presented, relating to the case where fluid and particles are not in relative motion, corresponds to a magnetohydrostatic situation labeled MHS (although often the fluid will not be strictly at rest as theoretical and experimental results will show). The second study relates to the case where fluid and particles are in relative motion with an average velocity of U and corresponds to the magnetohydrodynamic (or MHD) situation.

From a theoretical point of view, research was above all directed toward the influence of shape and electric conductivity of the particle and also their orientation with regard to the electromagnetic force field in magnetohydrostatic situations. The research brought two distinct cases into view.

The first case appears in certain situations (Δ revolution axis of insulating and conducting cylinders parallel to $\vec{J}_0$ or $\vec{B}_0$, for example) where the voluminal density of electromagnetic forces is equilibrated by a pressure gradient such as grad $p = \vec{J} \times \vec{B}$. A very simplified theory relative to cylindrical obstacles is founded on the following assumptions: the length and the electrical conductivity of the conducting cylinders, as well as the volume of the fluid, are all considered infinite and the magnetic field $\vec{b}_0$ generated by the imposed current density $\vec{J}_0$ is neglected. Solutions of Laplace equations yield the current lines of vector $\vec{J}$ around particles and, by integrations, the resulting pressure (F_{RP}), volume (F_{RV}), and total (F_{RT}) electromagnetic forces applies to the inclusion are found.

The second distinct case occurs when the electromagnetic force density $\vec{J} \times \vec{B}$ is not equilibrated by a pressure gradient (as is the case of cylinders and ellipsoids Δ axis of which is parallel to $\vec{J}_0 \times \vec{B}_0$ and of spheres). Convection flows with four cells appear around the inclusion and the magnetohydrostatic equilibrium is broken (Ricou 1975).

II. Experimental Studies of Magnetohydrostatic Situations

Experimental Method

The experimental method adopted is used to determine the pressure resultant electromagnetic force F_{RP} and the total resultant electromagnetic force F_{RT} (including the volume force F_{RV}). Particles are placed at the center of a vessel of dimensions 4 x 4 x 4 cm containing mercury. In order that the electric current density in the absence of inclusions ($J_0 = I_0/S$) is nearly uniform, the electric current of intensity I_0 arrives at each of the stainless steel electrodes of surface area S by means of 16 conducting wires. A magnetic field $\vec{B}_0$, uniform and perpendicular to $\vec{J}_0$, is imposed by means of an electromagnet, the poles of which have a 16 cm diam. The electromagnetic parameters vary between the following ranges:

$$6250 < J_0 < 215{,}000 \text{ A/M}^2$$
$$0.08 < B_0 < 0.8 \text{ T}$$
$$6875 < J_0 B_0 < 148{,}000 \text{ N/M}^3$$

Pressure profiles taken according to the classical method: a pressure tap of variable azimuth θ is situated upon the cylinder and is connected to an electromagnetic manometer (Branover 1978), as well as another pressure tap which is situated upon a wall of the vessel. Methodical experimentation permits a graph of $\Delta P_{(\theta)} = P_\theta - P_0$ as a function of θ for different values of $\vec{B}_0$ and $\vec{J}_0$. For cylinders, a graphical integration yields the pressure resultant force F_{RP} by unit length of the cylinder of diameter D,

$$F_{RP} = \int_0^\pi \Delta P_{(\theta)} D \cos\theta \, d\theta = \int_0^\pi G_{(\theta)} d\theta$$

where $G_{(\theta)} = \Delta P_{(\theta)} D \cos\theta$ represents the elementary pressure force distribution as a function of θ.

The total resultant force F_{RT} is directly measured by weighing on a precision balance with an unequal beam. The resultant of gravity and hydrostatic forces are equilibrated beforehand and then the electromagnetic force field is applied. Finally, several local velocity measurements were taken by means of a new electromagnetic conduction probe (Chambarel et al. 1981).

This systematic experimental study has given a large number of results, and thus we will limit ourselves to the presentation of only the most significant examples.

Spherical Inclusions

Results relating to insulating and conducting spheres of 1 cm diam are shown in Fig. 1 where $F^* = F_{RT}/J_0B_0v$ (J_0B_0v is the electromagnetic force exerted on the displaced volume of liquid metal). In the case of the insulating sphere for $B_0 = 0.2$ T, and when J_0B_0 exceeds 20,000 N/m³, measurements become impossible due to the appearance of strong instabilities. On the other hand, for $B_0 = 0.535$ T and $J_0B_0 = 60,000$ N/m³, measurements are again feasible; the stabilizing effect is entirely due to the imposed magnetic field. The total resultant force is proportional to J_0 and, given the product J_0B_0, the proportional factor becomes all the greater as B_0 becomes smaller.

In the case of the conducting sphere, observations are similar except experiments show that the voluminal forces are superior to area forces. It follows that, for a given direction of the magnetic field, insulating and conducting spheres have a tendency to move in the opposite direction.

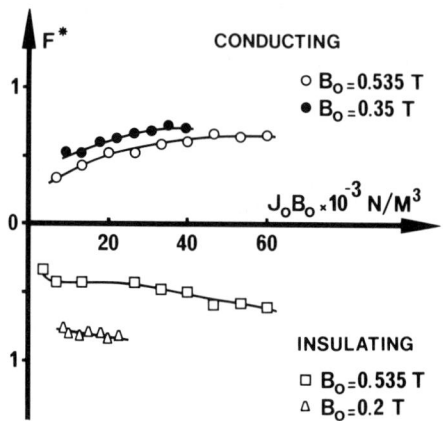

Fig. 1 F* as a function of the electromagnetic force density in the case of insulating and conducting spheres.

The media confinement can be represented by the parameter $V^* = V/v$, which is the ratio between the volumes V of spheres with diameters of 1 and 2 cm, respectively, V^* lying between 122 and 15, experimental results obtained for F^* are practically identical.

Electric field measurements were carried out upon the surface of an insulating sphere of 1 cm diam. To do this, two electrodes, A and A', of variable azimuth θ (Fig. 2) are placed upon the equatorial circle with a distance of 2 mm between them. For a given value of θ, the voltage $V_{AA'}$, between A and A', is measured for different values of the electric current I_0 and the magnetic field (or of the Hartmann number $M = B_0 a (\sigma/\eta)^{1/2}$. The electric field is given by $E_\theta = V_{AA'}/AA'$.

The increase of the electric field with B_0 shows evidence of the existence of flows provoked by the density of the electromagnetic force field not being balanced by a pressure gradient. For $B_0 = 0.45$ T, the increase of the electric field is 45% and, in all cases, for a given value of B_0, the electric field is directly proportional to I_0 (or to E).

Figure 3 shows the velocity distribution around an insulating sphere as a function of θ for different values of r, $\vec{B}_0$, and $\vec{J}_0$. The presence of four cells can be observed, the dimensions of which depend upon the distance r (Fig. 3a) and the Hartmann number (Fig. 3b) and are independent of $\vec{J}_0$.

Cylindrical Inclusions

The particles in this case are cylinders of 10 mm diam, are 13 mm long, and are made of copper, brass, or Plexiglass.

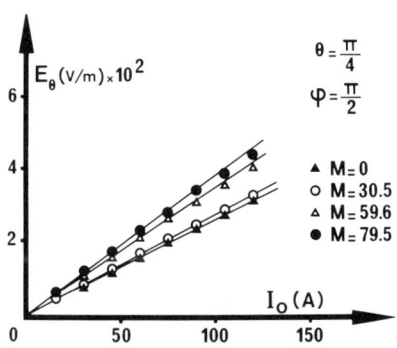

Fig. 2 Electric field upon an insulating sphere as a function of the imposed electric current, for different values of the Hartmann number.

Only the three directions in which the Δ revolution axis is parallel to $\vec{B}_0$, to $\vec{J}_0$, and to $\vec{J}_0 \times \vec{B}_0$ will be studied. Later scrutiny of a number of ingots will show that cylinders are practically always oriented in one of these directions.

Δ Revolution Axis is Parallel to $\vec{B}_0$. Figure 4 corresponds to an insulating cylinder with ΔP being the pressure difference for θ = 0. According to the theory (Vives 1980), the pressure is virtually constant upon the cylinder. The comparison of theoretical and practical results becomes all the more satisfying as the product $J_0 B_0$ becomes weaker. The total resultant of electromagnetic forces is not exactly of zero value, but it remains weak and is oriented in the opposite direction with respect to $\vec{J}_0 \times \vec{B}_0$.

Figure 5 relates to a conducting cylinder in accordance with the calculation which yields

$$\Delta P_{(\theta)} = -2 J_0 B_0 a \cos \theta$$

The profile of pressure as a function of θ is quite nearly of sine curve shape for the weaker values of $\vec{J}_0$. The gap between theoretical predictions and experimental results increases as the product $J_0 B_0$. The resultant of the pressure electromagnetic forces is, at first, proportional to $J_0 B_0$ and is not negligible (Fig. 6); however, this rule is no longer confirmed for the values of $J_0 B_0$ superior to 25,000 N/m^3, which also corresponds to the existence of strong instabilities. Weighing shows that $\vec{F}_{RT}$ is very weak and is oriented in the direction of $\vec{J}_0 \times \vec{B}_0$. The pressure

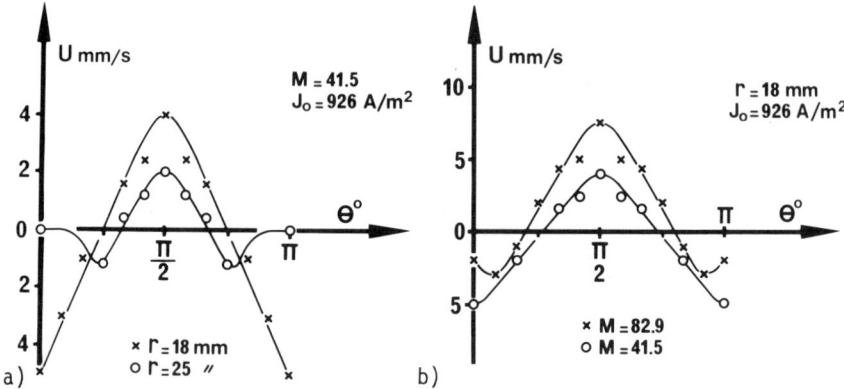

Fig. 3 Velocity profiles around an insulating sphere drawn up the equatorial plane for different values a) of r and b) of the Hartmann number.

forces are almost totally counterbalanced by voluminal forces which demonstrate a somewhat slight predominance.

In summation of Δ revolution axis as parallel to $\vec{B}_0$, this disposition of insulating and conducting cylinders with respect to J_0 and B_0 is unfavorable to their extraction from a molten metal.

Δ Revolution Axis is Parallel to $\vec{J}_0$. Figure 7 shows that $\overline{F_{RT}} = \overline{J_0 B_0 v}$ for an insulating cylinder, whereas for a conducting cylinder F* depends upon the particular values of J_0 and B_0 and, given $J_0 B_0$, increases as J_0. For instance, when the density difference between the cylindrical inclusion and the liquid metal is of the order of magnitude of one, phase separation is possible with an electromagnetic force density of 10,000 N/m³. This level of force density may be realized by a magnetic field of 0.2 T and an electric current density of 5 A/cm². A low electromagnetic force density of this nature provides two advantages: it does not

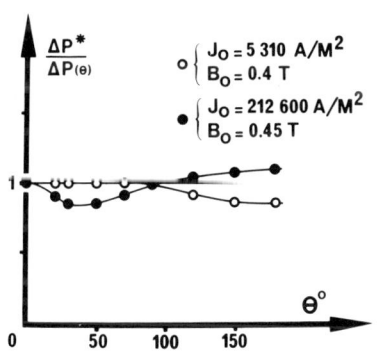

Fig. 4 Pressure distributions upon an insulating cylinder, the Δ axis of which is parallel to B_0.

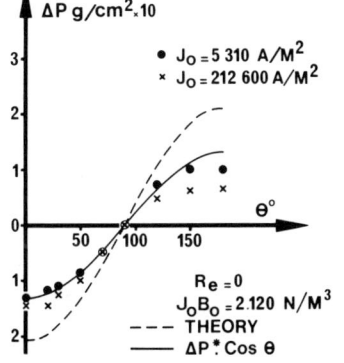

Fig. 5 Pressure distribution upon a conducting cylinder, the Δ axis of which is parallel to B_0.

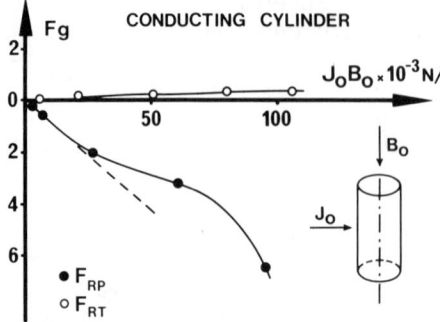

Fig. 6 Resultants of both the total and surface electromagnetic forces exerted upon a conducting cylinder, the Δ axis of which is parallel to B_0, as a function of the electromagnetic force density.

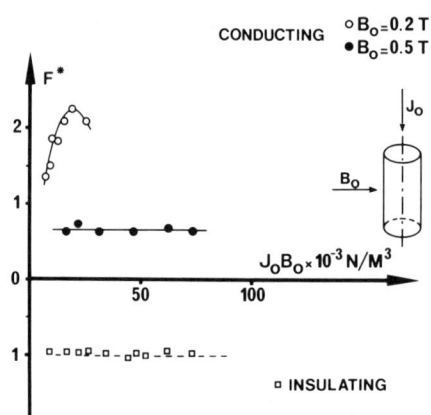

Fig. 7 F* as a function of the electromagnetic force density in the case of insulating and conducting cylinder, the Δ axis of which is parallel to B_0.

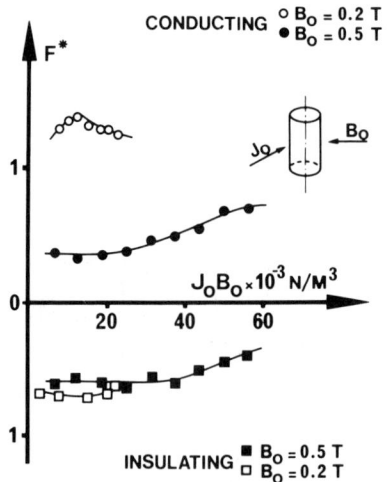

Fig. 8 F* as a function of the electromagnetic force density in the case of insulating and conducting cylinder, the Δ axis of which is parallel to B_0.

give rise to strong instabilities, plus it is easy to put into practice.

Δ Revolution Axis is Parallel to $\vec{J}_0 \times \vec{B}_0$. Figure 8 shows that, for an insulating cylinder and a given $J_0 B_0$ value, F_{RT} is independent of the particular values of J_0 and B_0. However, in the case of the conducting cylinder and given the product $J_0 B_0$, F_{RT} increases as J_0.

III. Experimental Studies in MHD Situations

Results obtained in magnetohydrostatic cases should be modified when particles and liquid metal are in relative motion. Therefore, for different situations, the variations of F_{RP} and of F_{RT} are studied in function of the Reynolds number $Re = UD/\nu$ for a particle of diameter D.

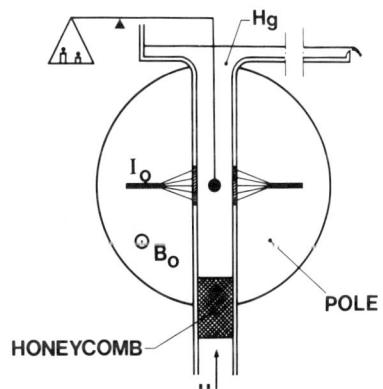

Fig. 9 Set up of one part of the study device for MHD situations.

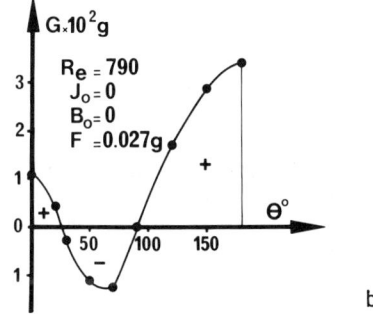
a)

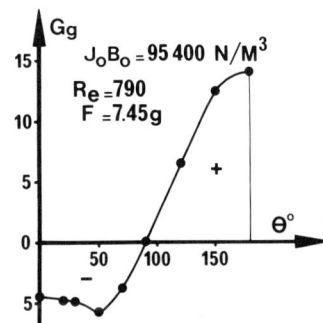
b)

Fig. 10 Pressure force distributions upon a conducting cylinder, the Δ axis of which is parallel to B_0: a) in absence and b) in presence of magnetic field.

Experimental Device

Particles are placed at the center of a vertical duct (Fig. 9) of cross section 4 x 4 cm which is included in a mercury loop containing a pump, an electromagnetic flowmeter, and a ball valve to govern the average velocity U. A honeycomb composed of 1 mm diam glass balls subdues the electromagnetic end effects.

For a given Reynolds number, the resultant of hydrostatic, hydrodynamic, and gravitational forces is equilibrated beforehand in the absence of an electromagnetic field. Afterward the effect of electromagnetic forces is measured.

Experimental Results

Provided that the Reynolds number of the particle is less or equal to 1500 or so, experimental results obtained in magnetohydrodynamic situations for F_{RT} are comparable to those obtained in the corresponding magnetohydrostatic cases. Experiments show, through the appearance of very strong instabilities, that separation may be effective only if the direction of $\vec{F}_{RT}$ and $\vec{U}$ are opposite or, put simply, when the particles move in counterflow.

Figure 10 shows the elementary pressure force distribution as a function of θ in the case of an electroconducting cylinder with the Δ axis parallel to $\vec{B}_0$ and the direc-

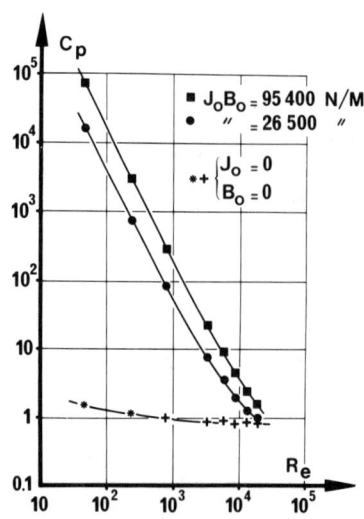

Fig. 11 Pressure drag coefficient of a conducting cylinder, the Δ axis of which is parallel to B_0 as a function of the Reynolds number for different values of the electromagnetic force density.

tion $\vec{J}_0 \times \vec{B}_0$ opposed to that of $\vec{U}$. In Fig. 10a, relating to $J_0B_0 = 0$, two zones of overpressure can be seen. When θ lies between $0 < \theta < 25$ deg and $90 < \theta < 180$ deg, separated by an underpressure zone, the pressure resultant force F_{RP} is 0.027 g. Figure 10b corresponds to the same case but the imposed electromagnetic force density is of 95,400 N/m³, the underpressure zone now extends between 0 and 90 deg, and the pressure resultant force is multiplied by approximately 300.

Figure 11 presents, still in the same situation, the variations of the pressure drag coefficient $Cp = 2F_{RP}/\rho DU^2$ as a function of the Reynolds number for different values of the product J_0B_0. The resultant electromagnetic force is multiplied by about 40,000 when Re = 50 and $J_0B_0 = 95,400$ N/m³. The relative effect of the electromagnetic field decreases when the Reynolds number increases and here the strong pressure forces are practically counterbalanced by the volume forces.

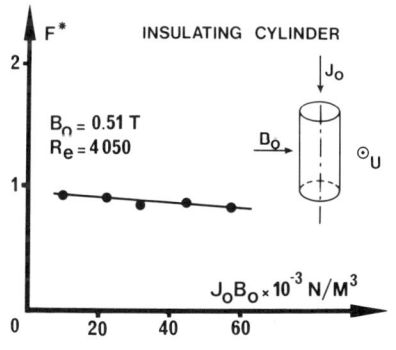

Fig. 12 F* as a function of the electromagnetic force density in the case of an insulating cylinder, the Δ axis of which is parallel to B_0.

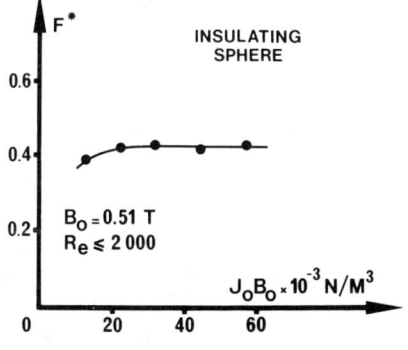

Fig. 13 F* as a function of the electromagnetic force density in the case of an insulating sphere.

Fig. 14 Copper electromagnetic separation from a lead bath.

For an insulating cylinder with Δ axis parallel to J_0 (Fig. 12), it can be observed that when $\vec{J}_0 \times \vec{B}_0$ is opposite to $\vec{U}$, F* is nearly one as in the magnetohydrodynamic case, even when Re = 4050.

In the case of an insulating sphere, when $J_0 \times B_0$ and U have the same direction (Fig. 13), it is found that F* = 0.4, when B_0 = 0.51 T and 0 < Re < 2000 (i.e., practically the same result as obtained in the analogous magnetohydrostatic situation). This finding is easily understood in light of the fact that the forces of hydrodynamic origin are negligible before the electromagnetic forces as long as the Reynolds number is not exceedingly elevated.

IV. Tests of Electromagnetic Decantation

Conclusions formed in this systematic study are confirmed by observations of ingots obtained after solidification of various molten metals (tin, lead, aluminum) into which artificial suspensions (copper, brass, silica) of diverse shapes had been introduced.

Two parallelopipedic crucibles of respective dimensions of 35 x 35 x 40 mm and 45 x 60 x 100 mm were fabricated from refractory materials (silica, cement) and provided electrodes of copper or stainless steel. Maximal values reached

Table 1 Results of study

	Δ // J_o	Δ // B_o	Δ // $J_o \times B_o$
Insulating cylinder	F* = 1	F* ≃ 0.1	F* ≃ 0.6
Conducting cylinder	0.6 < F* < 2.3	F* ≃ 0.1	0.4 < F* < 1.4
Insulating cylinder		0.45 < F* < 0.9	
Conducting cylinder		0.5 < F* < 0.7	

Fig. 15 Ingot of lead and copper particles including traces of iron, showing unsuccessful electromagnetic separation.

were of 400 A for the electrical current, 0.6 T for the magnetic field, and 88,000 N/m³ for the electromagnetic force density.

Notwithstanding a density difference of 2.5, copper and brass particles of diameters between 0.1 and 10 mm go to the bottom of an ingot of lead when $J_o \times B_o$ is directed vertically in a top to bottom direction (the ratios of electric conductivity are σ_{Cu}/σ_{Pb} = 57 and $\sigma_{brass}/\sigma_{Pb}$ = 7.5). Figure 14 shows a successful separation of copper particles from a lead bath.

Examination of the ingot in Fig. 15 shows that the orientation of the brass cylinders is the same as the orientation of the imposed magnetic fields. This is due to the presence of iron in extremely small concentrations contained within the particles. The Δ revolution axis is parallel to B_o, and, in agreement with theoretical and experimental results, the separation of solid and liquid phases is unsuccessful. Also of note is that the Δ revolution axis of the particle situated at the bottom of the ingot is parallel to $\vec{J}_0$ and this position sharply increases the apparent weight of the inclusion.

V. Conclusions

The results of this study are summarized in Table 1.

F* generally differs from one, and the analogy with Archimede's principle is legitimate only when the Δ revolution axis of an insulating cylinder is parallel to $\vec{J}_0$, because in this situation the inclusion does not significantly disturb the electromagnetic force field.

The F* values are generally valid for Reynolds numbers less than 1500 and decantation problems set before metallurgists are related to inclusions with characteristic dimensions less than 1 mm. Therefore, as long as the relative velocity of the particles does not exceed 20 cm/s, the results presented herein may be legitimately adopted.

To summarize, this first methodical study has shown that this process of electromagnetic decantation (which consumes little energy on account of the high electric conductivity of molten metals) could be successfully used in continuous casting with a vessel as large as 50 liters. This method could easily be applied to tin, lead, zinc, and aluminum as well as to their numerous alloys with fusion temperatures less that $800^\circ C$, beyond which temperature technological problems relating to the constitution of the necessary electrodes can arise.

Research is still in progress in our laboratory in an attempt to explain the causes of variations of parameter F* for a given product $J_0 B_0$ as a function of J_0 and B_0. Also we are working to better understand the appearance in certain cases of important instabilities prejudicial to the efficiency of this separation process and due particularly to electromagnetic end effects.

References

Andres U. (1976) <u>Magnetohydrodynamic and Magnetohydrostatic Methods of Mineral Separation</u>, Israel Universities Press, Jerusalem.

Branover H. (1978) <u>Magnetohydrodynamic Flow in Ducts</u>, John Wiley & Sons, New York and Israel Universities Press, Jerusalem, pp. 56-57.

Chambarel A., Ricou R., and Vives Ch. (1981) "Les méthodes de diagnostic en magnétodynamique des métaux fondus," J. Mech. Appl. (France), No. 4, p. 453.

Menchikov P., Anosov V., Pavlov V., and Guldin J. (1978) "Electromagnetic refining in aluminium production," Cvetn. Metally. SSSR, No. 1, pp. 35-37.

Ricou R. (1975) "Perturbations des paramètres electriques et mécaniques dues à l'introduction d'un porte-sonde, au sein d'un fluide electroconducteur, en présence d'un champ de force electromagnétique," Thèse d'Université, Univ. d'Aix-Marseille III, France, pp. 3-28.

Vives Ch., Bas J., and Ricou R. (1980) "Remarques sur les mécanismes de ségrégation electromagnétique des phases solide et liquide en métallurgie," C.R. Acad. Sci. (France), Série B, T 291, No. 6, pp. 165-167.

Pressure and Velocity Distribution around an Obstacle Immersed in Liquid Metal Subjected to Electromagnetic Forces

Ph. Marty* and A. Alemany†
Université de Grenoble, Grenoble, France
and
R. Ricou‡ and Ch. Vives§
Centre Universitaire d'Avignon, Avignon, France

Abstract

Studied is the perturbation due to the presence of an obstacle of any electrical conductivity, immersed in liquid metal subjected to a homogeneous magnetic field and an electric current which are perpendicular to each other. Only the case of an infinitely long cylinder is studied. From the theoretical point of view, calculations are made for three configurations: the cylinder is parallel to the electric current, to the magnetic field, and to the force field. In this last case, an induced flow around the obtacle appears. This flow is studied for any value of the Hartmann number M when the cylinder is insulating and only for small value of M in the other cases. In each case, the electromagnetic force acting on the cylinder is expressed as a function of the electrical conductivities of the fluid and of the obstacle. From the experimental point of view, force and velocity measurements are presented and compared with

---

Paper presented at Third Beer-Sheva International Seminar on Magnetohydrodynamic Flows and Turbulence, Ben-Gurion University of the Negev, Beer-Sheva, Israel, March 23-27, 1981. Copyright © American Institute of Aeronautics and Astronautics, Inc., 1982. All rights reserved.
*Ingénieur au C.N.R.S., Institut de Mécanique de Grenoble.
†Chargé de Recherche au C.N.R.S., Institut de Mécanique de Grenoble.
‡Assistant de l'Université, Faculté des Sciences.
§Professeur de l'Université, Faculté des Sciences.

the theoretical results. The proposed mechanism may be applied efficiently to the purification process of liquid metals during the casting.

I. Introduction

Elimination of impurities in the castings of alloys is a very difficult problem to solve in the metallurgical industry. Fragments of sand or oxide cannot be completely suppressed by classical procedures such as decantation, centrifugation, etc. (the decantation time often being too long).

We intend to solve this problem by an electromagnetic method (Andres 1976, Marty 1980). The principle of this method is as follows (Fig. 1): a liquid metal is subjected to the influence of a magnetic field $\vec{B}$ perpendicular to an electric current of density $\vec{J}$. If there are walls which are perpendicular to the electromagnetic forces

$$\vec{F} = \vec{J} \times \vec{B}$$

then, the pressure gradient distribution exactly balances those forces which are irrotational and there is no induced flow. This field of forces can be looked at as an electromagnetic gravity which will accelerate the decantation of impurities (Shilova 1975). However, when an obstacle is immersed in this liquid metal, the streamlines of the electric current are modified and the forces field distribution is able to become rotational (Fig. 2).

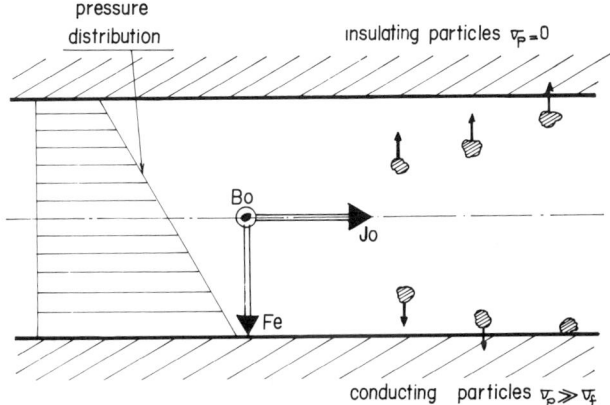

Fig. 1 Principle of the method.

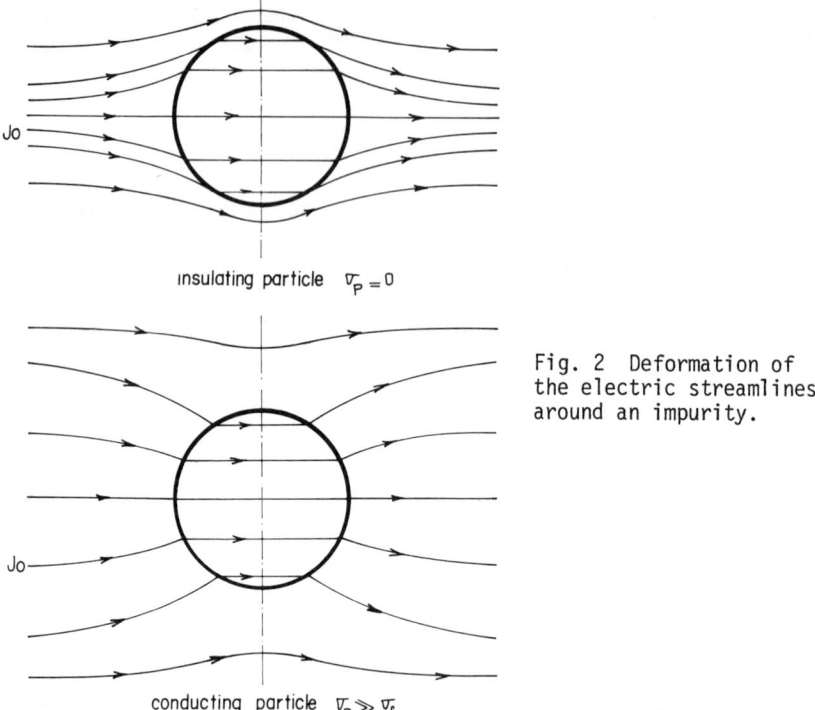

Fig. 2 Deformation of the electric streamlines around an impurity.

Our purpose is to calculate the modification of the local pressure distribution in the liquid and the possible flow induced by the perturbation of the field of forces.

II. Theoretical Model

Modelization

To take into account the different shapes of the obstacles, we suppose them to have either no preferred direction or to be elongated. In the first case, we consider the particles to be spherical and in the second case to be an infinitely long cylinder. For the cylindrical shape, three configurations have been considered: case A in which the cylinder axis is in the direction of the electric current, case B in which the cylinder axis is in the direction of the magnetic field, and case C in which the cylinder axis is perpendicular both to the current and the magnetic field (see Fig. 3).

In this study, we shall consider only the cases of obstacles with cylindrical shapes. More complete results in this case will be found in (Marty and Alemany 1983). Another article will study spherical shapes.

Equations

Let us suppose an unbounded and electrically conducting fluid, with a conductivity σ_f, a kinematic viscosity ν, and a magnetic permeability μ_0, is submitted to an electric current of uniform density $\vec{J}_0$, and an applied magnetic field $\vec{B}_0$ perpendicular to J_0.

In most cases, the typical size d of the particles is very small. Therefore, the Reynolds number

$$Re = V \cdot d/\nu \tag{1}$$

is low enough to neglect the importance of the inertial terms in the Navier-Stokes equations (V is the relative velocity between the particle and the fluid).

The concentration of the particles is assumed to be small enough to consider the obstacle alone in the fluid. So, far away from it, the field of electromagnetic forces ($\vec{F}_e = \vec{J}_0 \times \vec{B}_0$) is uniform.

Let us use a reference frame which is fixed with respect to the cylinder and let us search for a steady-state solution. If we call z the axis of this infinitely long cylinder ($\partial/\partial z = 0$), the equations are

$$\text{curl } \vec{E} = 0 \tag{2}$$

$$\text{div } \vec{B} = 0 \tag{3}$$

$$\text{curl } \vec{B} = \mu_0 \cdot \vec{J} \tag{4}$$

$$\vec{J} = \sigma_f(\vec{E} + \vec{V} \times \vec{B}) \tag{5}$$

$$\text{div } \vec{J} = 0 \tag{6}$$

$$-\text{grad } p + \vec{J} \times \vec{B} + \rho\nu\nabla^2\vec{V} = 0 \tag{7}$$

where $\vec{E}$, $\vec{B}$, $\vec{J}$, $\vec{V}$, and p are respectively the electric field, magnetic field, electric density, velocity, and pressure. ρ is the density of the liquid metal.

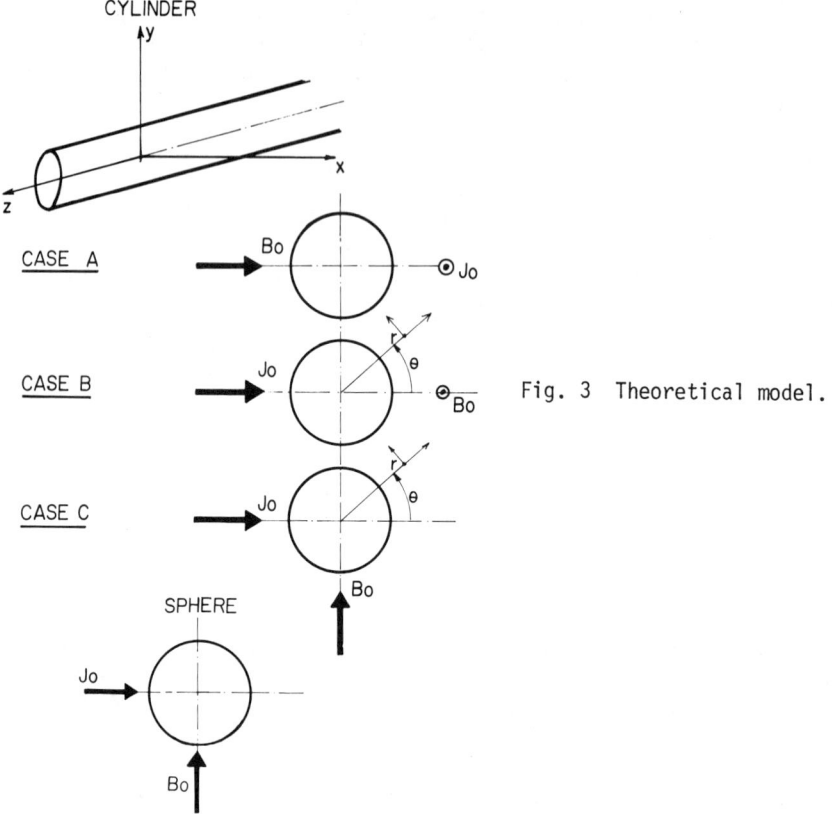

Fig. 3 Theoretical model.

III. Solutions

Axis of the Cylinder is in the Direction of the Electric Current (Case A)

This is the most simple case because J_0 stays unchanged outside the cylinder. The current density $\vec{J}_0'$ inside the cylinder is given by Eq. (2)

$$\vec{J}_0' = (\sigma_p/\sigma_f) \cdot \vec{J}_0 \qquad (8)$$

where σ_p is the electrical conductivity of the particle.

Then the resultant force $\vec{F}_R$ per unit of volume, which is equal to the sum of a pressure term $\vec{F}_S$ and a volume term $\vec{F}_V$, is

$$\vec{F}_R = \vec{F}_S + \vec{F}_V = [1-(\sigma_p/\sigma_f)]J_0 B_0 \qquad (9)$$

In this case, there is no induced flow because the field of forces is not rotational.

Axis of the Cylinder is in the Direction of the Magnetic Field (Case B)

First, we calculate in cylindrical coordinates (r,θ), the distribution of the electric current $\vec{J}_f$ in the fluid, $\vec{J}_p$ in the particle, and the magnetic field around the obstacle with the supposition that there is no induced flow. Then Eqs. (2-5) give

$$\text{div } \vec{J} = 0$$

$$\text{curl } \vec{J} = 0 \tag{10}$$

The solution which satisfies the boundary conditions at infinity

$$J_{fr} = J_0 \cdot \cos\theta$$

$$J_{f\theta} = -J_0 \cdot \sin\theta \tag{11}$$

and at the surface of the cylinder $(r = a)$

$$J_{fr} = J_{pr}$$

$$J_{f\theta}/J_{p\theta} = \sigma_f/\sigma_p \tag{12}$$

is in the fluid,

$$J_{fr} = J_0 (1 - \frac{\sigma_f - \sigma_p}{\sigma_f + \sigma_p} \cdot \frac{a^2}{r^2}) \cdot \cos\theta$$

$$J_{f\theta} = -J_0 (1 + \frac{\sigma_f - \sigma_p}{\sigma_f + \sigma_p} \cdot \frac{a^2}{r^2}) \cdot \sin\theta \tag{13}$$

In the cylinder, the uniform current density is

$$J_{pr} = \frac{2\sigma_p}{\sigma_p + \sigma_f} \cdot J_0 \cdot \cos\theta$$

$$J_{p\theta} = -\frac{2\sigma_p}{\sigma_p + \sigma_f} \cdot J_0 \cdot \sin\theta \tag{14}$$

Then, the field of forces is irrotational

$$\text{curl } (\vec{J} \times \vec{B}) = 0 \qquad (15)$$

and can be balanced by a pressure gradient.

As was supposed, there is no induced flow around the cylinder. The resulting force $\vec{F}_R$ is equal to zero because the surface force

$$F_S = -\frac{2\sigma_p}{\sigma_p + \sigma_f} \cdot J_o B_o \qquad (16)$$

exactly balances the volume force.

This result can be extended to the case of an infinitely long shape of any uniform cross section. The electromagnetic force can generally be written

$$\vec{J} \times \vec{B} = \frac{1}{\mu_o} (\vec{B} \cdot \vec{\nabla}) \vec{B} - \text{grad } \frac{B^2}{2\mu_o} \qquad (17)$$

with the first term equal to zero because of the hypothesis that $\partial/\partial z = 0$. The resultant force $\vec{F}_R$, which can be written

$$\vec{F}_R = \iint_S \frac{B^2}{2\mu_o} \vec{n} \cdot d\vec{S} - \iiint_V \text{grad } \frac{B^2}{2\mu_o} \, dv \qquad (18)$$

(where $\vec{n}$ is the unit surface vector), vanishes because of the gradient theorem (Durand 1966), see Fig. 4.

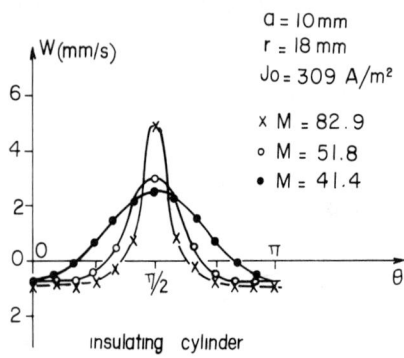

Fig. 4 Gradient theorem.

Axis of the Cylinder is Perpendicular both to the Current and the Magnetic Field (Case C)

By comparison with the previous case, the direction of the external magnetic field has changed. Using the expression of $\vec{J}$ in Eq. (13) in the fluid, we find

$$\text{curl }(\vec{J} \times \vec{B}) \neq 0 \qquad (19)$$

The electromagnetic forces cannot be balanced by the pressure gradient, so it then appears an induced force around the obstacle. These forces are directed on the z axis. So, we suppose the velocity W to be directed on the same axis. Then, this flow induces an extra distribution of current $\vec{J}'$ in the x-y plane, compatible with the hypothesis used for the velocity direction.

In the case of insulating particles (the magnetic field vanishes at the wall), Ricou (1975) has found an exact solution for the velocity distribution and the magnetic field

$$W^* = -\frac{1}{2M}[I_{-1}(M/2)+I_1(M/2)]\left[\frac{K_0(Mr^*/2)}{K_0(M/2)}\right]\cdot\cosh\left(\frac{Mr^*\sin\theta}{2}\right)$$

$$+ \sum_{p=1}^{\infty}\frac{(-1)^{p+1}}{M}[I_{2p-1}(M/2)+I_{2p+1}(M/2)]\left[\frac{K_{2p}(Mr^*/2)}{K_{2p}(M/2)}\right]\cos 2p\theta$$

$$\cdot\cosh\left(\frac{Mr^*\sin\theta}{2}\right) - \sum_{p=0}^{\infty}\frac{(-1)^{p+1}}{M}[I_{2p}(M/2)+I_{2p+2}(M/2)]$$

$$\cdot\left[\frac{K_{2p+1}(Mr^*/2)}{K_{2p+1}(M/2)}\right]\sin(2p+1)\theta\cdot\sinh\left(\frac{Mr^*\sin\theta}{2}\right) \qquad (20)$$

$$B^* = \frac{1}{2M}[I_{-1}(M/2)+I_1(M/2)]\left[\frac{K_0(Mr^*/2)}{K_0(M/2)}\right]\cdot\sinh\left(\frac{Mr^*\sin\theta}{2}\right)$$

$$- \sum_{p=1}^{\infty}\frac{(-1)^{p+1}}{M}[I_{2p-1}(M/2)+I_{2p+1}(M/2)]\left[\frac{K_{2p}(Mr^*/2)}{K_{2p}(M/2)}\right]\cos 2p\theta$$

$$\cdot\sinh\left(\frac{Mr^*\sin\theta}{2}\right) + \sum_{p=0}^{\infty}\frac{(-1)^{p+1}}{M}[I_{2p}(M/2)+I_{2p+2}(M/2)]$$

$$\cdot\left[\frac{K_{2p+1}(Mr^*/2)}{K_{2p+1}(M/2)}\right]\sin(2p+1)\theta\cdot\cosh\left(\frac{Mr^*\sin\theta}{2}\right) + \frac{r^*\sin\theta}{M} \qquad (21)$$

where

$$W^* = \frac{W}{J_0 B_0 a^2/\rho\nu} \quad , \quad B^* = \frac{B_z}{Ma\mu_0 J_0} \quad , \quad r^* = r/a$$

and M is the Hartmann number, $B_0 \cdot a \cdot \sqrt{\sigma_f/\rho\nu}$.

I and K are the modified Bessel functions of the first and second order, respectively. The flow is arranged in cells, symmetrical with respect to the axis, $\theta = \pi/2$ (Fig. 5).

In the case of particles of any conductivity, we have found a solution only in the case of small Hartmann numbers. Then, the magnetic field is not modified by the induced current due to the velocity. If we assume the rotational part of the electromagnetic forces to be balanced by the viscosity

$$\rho \cdot \nu \cdot \nabla^2 W \sim J_0 B_0 \Rightarrow \rho \cdot \nu \cdot (W/a^2) \sim J_0 B_0 \tag{22}$$

and because of Ohm's law

$$J' \sim \sigma_f \cdot W \cdot B_0 \tag{23}$$

Then,

$$J'/J_0 \sim M^2 \tag{24}$$

In the case of small cylinders, M is very much less than unity and the distribution of electric current will be the same as Eqs. (13) and (14). The forces can be expressed independently of the velocity field. With this hypothesis, the Navier-Stokes equations become in cylindrical coordinates

$$\nabla^2 W^* = \frac{\sigma_f - \sigma_p}{\sigma_f + \sigma_p} \cdot \frac{\cos 2\theta}{r^{*2}} \tag{25}$$

In these equations, the irrotational part of the electromagnetic forces has been included in the pressure term

$$p = p_0 + J_0 B_0 z - (B_z^2/2\mu_0) \tag{26}$$

where p_0 is the pressure value without any electromagnetic forces.

The solution which satisfies the boundary condition at the wall ($r^* = 1$)

$$W^* = 0 \tag{27}$$

and which has a finite value far from the cylinder is

$$W^* = \frac{\sigma_f - \sigma_p}{\sigma_f + \sigma_p} 1/4(\frac{1}{r^{*2}} - 1) \cos 2\theta \qquad (28)$$

The resultant force per unit of volume can be easily expressed

$$F_R = F_S + F_V = -J_o B_o + \frac{2\sigma_p}{\sigma_p + \sigma_f} J_o B_o = \frac{\sigma_p - \sigma_f}{\sigma_p + \sigma_f} J_o B_o \qquad (29)$$

The case of great values of Hartmann numbers is more difficult because the field of forces depends on the velocity distribution. The resolution of this problem has been nearly achieved.

IV. Experimental Results

The apparatus we used to measure the acting force on different cylinders and the field of velocity around it is shown in Fig. 6.

It is a parallelepiped, filled with mercury and placed in the gap of a magnet. Two electrodes supply homogeneous electric current (dc) perpendicular to the direction of the

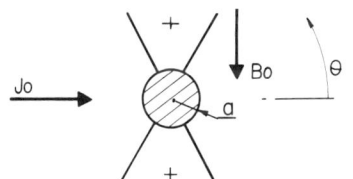

Fig. 5 Flow around an insulating cylinder (experimental results).

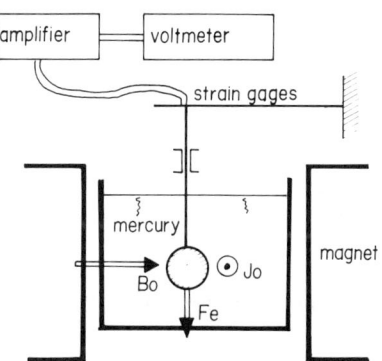

Fig. 6 Experimental apparatus.

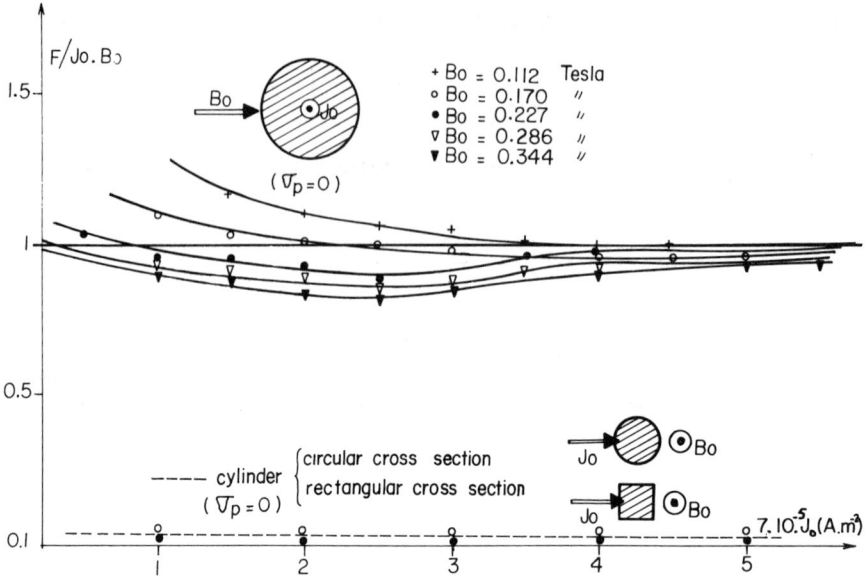

Fig. 7 Force acting on cylinders of different orientation (experimental results).

magnetic field. The cylinders, immersed in the mercury, are hung by a thin vertical rod to a flexion bar. The deformation of the bar, due to the force F_R, is recorded by strain gages. The results are shown in Fig. 7.

The velocity profiles were measured with special probes (Chambarel et al. 1979) (Fig. 5) developed in the MHD laboratory in Avignon. The only results presently achieved concern the cases A and B. They satisfy the theoretical results of Eqs. (9) and (16). In the case of conducting cylinders, the contact resistance at the surface modifies the apparent conductivity σ_p. The surface force F_S remains unchanged but the volume force F_V is less than its predicted value.

V. Conclusion

This first work is about cylindrical obstacles. The theoretical results concern only the three main configurations:

Case A: the cylinder axis is in the same direction as the electric current.

Case B: the cylinder axis is in the same direction as the magnetic field.

Case C: the cylinder axis is perpendicular both to the current and the magnetic field.

The deformation of the electric streamlines is a very important point and can completely change the intensity of the resultant decantation force. This point is especially dramatic in the case B where the total force vanishes.

The experimental results agree with the theory. Nevertheless, in the case of conducting particles (which have not been presented in this paper), the contact resistance with the mercury modifies the apparent conductivity σ_p.

Our future work will concern the case of spherical impurities, which is the most important in the industrial context. A final objective is the industrial application of this process to a real casting of liquid metal.

References

Andres U. (1976) <u>Magnetohydrodynamic and Magnetohydrostatic Methods of Mineral Separation</u>, Israel Universities Press, Jerusalem.

Chambarel A., Ricou R., and Vives Ch. (1979) "Procédés et dispositifs pour mesurer localement des vitesses instantanées d'un fluide électroconducteur," <u>Brevet Anvar</u>, No. 79, 19818, Juillet.

Durand E. (1966) <u>Electrostatique,</u> Masson (Ed.), 120, Bd. St. Germain, Paris, Vol. 1, p. 11.

Marty Ph. (1980) "Séparation électromagnétique continue," <u>D.E.A. of Fluid Mech.</u>, Institut de Mécanique de Grenoble, France.

Marty Ph. and Alemany A. (1983) "Ecoulement dû à des champs magnétique et électrique croisés autour d'un cylindre de conductivité quelconque," <u>J. de Méc.</u> (in press).

Ricou R. (1975) "Perturbations des paramètres electriques et mécaniques dues à l'introduction d'un porte-sonde, au sein d'un fluide electroconducteur, en présence d'un champ de force electromagnétique," Thèse d'Université, Univ. d'Aix-Marseille III.

Shilova E.I. (1975) "Removal of nonconducting impurities from liquid metals in the self magnetic field of an electric current," <u>Magni. Gidrodin.</u>, No. 2, April-June, pp. 142-144.

Liquid-Metal Columns Confined by External Parallel Conductors and Surface Tension Part I: Two-Dimensional Theory

J.A. Shercliff*
University of Cambridge, Cambridge, England

Abstract

High-frequency magnetic fields, produced by vertical conductors near the falling liquid column, exert a magnetic pressure at the surface. We ignore stirring of the fluid by the forces in the skin and also vertical acceleration under gravity. It is possible to treat the problem two-dimensionally by complex variable methods combined with numerical integrations. The method is a generalization of the Schwarz-Christoffel transformation to doubly infinite products, combined with an integral equation expressing the surface tension boundary condition and solved by iteration. The cases considered comprise: 1) the far field is uniform, squashing the column away from its naturally circular shape under surface tension; 2) the far field is of quadrupole form, tending to generate sharp corners and concave flanks on the column; and 3) the external field is due to four line conductors with currents in alternate directions. In case 3, the problem of zero surface tension can be solved analytically by the hodograph method. For strong enough surface tension, the concavity of the flanks is so great that protuberances develop at the corners until they are "pinched off" by the concave flanks coming together.

Paper presented at Third Beer-Sheva International Seminar on Magnetohydrodynamic Flows and Turbulence, Ben-Gurion University of the Negev, Beer-Sheva, Israel, March 23-27, 1981. Copyright © American Institute of Aeronautics and Astronautics, Inc., 1982. All rights reserved.
*Professor of Applied Thermodynamics.

I. Introduction

Continuous casting is becoming standard practice in industrial metallurgy. Some subsequent processes could perhaps be saved and the life of molds extended if magnetic forces were exploited so as to confine and shape the vertically descending molten liquid column as its outer layers begin to solidify. The necessary magnetic pressure can be generated by using external horizontal magnetic fields, alternating at around 100 kHz so as to be excluded from the metal by skin effect. In this treatment vertical variations due to gravity are ignored so as to allow a two-dimensional analysis of the competition between magnetic pressure and surface tension (tending to make the column circular). Any stirring of the liquid by the $\bar{j} \times \bar{B}$ forces in the skin is also ignored. The problem becomes quasimagnetohydrostatic, i.e., virtually the equilibrium of a perfect conductor under a steady field equal to the rms value of the actual field.†

The two-dimensional results which relate the fluid pressure p to the surface tension τ, the external magnetic field, and the size of the column cross section can be approximately applied to the true three-dimensional problem where, because of gravity g, the cross section is liable to change and the associated changes in vertical velocity can be related to pressure by means of Bernoulli's equation, including the gravity term. If the velocities are low enough, the vertical distribution of pressure is simply related to height by hydrostatics. The external magnetic field can be varied with height to give a further degree of freedom in shaping the column. For simplicity, all of the cases considered here have high symmetry, but this is not necessary in practice, e.g., if more complicated shapes are to be continuously cast.

II. Mathematical Method

To illustrate the mathematical method consider the simple two-dimensional case where the far-field B_∞ is uniform. In Fig. 1a the cross section is shown cross-hatched. The line ψ = const is a typical field line, for the magnetic field is described by a complex potential $\Omega = \phi + i\psi$, an analytic function of $z = x + iy$. If we nondimensionalize

---
†This paper gives an abbreviated account of the work, which is more fully described in Shercliff (1981), although some new information is presented here.

in terms of a length ℓ and typical field B_s, the condition for equilibrium between true and magnetic pressure and surface tension becomes

$$\frac{d\theta}{d\Omega} = K\left(\frac{1}{r} - ar\right) \qquad (1)$$

where θ = surface slope (see Fig. 1), r = dimensionless field strength, $K = p\ell/\tau$, and $a = B_s^2/2\mu_0 p$.

The conformal transformation of the upper half of the z plane (Fig. 1a) outside the liquid into the upper half of the Ω plane (Fig. 1b) may be expressed as

$$\log \frac{d\Omega}{dz} = \frac{1}{2}\log(1-\Omega^2) - \frac{1}{\pi}\int_{-1}^{1} \log(S-\Omega)\frac{d\theta}{dS}dS + C_r - \frac{i\pi}{2} \qquad (2)$$

if $\Omega = 0$ at C and $\Omega = 1$ at D, C_r being a constant equal to $\log(B_\infty/B_s)$ and θ a function of S (or Ω) in the range $|S| \le 1$. On the surface BCD, with Ω real, Eq. (2) becomes

$$\log r = \frac{1}{2}\log(1-\Omega^2) - \frac{1}{\pi}\int_{-1}^{1} \log|S-\Omega|\frac{d\theta}{dS}dS$$
$$+ \frac{2}{\pi}\int_0^1 \log S \frac{d\theta}{dS}dS \qquad (3)$$

if we take B_s as the field at C, so that $r = 1$ there. The problem is solved iteratively from an initial distribution of $d\theta/dS$ or $d\theta/d\Omega$ against S or Ω, which yields an r distribution from Eq. (3) and thence a net set of values of $d\theta/d\Omega$ from Eq. (1), K being updated so that θ always rises by $\pi/2$

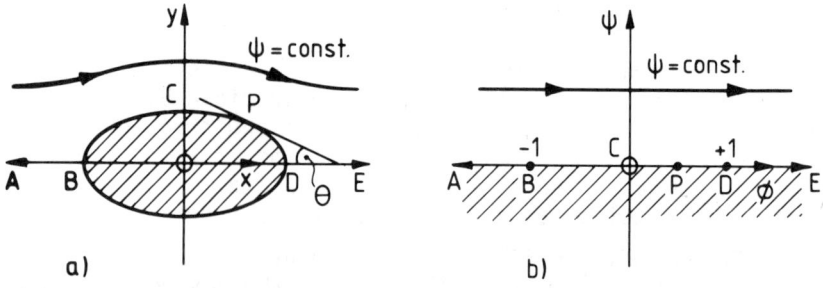

Fig. 1 Liquid column in uniform far field: a) z plane; b) Ω plane.

between C and D. To start the calculation, the known analytical solution for a = 0 (a circular column) is used and, subsequently, a is stepped to yield other cases. Actually, new independent variables ω, s are used (where Ω = sin ω, S = sin s) to give smoother plots and better conditioned numerics. To avoid undefined integrands, integrals such as the last in Eq. (3) are rearranged as

$$\int_0^1 \log S \frac{d\theta}{dS} dS = \int_0^{\pi/2} \log S \left[\frac{d\theta}{ds} - \left(\frac{d\theta}{ds}\right)_o\right] ds - \left(\frac{d\theta}{ds}\right)_o \frac{\pi}{2} \log 2$$

Finally the shape of the column is found from integrating

$$\frac{dz}{d\Omega} = \frac{e^{-i\theta}}{r}$$

As a rises from 0 to 1 the column is squashed into an increasingly slender oval, long in the field direction. However, it should be noted that these cases would be unstable in three-dimensions in the "sausage-mode," just as in the nonmagnetic case a = 0.

In the subsequent cases discussed, the method is very similar except that to allow a to be stepped monotonically across all the cases, we take K = const and allow r at C to float.

The next case solved is that in which the external field is of quadrupole form and $\Omega = iQz^2$ for large $|z|$. Such a field can be produced by having two or more probably four vertical conductors, symmetrically placed far from the liquid column. This is the limit of Fig. 4 as OQ becomes large, cf., OC. When surface tension is weak (a → ∞) the column approaches the four-cusped hypocycloid ($x^{2/3} + y^{2/3}$ = const to suitable axes) as shown by Berkowitz et al. (1981). Figure 2 shows the shapes which result as a rises from low values (magnetic pressure negligible so that from the circular solution pOC/τ = 1) to high values (surface tension negligible so that from the hypocycloid solution $Q^2 OC^2/\mu_0 p$ = 2/9 and OD = 2 OC). Note that putting τ = 0 gives good estimates of OC up to quite high values of τ, whereas OD is greatly in error because the surface tension strongly truncates the cusps, while not affecting the solution elsewhere very much. The ratio OD/OC measures the departure from circularity. Figure 3, which is not given in Shercliff (1981), plots the results of the calculation in the

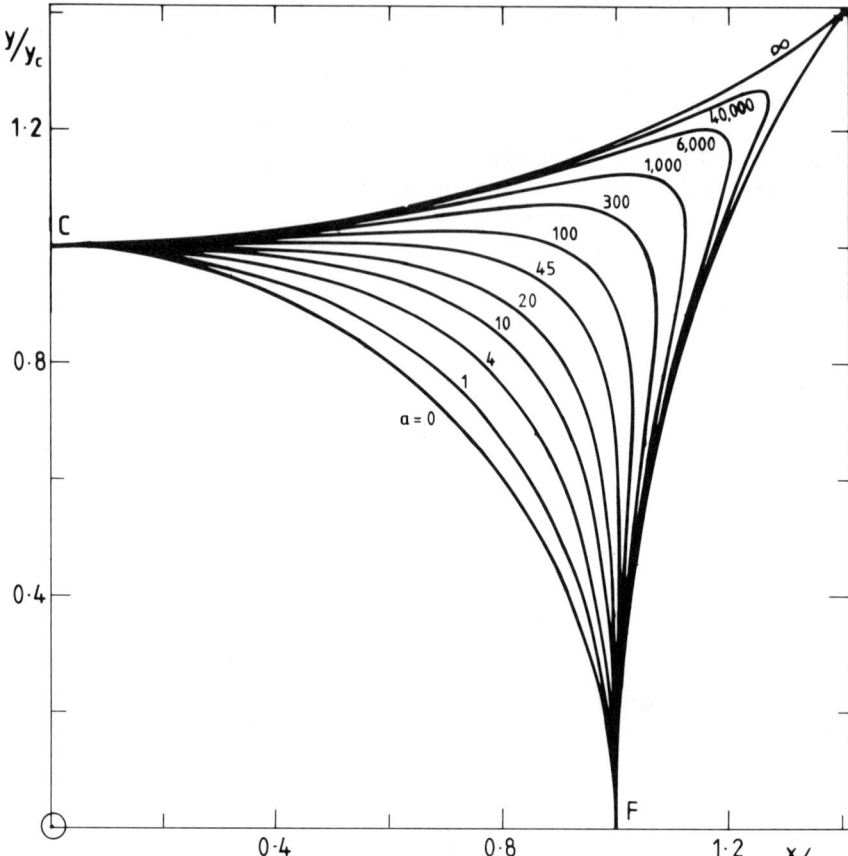

Fig. 2 Liquid column in quadrupole field (first quadrant only).

most useful form. The natural length scale, for given values of the far field (measured by Q) and the surface tension τ, is the quantity $L = (Q^2/\mu_0\tau)^{-1/3}$. The abcissa in Fig. 3 is radial (horizontal) distance to either C or D, nondimensionalized by dividing by L. Meanwhile the pressure is nondimensionalized by dividing by $(Q^2\tau^2/\mu_0)^{1/3}$ and is plotted downward as ordinate in view of the fact that under hydrostatic conditions p increases with depth. Figure 3 also shows how the curve for OC fairs smoothly into the approximate curves (shown broken) corresponding to the cases where either B or τ is neglected. The minimum value of dimensionless pressure is 2.95.

In the hydrostatic case $p = \rho gz$, if z is depth measured below a suitable datum, and then the ordinate in Fig. 3

LIQUID-METAL COLUMNS: TWO-DIMENSIONAL THEORY   419

is equal to $\alpha(z/L)$, where α is the dimensionless number $\rho g(\mu_0^2/\tau Q^4)^{1/3}$ which measures the vertical/horizontal scale distortion if Fig. 3 is regarded as a geometric diagram showing the actual side views of a liquid column confined by a quadrupole field which does not vary with height, on the assumption that the two-dimensional treatment suffices. If we take ρg = 50,000 N/m^3, τ = 1 N/m, and Q = 1 (for which the magnetic field is 0.2 T when $|z|$ = 0.1 m), then $\alpha \simeq 5$ and Fig. 3 is undistorted. (Note that the horizontal and vertical numerical scales differ by a factor of 5.) Obviously, the two-dimensional treatment is very questionable near the top.

Equilibrium is impossible above the point marked T. If the mold delivers a nearly circular column with a pressure corresponding to T there is a bifurcation into two alternative branches. The left-hand τ-dominated one is obviously unstable. (A local increase in radius would be catastrophic as the surface tension's ability to confine a given pressure falls off with increased radius.) The increasingly cusp-shaped right-hand branch is probably stable however (on the analogy with similar fusion problems).

Alternately the mold could deliver a column which descended from any level on the curves to the right of T.

Finally we turn to the slightly more realistic case where a quadrupole-like field is produced by four slender vertical conductors arranged symmetrically in a square around the liquid column, with equal currents alternately in opposite directions (see Fig. 4).

In the case where surface tension is negligible the corners such as D become cusps and an analytical solution using the complex hodograph method becomes possible (for details, see Shercliff 1981). These results fair smoothly into the sequence of results found numerically when $\tau \neq 0$ by an extension of the iterative method described earlier. These cases form a two-parameter family because now there is the additional length scale OQ, the field strength being characterized by the current I in each conductor. The cases fall into two classes:

1) Those where, as $\tau \to 0$, four cusps form at points such as D.

2) Those where, as τ falls, the four protuberances such as D extend progressively outward and the sequence is termi-

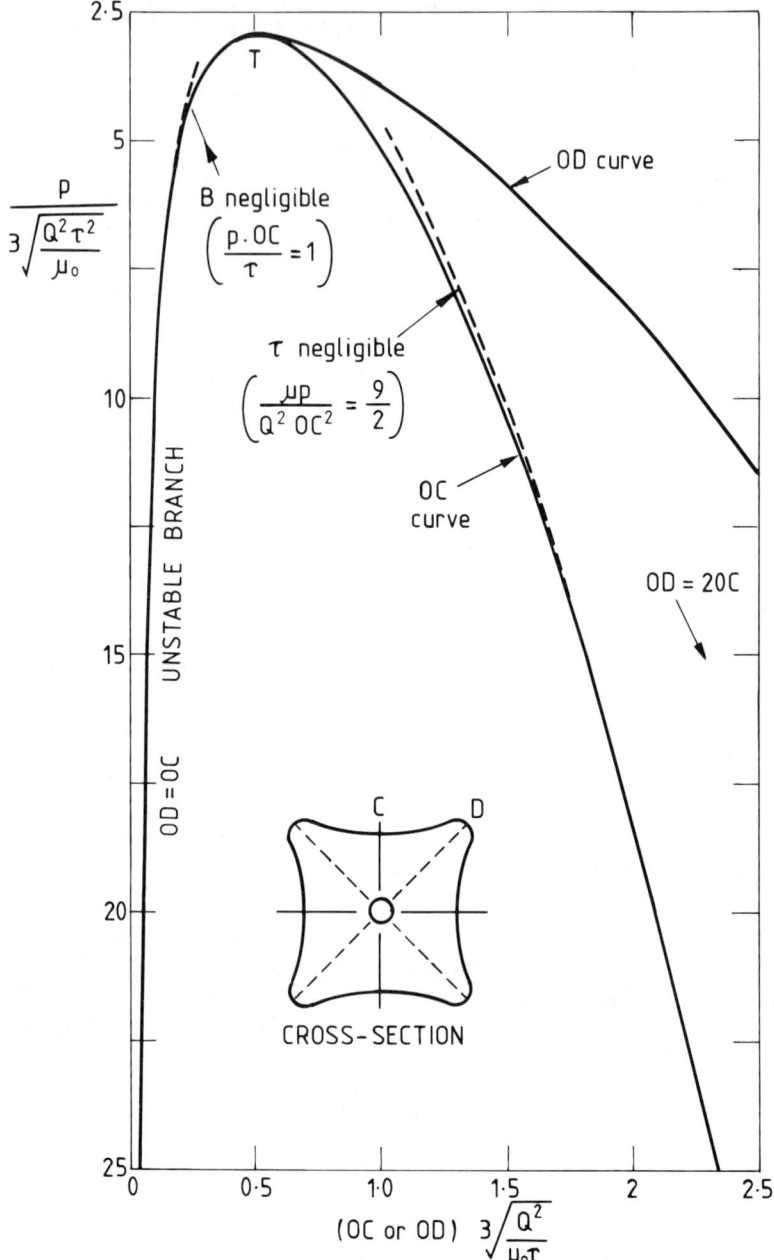

Fig. 3 Liquid column in quadrupole field: dimensionless plot of pressure (or depth in the hydrostatic case) against radial lengths OC and OD.

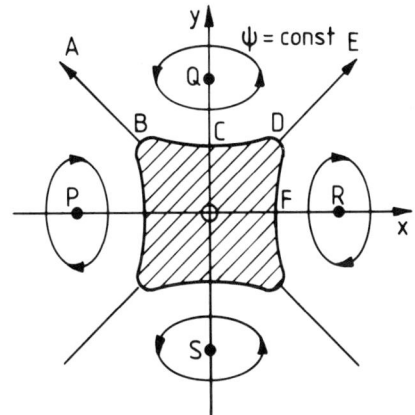

Fig. 4 Liquid column confined by four conductors (P, Q, R, and S).

nated when the two concave flanks of each protuberance come together somewhere between the two nearest conductors and the protuberance "necks" off and the main column sheds liquid. How fully such behavior could be realized in an experiment or in practice remains to be seen.

In addition to the obvious shortcomings of the two-dimensional model used here in the face of the vertical variations that are inevitable, due to gravity, another idealization is that the external conductors PQRS are treated as line currents, whereas in any real apparatus conductors of large cross section would be used to carry the high currents and coolant flow.

Acknowledgment

The author is grateful to R. Moreau and his co-workers in Grenoble for introducing him to the magnetic shaping problem.

References

Berkowitz J., Friedrichs K.O., Goertzel H., Grad H., Killeen J., and Rubin E. (1953) "Cusped geometries," Proc. 2nd Int. Conf. on Peaceful Uses of Atomic Energy, Vol. 31, United Nations, Geneva, p. 171.

Shercliff J.A. (1981) "Magnetic shaping of molten metal columns," Proc. Roy. Soc. London, Ser. A, Vol. 375, pp. 455-473.

Liquid-Metal Columns Confined by External Parallel Conductors and Surface Tension Part II: Experiment

J. Etay* and M. Garnier†
Université de Grenoble, Grenoble, France

I. Introduction

Experimental simulations of magnetic shaping devices are presented in this paper. The experiments are done with mercury columns exposed to electromagnetic forces induced by high-frequency (f ≥ 100 kHz) magnetic fields. The skin effect excludes the magnetic field from the liquid-metal interior and we analyze the competition between the magnetic pressure and the surface tension which tends to give the column a circular shape. The experimental conditions are then very close to those considered in Shercliff's theoretical analysis (1981) which assumes that the frequency is so high that the current-bearing skin depth is zero and that the stirring tendency due to the curl of magnetic forces is negligible. We particularly consider the case which allows the best comparison with the theory, where the external magnetic field is a quadrupole-like field on the liquid-metal column axis. The theoretically predicted equilibrium-free surface imposed by the magnetic field is a four-cusp hypocycloid-like shape with blunted cusps because of surface tension. It will be demonstrated that such an inductor is able to guide a liquid-metal flow against gravity with no contact between the liquid metal and a wall.

Paper presented at Third Beer-Sheva International Seminar on Magnetohydrodynamic Flows and Turbulence, Ben-Gurion University of the Negev, Beer-Sheva, Israel, March 23-27, 1981. Copyright © American Institute of Aeronautics and Astronautics, Inc., 1982. All rights reserved.
  *Attachée de recherche, Institut de Mécanique.
  †Chargé de recherche, Institut de Mécanique.

II. The Basic Mechanisms

The Magnetic Shaping

Let us consider an initially circular vertical column of molten metal centered on the axis of the four parallel conductors generating a quadrupole field. The inductor is provided with high-frequency currents in such a way that the skin depth δ is very small compared to the radius R of the liquid-metal column (Fig. 1). In the skin depth the induced currents $\bar{j}$ (which react against the flux variations experi-

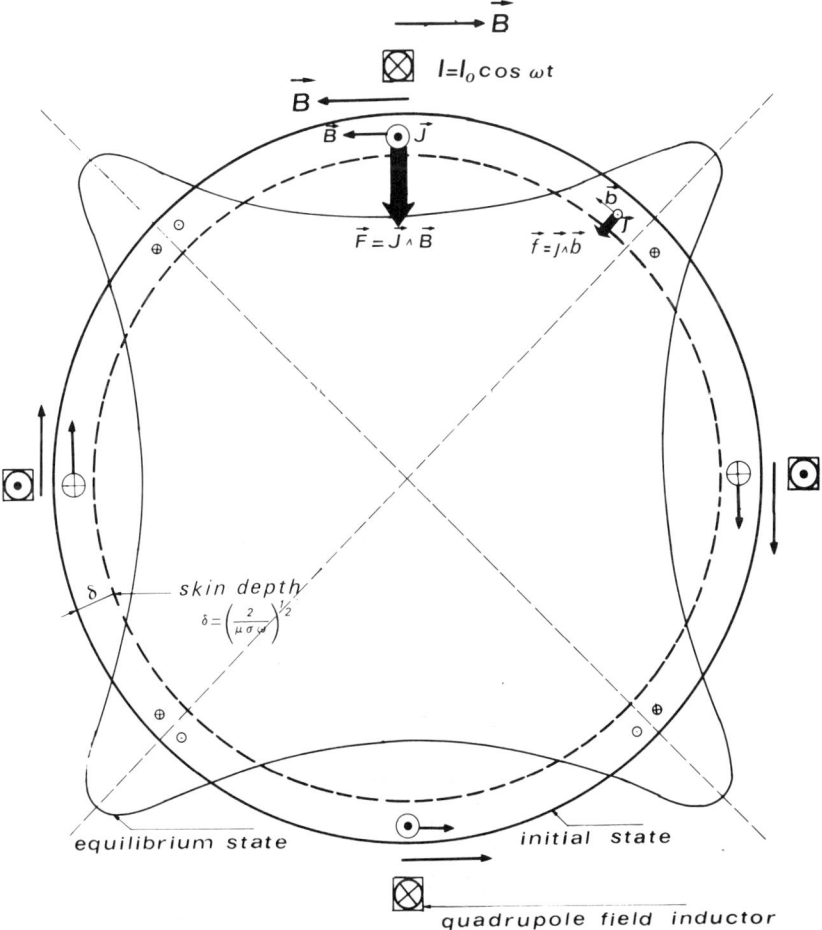

Fig. 1 Effect of a quadrupole field on an initially circular liquid-metal column.

enced by the metal) are opposed to the inducting currents. They interact with the magnetic field $\overline{B}$ generated by the four parallel conductors, giving rise to a Lorentz force $\overline{j} \times \overline{B}$ always directed toward the axis of the liquid column. The intensity of the electromagnetic force is nonuniform around the initially circular column. In particular, because of the symmetry, the magnetic field and the induced currents are very weak in the neighborhood of the generating lines

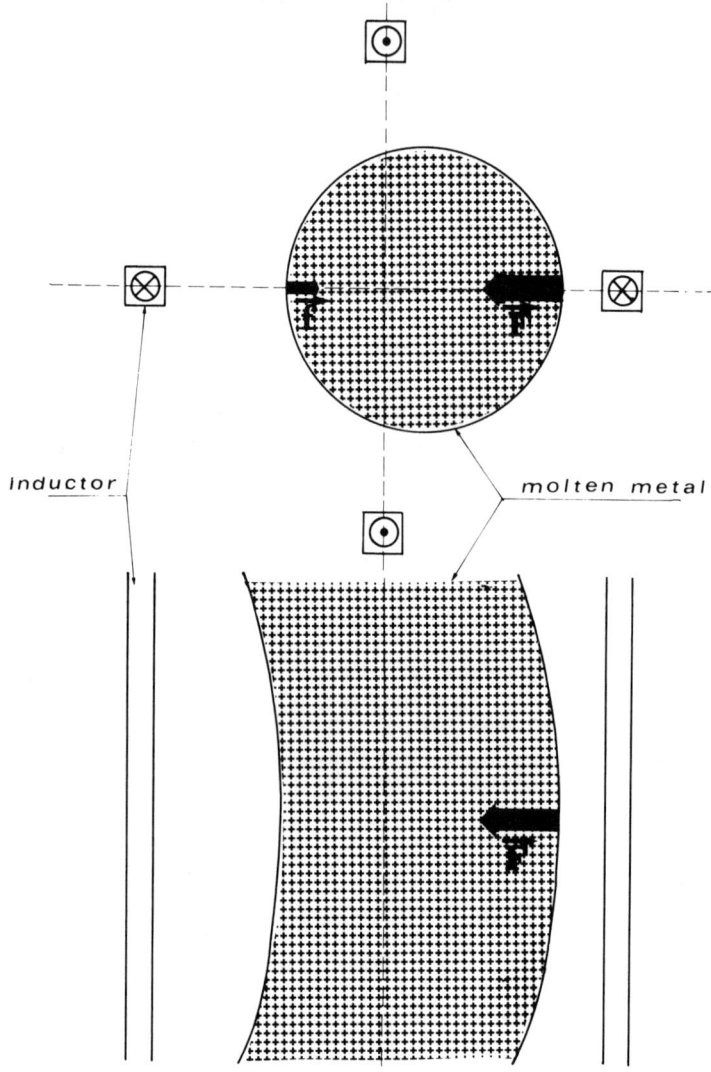

Fig. 2 Mechanism of magnetic guiding.

lying at equal distance of two conductors. The resulting electromagnetic force is therefore very weak in these regions. On the other hand, near the conductors the induced currents and magnetic field are strong, as is the electromagnetic force. The liquid metal tends, under the action of these repulsive forces, to escape from the region where the magnetic field is very intense and to flow into the region where the repulsive force is weak. The equilibrium shape of the free surface is a four-cusp blunted hypocycloid governed by the competition between magnetic pressure and surface tension.

The Magnetic Guiding

The magnetic guiding is based on the same principle as the magnetic shaping: the liquid metal tends to escape from the regions where it experiences strong magnetic flux variations. Thus, if a magnetic field configuration is imposed, in which the magnetic field intensity is very weak along a given line, a liquid-metal column initially falling in the neighborhood of this line will be flowing in such a way that its axis will coincide with this particular line. The inductor generating a quadrupole-like field produces such a configuration. Indeed, let us consider a disturbance of the liquid-metal column which brings it near one of the four conductors (Fig. 2). The magnetic field and the induced currents locally increase near this conductor and decrease in the diametrically opposed region. A strong restoring force then appears which tends to bring the metal column back to its initial position. To have a good guiding effect two conditions must be respected: the diameter of the metal column must be much larger than the skin depth computed with the frequency of the inducting currents and must be small enough, for a given electric power, to insure a balance between the momentum of the deflected column and the induced magnetic pressure.

III. Experiments and Results

Description of the Experiment

The experiment is composed of a hydraulic circuit and an oscillating electric circuit (Fig. 3) (Etay, 1980). The hydraulic mercury circuit includes a mercury tank, a pump, a constant head tank, and an experiment vein where the mercury flows vertically under the simple effect of the gravity. In order to improve the hydraulic quality of the liq-

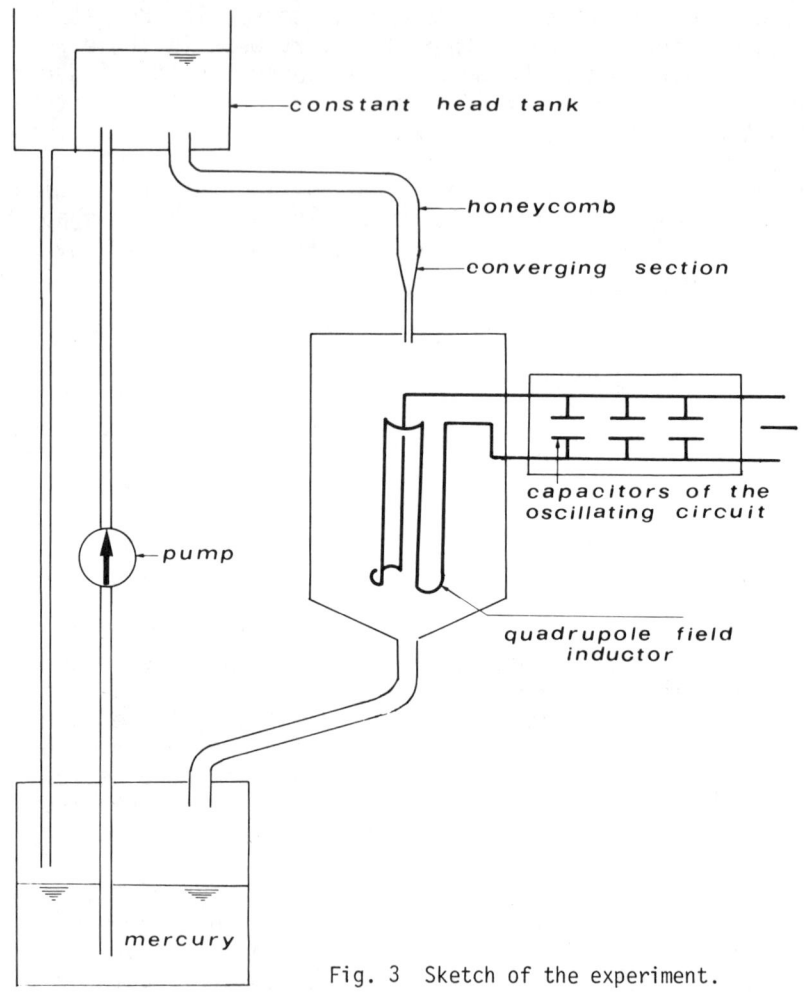

Fig. 3 Sketch of the experiment.

uid-metal column, a honeycomb and a coverging section are placed in the nozzle at the top of the vein. The oscillating circuit generates an alternating current through adjustable capacitors and the self-inductance of the quadrupole field inductor. Copper shields are placed at the top of the inductor in order to avoid any fringe effects caused by the connections between the inductor and the capacitors (Fig. 4). The induced currents generated in the shields are opposed to the currents in the connections and the resulting magnetic field within the two shields is very weak. The electric power supply is a generator which remains at all

LIQUID-METAL COLUMNS: EXPERIMENT 427

Fig. 4 Inductor with copper shields.

Fig. 5 Magnetic shaping in zero magnetic field.

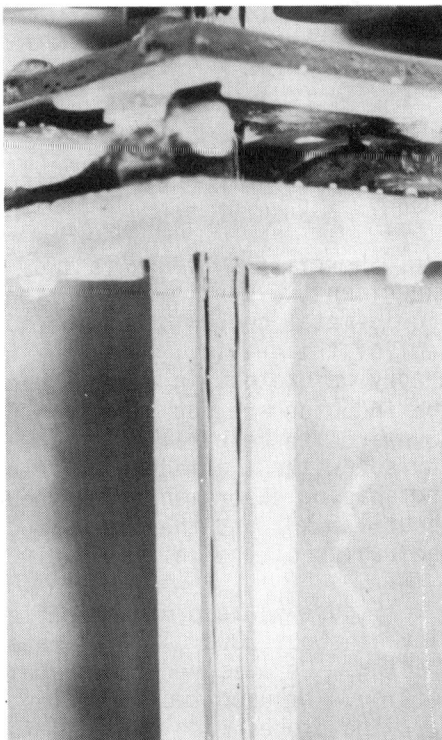

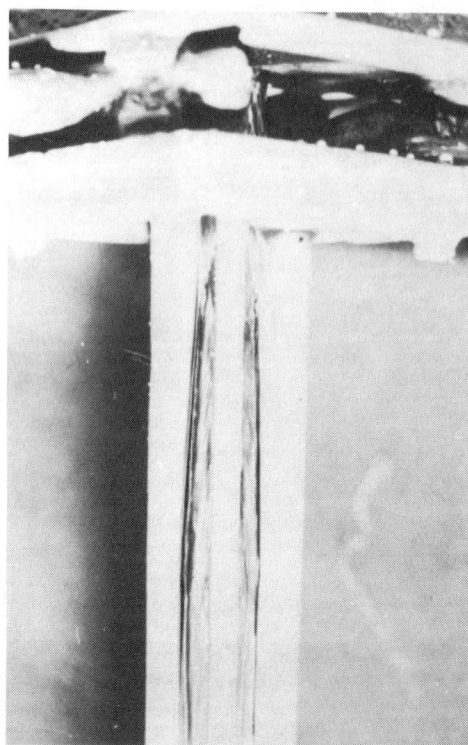

Fig. 6 Magnetic shaping in magnetic field of 3000 G.

times at the resonance by suitable modifications of the frequency.

Shercliff's theory is two-dimensional. However, in our experiment, a vertical reduction of the cross section of the liquid-metal column occurs because of the accelerating effect of the gravity. To get a better comparison with the theory and to obtain a good shaping, the four conductors of the inductor are not parallel but they converge slightly downwards to be parallel to the generating lines of the free surface without a magnetic field. In the magnetic guiding and shaping experiments the inducting current frequency is about 175 kHz and the corresponding skin depth is 1.2 mm. The radius of the initially circular column is 5 mm.

Figure 5 (zero magnetic field) and Fig. 6 (field intensity of about 3000 G) illustrate the possibility of magnetic shaping. We measured the shapes of the mercury column cross section. We used two XY tables, whose precision is 10^{-2} mm, carrying probes made of glass tubes containing a tungsten

LIQUID-METAL COLUMNS: EXPERIMENT

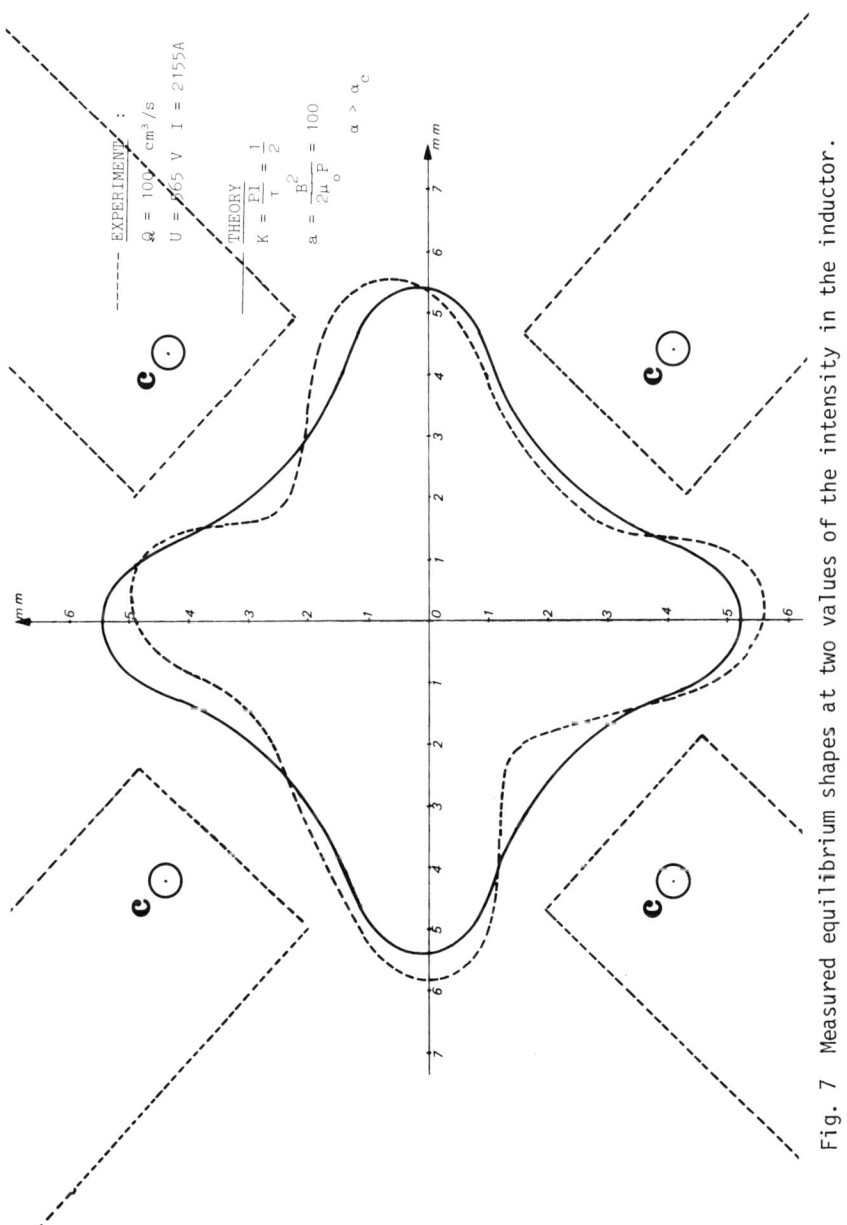

Fig. 7 Measured equilibrium shapes at two values of the intensity in the inductor.

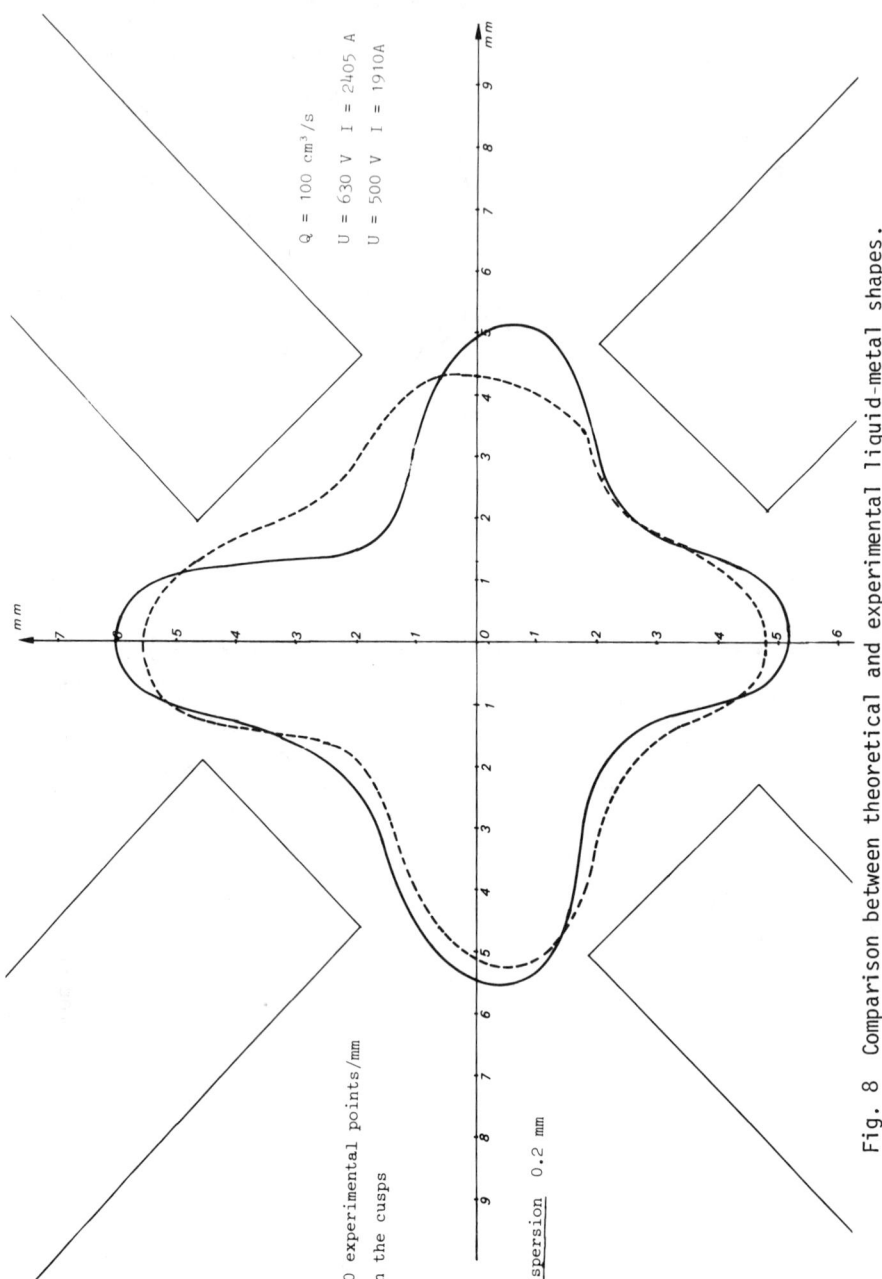

Fig. 8 Comparison between theoretical and experimental liquid-metal shapes.

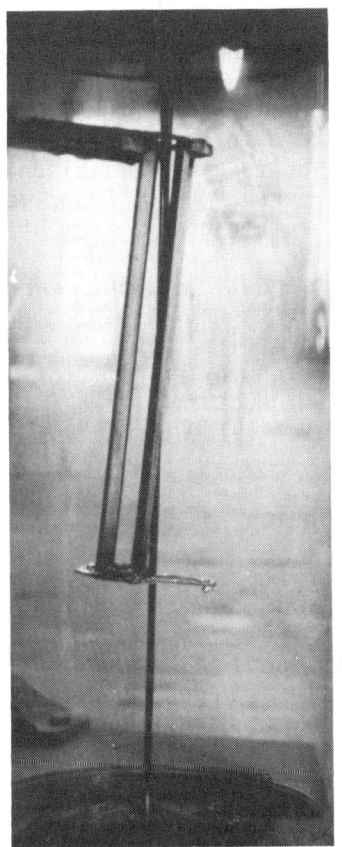 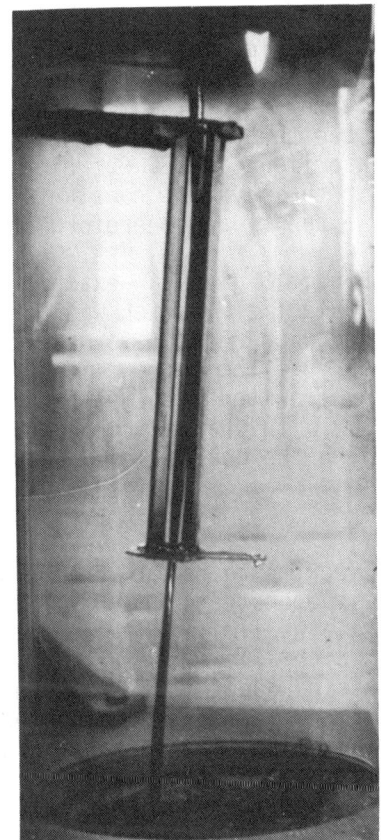

Fig. 9 Example of magnetic guiding.

wire. The measurement apparatus is very simple: when the tungsten wire touches the mercury column, it creates an electric circuit including the mercury column, a lamp, a dc generator, and the probe. The measured equilibrium shapes are plotted on Fig. 7 for various values of the intensity in the inductor and a given value of the flow rate. The shapes are not symmetrical with respect to the axis of the inductor, a result of the difficulty in precisely centering the initial circular flow in the inductor. Even with a very small default in the initial centering, the liquid-metal column weakly oscillates in space in order to bring its axis in coincidence with the axis of the inductor. A lack of symmetry results.

We compared our experimental results with Shercliff's theoretical analysis. In Fig. 8 two shapes are superimposed on one another: a theoretical one (which is symmetrical) and an experimental one which has the same area. The computed free surface corresponds to the values 0.5 and 100, respectively, for the two characteristic parameters, k and a, of the theory. The magnetic field intensity measured in our experiment is 980 G, which leads to a = 79, taking into account the physical constants of mercury. The experiments and the theory are therefore in a rather good agreement.

Figure 9 gives an example of magnetic guiding. The quadrupole field, which was used for magnetic shaping, is inclined with respect to the vertical direction. When the intensity of the current is increasing in the inductor, the mercury flow tends to get into the inductor to make its axis coincide with that of the quadrupole field without any contact with the conductors.

IV. Conclusion

Experimental simulations using mercury in industrial devices demonstrate the feasibility of magnetic shaping or guiding. We limit our interest in the study of a particular shape which allows comparison with Shercliff's theory. There is no doubt that the principle of magnetic shaping is valid for other more complicated shapes resulting from an equilibrium between magnetic pressure and surface tension. Our hope is that these experimental simulations lead to industrial applications in the metallurgical industry for continuous casting of rough metal sections.

References

Etay J. (1980) "Formage et guidage de métaux liquides sous l'action de champs magnétiques alternatifs," Rept. DEA, Univ. of Grenoble, France.

Shercliff, J.A. (1981) "Magnetic shaping of molten metal columns," Proc. Roy. Soc. London, Ser. A, Vol. 375, pp. 455-473.

Electromagnetic Devices for Molten Metal Confinement

M. Garnier*
Université de Grenoble, Grenoble, France

Abstract

Some examples of devices based upon the use of alternating magnetic fields, in which magnetic field lines play the part of walls, are presented. The aim of these devices is to achieve electromagnetically convergent or divergent flows by imposing a given area on the cross section of a liquid-metal column. Two patented devices, which mainly differ in the frequency of the applied magnetic field produce such an effect: the high-frequency device (characterized by a very strong skin effect in the liquid metal) composed of a coil of a cylindrical copper shield; and the low-frequency device, a simple coil of a suitable geometry, surrounding the liquid-metal column, provided with alternating currents leading to a skin depth as large as the radius of the column. Some experiments made with mercury to illustrate possible applications such as flow rate regulation or electromagnetic junction between two ducts are described.

I. Introduction

In classical metallurgical processes the contact between the molten metal and the walls is always undesirable and sometimes intolerable. Suppression of this contact and, with it, of chemical or physical contamination of the metal arising from the walls would lead to a better quality of

---

Paper presented at Third Beer-Sheva International Seminar on Magnetohydrodynamic Flows and Turbulence, Ben-Gurion University of the Negev, Beer-Sheva, Israel, March 23-27, 1981. Copyright © American Institute of Aeronautics and Astronautics, Inc., 1982. All rights reserved.
*Chargé de Recherches, CNRS.

the solidified ingots and to lower costs of metallurgical production.

The possibility of inducing electromagnetic forces in an electrical conducting medium, without any contact between this medium and a wall, by using alternating magnetic fields is the origin of the solution of many problems related to the presence of the walls. Some recent processes demonstrate that electromagnetic forces are able to stably support liquid-metal volumes against gravity: Okress et al. (1952) and Sagardia and Segsworth (1977) achieved the levitation melting of significant masses of metal; Getselev (1971) eliminated the ingot mold in the usual continuous casting of aluminum devices.

We present other examples of devices, based upon the use of alternating magnetic fields, in which magnetic field lines play the part of the usual walls. The aim of these devices is to achieve electromagnetically convergent or divergent flows by imposing a given area on the cross section of a liquid-metal column. Two patented devices which differ mainly in the frequency of the applied magnetic field produce such an effect.

II. The "High-Frequency" Device for Molten Metal Confinement

The device is composed of a cylindrical coil put around a vertically falling liquid-metal column and of a tubular copper shield, concentric with both the coil and the liquid-metal column, which penetrates partially inside the coil. The frequency is chosen high enough to insure that the skin depth in the liquid metal will be small compared to the radius of the molten metal column. If the thickness of the copper shield is larger than the skin depth in the copper for the chosen frequency, the magnetic field lines are extracted out of the liquid metal to penetrate into the shield. Thus the shield separates two regions: an outer region where the magnetic field is imposed by the coil and an inner region where there is no magnetic field. In the outer region the liquid metal experiences flux variations against which it reacts by generating induced currents, whose phase is opposite to that of the inducting currents. In the skin depth where they are located, the induced currents interact with the applied magnetic field to give rise to a Lorentz force directed toward the liquid-metal column axis where an overpressure arises. This overpressure vanishes with the mag-

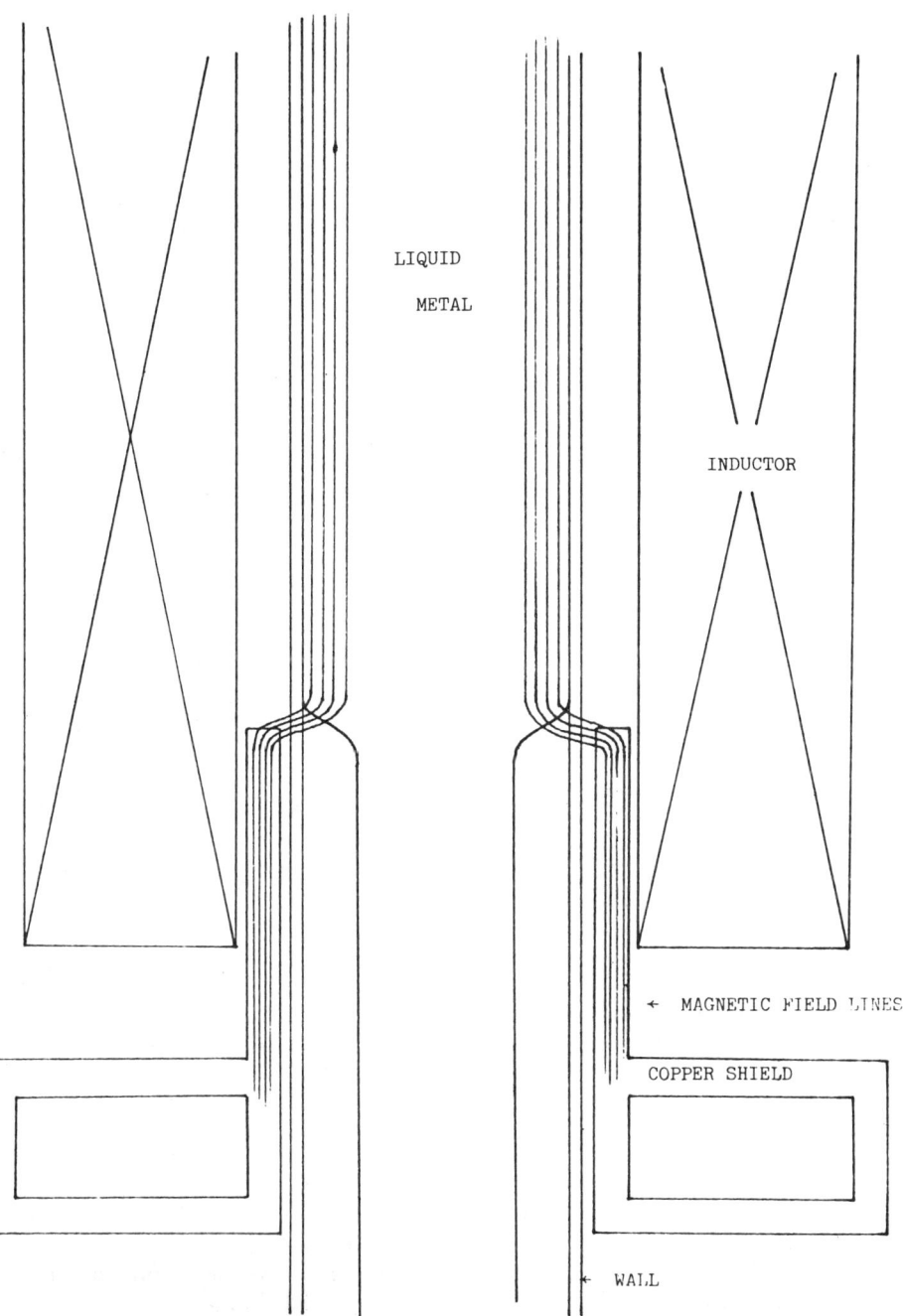

Fig. 1 "High-frequency" device for molten metal confinement.

netic field in the inner region of the shield. The momentum equation, taking into account the electromagnetic pressure, indicates that if the magnetic field vanishes, the velocity of the flow must increase. Because of the flow rate invariance, the diameter of the liquid-metal vein has to decrease. The ratio α between the diameter d of the liquid-metal column inside the shield and the diameter D outside is easily obtained

$$\alpha = \frac{d}{D} = (1 - \frac{B^2}{2\mu\rho gH})^{1/4}$$

where μ and ρ denote the magnetic permeability and the density of the metal, g the gravity, H the head of liquid metal above the constricted cross section, and B the root square value of the applied magnetic field. The separation line of the liquid metal from the wall is situated in the plane of the upper edge of the copper shield (Fig. 1).

III. The "Low-Frequency" Device for Molten Metal Confinement

This device is, from a technical point of view, simpler than the preceding one, since the copper shield is eliminated: a suitable choice of the frequency of the inducting currents leads to the same effect. The role of the shield was to separate a region where there is a strong magnetic field from a region where the magnetic field is zero. When the frequency is chosen in such a way that the skin depth δ is equal to the radius R of the liquid-metal column, two regions naturally appear: the magnetic field is nowhere zero but becomes negligible in a region where the flow must accelerate. If we suppose that $\delta = R$, we may consider that the magnetic field is zero along the liquid-metal column axis where an overpressure appears as in the preceding case. But when the diameter of the column locally decreases, the ration δ/R becomes larger than unity and the magnetic field is no longer zero along the axis: the overpressure is then weaker and the efficiency of the magnetic field too.

Consider a liquid metal flowing vertically in a duct whose end coincides with the lower edge of a coil. The magnetic field vanishes far downstream from the coil; then, to verify the momentum equation between a cross section in the coil and another where the magnetic field is zero, the diameter in the latter case must be smaller than it was when only the gravity acts. A general contraction must then appear which leads the efficiency of the magnetic field to decrease everywhere in the liquid-metal column. A new contraction arises then and the process goes on until the momentum equation is verified in all cross sections. The only stable position for the separation line locates in the plane of the upper edge of the coil; if this line moves upstream, the Lorentz force decreases and the diameter increases. On the contrary, if this line moves downstream, the Lorentz force increases and the diameter decreases. Figure 2 illustrates this device: the molten metal is simulated with mercury and the duct by a glass tube whose inner diameter is 1 in. The velocity of the flow is about 2 m/s and a mag-

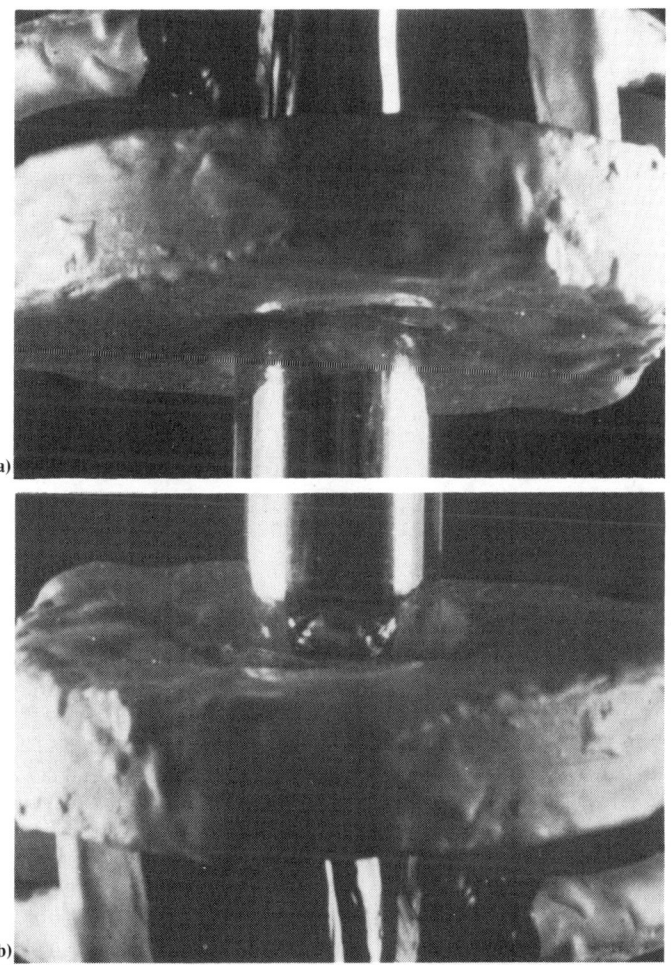

Fig. 2 "Low-frequency" device for molten metal confinement.

netic field of 1500 G leads to a reduction of the diameter of 4 for a frequency of 2500 Hz.

The formula giving the value of the ratio α between the diameters d and D after and before contraction with high-frequency fields does not hold with low frequencies. The averaged momentum in a cross section leads to the definition of the true magnetic pressure

$$B^2 e/2\mu = C_f(B^2/2\mu)$$

The coefficient C_f may be approximated by

$$C_f = 1 - R_\omega^{-\frac{1}{2}}[1 - \exp(-R_\omega^{\frac{1}{2}})]$$

where $R_\omega = 2\pi\mu\sigma f R^2$. The contraction ratio α is then given by

$$\alpha = (1 - C_f \frac{B^2}{2\mu\rho gH})^{1/4}$$

IV. Industrial Applications

The principal application of the two devices is flow rate regulation. The velocity of the liquid metal does not depend, in the constricted section, on the magnetic field and is imposed only by the head of liquid metal. To a given intensity of the current in the coil corresponds a given cross section and then a given flow rate. The ratio β of the flow rates with and without magnetic field is equal to

$$\beta = \alpha^2$$

The devices are quite reversible. It is possible to obtain the expansion of a free liquid-metal vein. When a given slice of liquid metal flows near the coil, its inner pressure and diameter increase (Fig. 3). It is possible to take advantage of these two opposite possibilities to achieve an electromagnetic joint between two ducts which are not put edge to edge to prevent leakage of the liquid metal. For this application two coils with opposed connections are to be used: the magnetic fields generated by the coils are in opposition and impose a zero magnetic field in the plane lying at equal distance from each coil where we wish to achieve the junction. The liquid metal separates from the

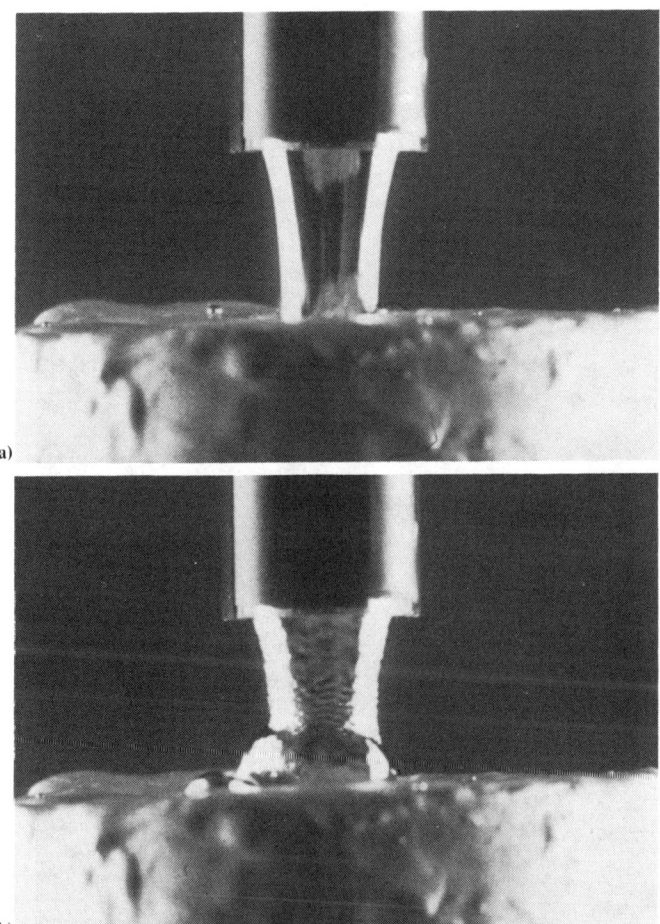

Fig. 3 Effect of the magnetic field on a free liquid-metal jet.

wall at the upper edge of the first coil and the free flow reattaches the wall at the upper edge of the second coil. Experiments of electromagnetic joints were made in our laboratory with mercury. To simulate the two ducts, we used a glass tube through which two holes were drilled. Without any magnetic field, a leak of mercury arises (Fig. 4a) which can be stopped by increasing the intensity of the current in the coil (Fig. 4b).

The possibility of suppressing the contact between a liquid-metal flow and a wall by using an alternating magnetic field may be very useful to reduce the erosion and increase the life of very expensive nozzles built of refrac-

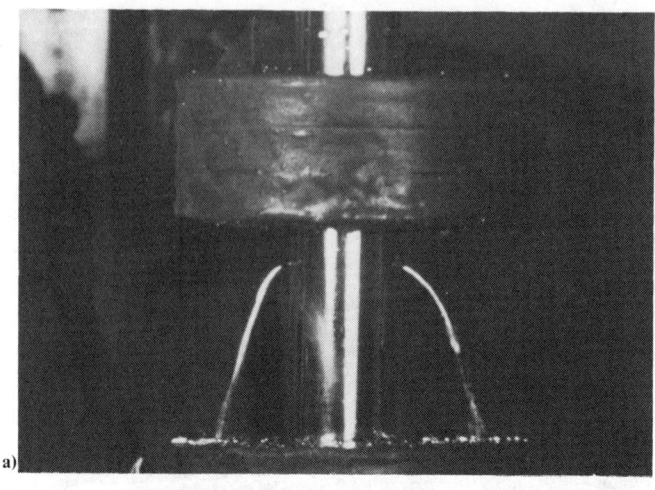

Fig. 4 Electromagnetic joint: a) without magnetic field; b) with magnetic field.

tory materials. Another application is the use of nozzles of large diameter to obtain small-diameter liquid-metal columns and to prevent any risk of obstruction.

V. Conclusion

The experiments achieved in our laboratory with mercury give new examples of the possibilities of suppressing the walls usually used to shape or to contain liquid-metal volumes and of replacing them by magnetic field lines of suitable geometry. In particular, we perfected two electromagnetic devices capable of confining a liquid-metal flow of circular or rectangular cross section. By using these devices, flow rate regulation and electromagnetic junction between two ducts are possible without any contact between the liquid metal and the wall.

References

Garnier J., Garnier M., and Moreau R. (1979) "Procédé et dispositif pour réaliser le confinement des métaux liquides par mise en oeuvre d'un champ électromagnétique," Brevet Francais No. 79-14-001.

Garnier M. and Moreau R. (1975) "Dispositif électromagnétique de confinement des métaux liquides," Brevet Francais No. 75-21-075.

Garnier M. and Moreau R. (1977) "Dispositif électromagnétique de confinement des métaux liquides pour réaliser une régulation de débit," Brevet Francais No. 77-21-121.

Getselev Z.N. (1971) "Dispositif de coulée continue et semi continue de métaux et installation pour sa mise en oeuvre," Brevet No. 2.160.281 - B 22d.

Okress E.C., Wroughton D.M., Comenetz G., Brace P.M., and Kelley J.C.F. (1952) "Electromagnetic levitation of solid and molten metals," J. Appl. Phys., Vol. 23, No. 5, p. 545.

Sagardia S.R. and Segsworth R.S. (1977) "Electromagnetic levitation melting of large conductive loads," I.E.E.E. Trans. Ind. Appl., Vol. IA 13, No. 1, p. 49.

Magnetic Levitation of Liquid Metals

J. Mestel*
University of Cambridge, Cambridge, England

This paper (see Mestel 1982) consists of an analytical and numerical study of the metallurgical process known as "levitation melting," in which a piece of metal is simultaneously levitated and heated by means of a high-frequency alternating magnetic field. This technique is of particular use when handling metals which would otherwise react with the crucible material at high temperatures. The field is taken to be generated by toroidal current loops and is thus axisymmetric and poloidal. With this geometry, the fluid tends to drip down the axis of symmetry along which the Lorentz force vanishes, and thus surface tension must be relied upon to provide the necessary support there. The magnetic Reynolds number is assumed low and thermal effects are neglected.

The governing equations for the mean velocity field and the unknown surface shape are derived in terms of three dimensionless parameters: the Reynolds number $R_e \gg 1$, the magnetic field penetration depth $\delta \ll 1$, and a Weber number W which measures the strength of surface tension. The fluid behavior is discussed in general, and numerical solutions are found in the limiting cases of high surface tension $W \to \infty$ and high frequency $\delta \to 0$.

In the limit of high surface tension the adopted shape is almost spherical, and a perturbation analysis is present-

---

Paper presented at Third Beer-Sheva International Seminar on Magnetohydrodynamic Flows and Turbulence, Ben-Gurion University of the Negev, Beer-Sheva, Israel, March 23-27, 1981. Copyright © American Institute of Aeronautics and Astronautics, Inc., 1982. All rights reserved.
*Department of Applied Mathematics and Theoretical Physics.

MAGNETIC LEVITATION OF LIQUID METALS 443

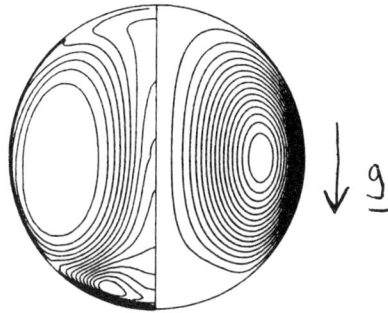

Fig. 1 Coil at $(R,z) = (1.732, -1)$, flow solution for $R_e = 400$ and $\delta = 0.05$ (right: streamlines; left: contours of potential vorticity).

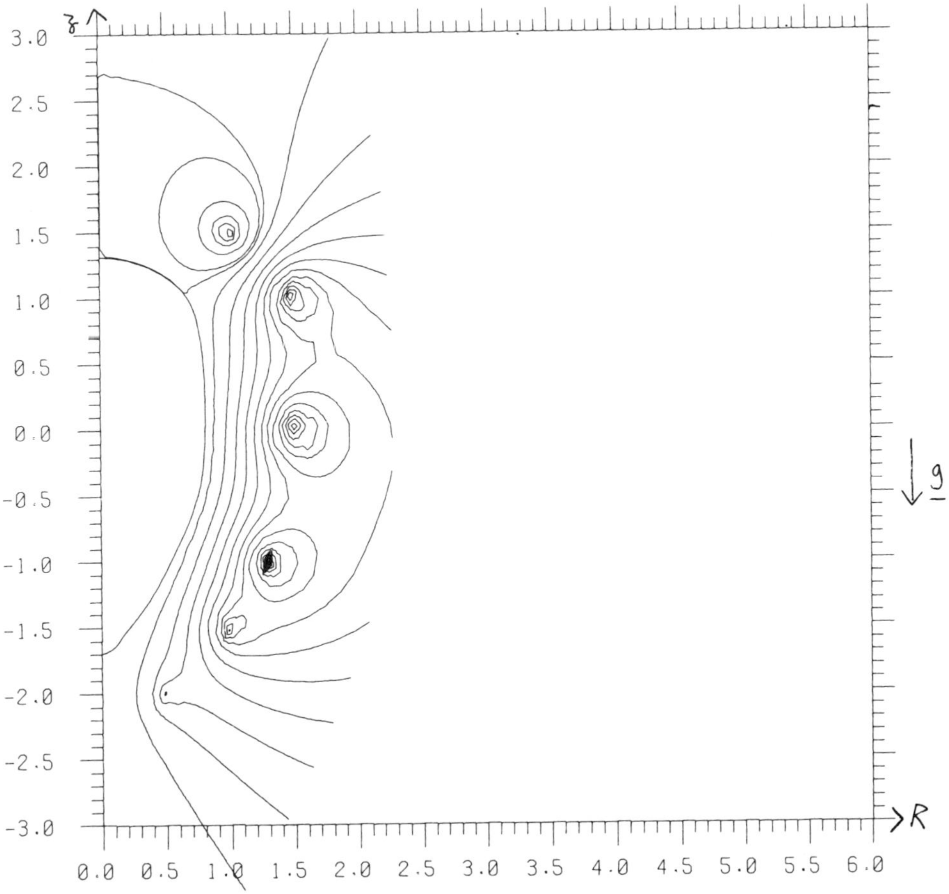

Fig. 2 Coils at $(R,z) = (0.5, -2)$, $(1, -1.5)$, $(1.3, -1)$, $(1.5, 0)$ and $(1.5, 1)$ and counterwound coil at $(R,z) = (1, 1.5)$ (magnetic field lines and equilibrium shape for $W = 0.2$).

ed. Finite-difference techniques are used to solve the Navier-Stokes equations inside a sphere and the surface perturbation then calculated. For the case of a single current loop a solution for the unperturbed velocity field is shown in Fig. 1. The streamlines are plotted for comparison alongside the contours of potential vorticity (i.e., the vorticity divided by the distance from the axis of symmetry). It is found that the potential vorticity is approximately constant along the streamlines and that outside the boundary layer there exists a large plateau region, in keeping with theoretical predictions.

In the high-frequency limit, it is shown that the dynamic pressure is negligible with respect to the magnetic pressure and gravity. The resulting magnetostatic problem is solved for the free surface shape by time-stepping techniques. For a particular value of W, Fig. 2 shows the equilibrium shape and magnetic field lines for a coil configuration consisting of five loops wound in one sense with one counterwound on top. In this reasonable model of a practical levitation device, the counterwinding is required to insure stability of the levitated metal. Further to this paper, attempts are being made to incorporate into the model the dynamic effects which are present at finite frequencies.

Reference

Mestel A.J. (1982) "Magnetic levitation of liquid metals," *J. Fluid Mech.*, Vol. 117, pp. 27-43.

New Electromagnetic Measuring Methods in Continuous Casting

F.R. Block*
Technical University, Aachen, Federal Republic of Germany

Abstract

The bath level in the mold of a continuous casting machine and the skin thickness of the billet can be measured using alternating axial magnetic fields. These magnetic fields induce azimuthal electromotive forces in the billet. The resulting electrical currents depend on the conductivity distribution. The magnetic fields of these electrical currents are determined from the induced voltage in a secondary coil. In the bath level measurement zone the conductivity is dependent on the presence of molten steel. The billet skin thickness measurement makes use of the fact that the conductivity is a monotonic function of temperature and exhibits a discontinuity between the liquid and solid phases. The conductivity distribution is determined by applying primary currents of different frequencies f_n. Three methods used to find the relation between the measured voltage $U(f_n)$ and the conductivity $\sigma(r)$ are described.

I. Introduction

In the field of continuous casting, two important measuring problems are encountered: the determination of the bath level in tubular molds and the nondestructive measurement of the skin thickness of the billet.

The methods to determine the bath level in the mold have to be reliable and should not disturb the continuous operation of the plant. Measuring techniques must also be

---

Paper presented at Third Beer-Sheva International Seminar on Magnetohydrodynamic Flows and Turbulence, Ben-Gurion University of the Negev, Beer-Sheva, Israel, March 23-27, 1981. Copyright © American Institute of Aeronautics and Astronautics, Inc., 1982. All rights reserved.
*Privatdozent, Dr. rer. nat.

used if the casting occurs in a closed system or if the molten steel surface is obscured by a powder layer. The installation should not disturb access to the mold and must be protected against damage caused by the overflowing steel. However, the only measuring method that fulfills all of these requirements uses γ rays, which entails a certain radiation risk.

II. Alternative Methods

Therefore, as an alternative an electromagnetic method has been developed in cooperation with Aciérie Réunies de Burbach-Eich-Dudelange, S.A. (ARBED). It is applicable to tubular molds of any cross section and uses axial alternating electromagnetic fields. The axial magnetic fields are generated by coils which surround the mold (Fig. 1). The axial magnetic fields are less strongly screened by the copper walls of the mold than the radial fields.

The reason for the stronger penetration can be deduced from Faraday's law: Normal alternating magnetic fields induce large-scale eddy currents in the walls of the copper mold and are therefore screened out of the liquid steel re-

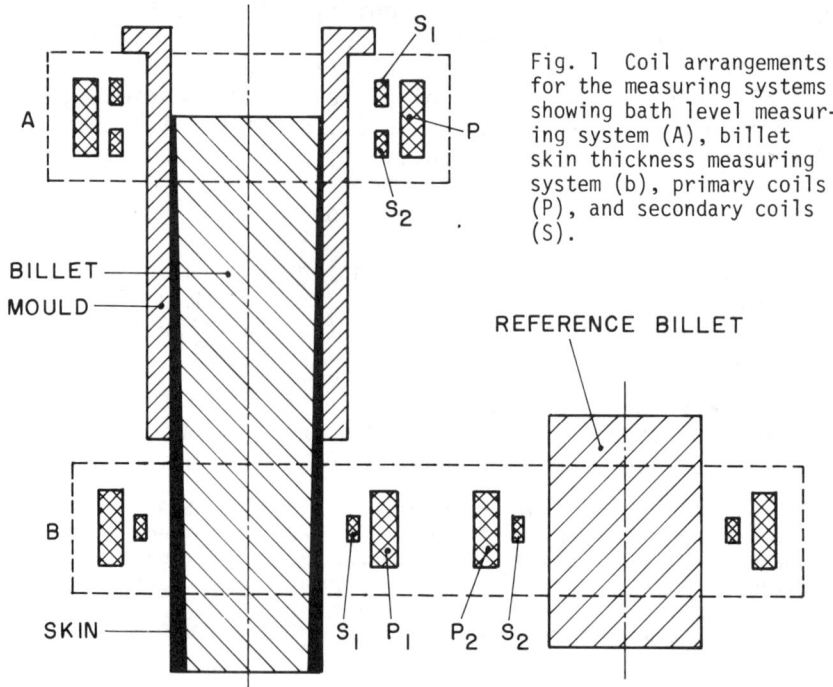

Fig. 1 Coil arrangements for the measuring systems showing bath level measuring system (A), billet skin thickness measuring system (b), primary coils (P), and secondary coils (S).

gion. On the other hand, if the fields are directed parallel to the wall of the mold, eddy currents are limited by the thickness of the wall. The parallel fields penetrate the copper wall more easily and induce in the liquid steel significant currents which can be detected by a suitable arrangement of secondary coils. The voltage of the secondary coils allows one to determine the level of the bath to within 1 mm over a range of 150 mm, if the bath level is undisturbed.

This new mold level control system has been proved in industrial plants and will be applied in other continuous casting machines.

For measuring the thickness of the solidified skin of billets leaving the mold (Fig. 1), another electromagnetic method has been proposed and tested in the laboratory. A precise knowledge of the solidification helps to determine the optimal withdrawal speed. Until now the growth of the skin thickness has been measured by drilling into the billet after certain periods of cooling and allowing the liquid steel to flow out. This has meant the destruction of the billet.

The new method makes use of the temperature dependence of the electrical conductivity $\sigma = \sigma(T)$, and the fact that a discontinuity exists in the conductivity when changing from the liquid to the solid phase.

Figure 2 shows a typical example for the resistivity of steel as a function of the temperature (Landolt-Börnstein 1972). Because the graph is monotonic, the temperature dependence on the radius and especially the solid/liquid boundary can be derived by determining the electrical conductivity as a function of the radius of the billet.

A primary axial magnetic field is produced by a long coil or a Helmholtz coil placed around the billet. The magnetic fields of the induced currents circulating in the billet are superposed onto the primary field. The resulting field is measured by a second coaxial coil placed in the center of the primary coil in order to be in the homogeneous region of the field.

Since the measurement of the electromotive force in the secondary coil uses virtually no power, the disturbance of the fields by the measuring coil can be neglected if the windings are made from a nonmagnetic material such as copper.

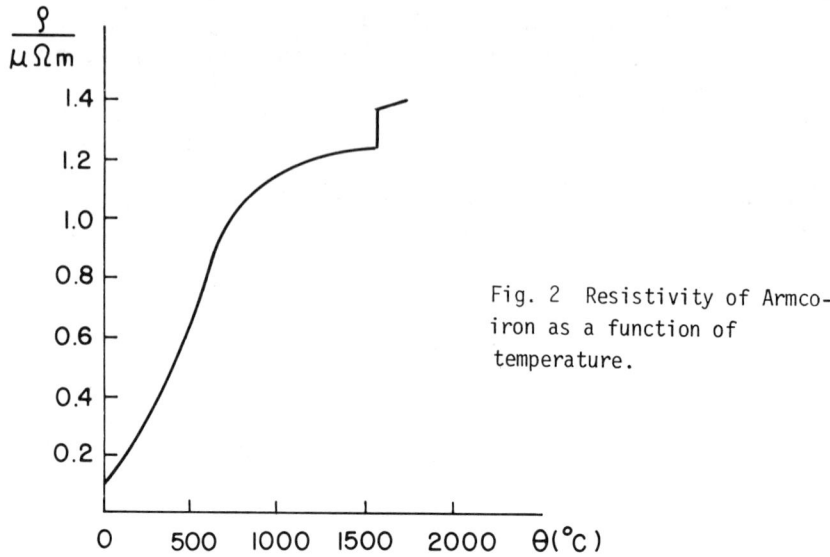

Fig. 2 Resistivity of Armco-iron as a function of temperature.

In order to reach zones of different depth in the billet, fields of different frequencies are used. High-frequency fields give information primarily on the outer layers of the rod; low-frequency fields provide information about the internal as well as external layers.

The results obtained with differing frequency fields are used to deduce the conductivity as a function of the radius and, therefore, the temperature distribution across the rod. The conductivity distribution can be considered as the superposition of a large temperature-independent term and a small temperature-dependent part. The desired signal forms only a small part of the total electromotive force in the secondary coil.

In order to be able to increase the measuring sensitivity, it is advisable to subtract the nontemperature-dependent component from the signal. This is achieved by using a comparator circuit in which a standard of known constant conductivity is placed. Figure 3 shows diagrammatically the arrangement of the complete circuit.

Theoretical considerations start with Maxwell's equations. In the applied frequency range up to 10 kc/s, the rate of change with time of the electrical displacement is small compared with the electrical current in the steel.

Therefore, the equation governing the magnetomotive force can be simplified to Ampere's law.

The relativistic invariant material equations (Pauli 1963) can be reduced to those valid for a local center of mass system, because:

1) The speed of the bar is limited to $v \leq 0.07$ m/s; therefore, all terms of the order v^2/c^2 (where c is the light velocity) can be neglected.

2) The main part of the field lies parallel to the velocity of the billet.

3) The term $(\varepsilon - \varepsilon_0 \mu_0/\mu)$ is much smaller than the permittivity of vacuum ε_0.

4) Steel is a good conductor.

From Ampere's, Ohm's, and Faraday's laws the equation

$$\text{curl } \frac{1}{\sigma} \text{curl } \overline{H} = -\mu \frac{\partial \overline{H}}{\partial t} \qquad (1)$$

is deduced.

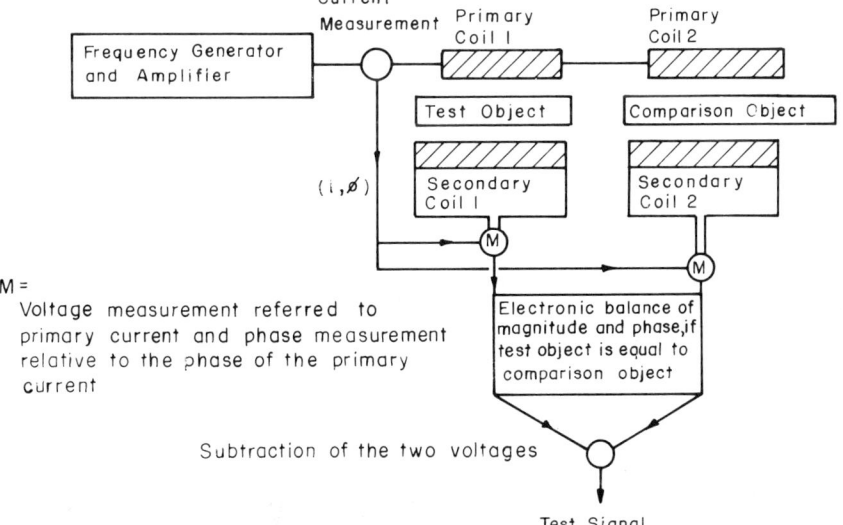

Fig. 3 Principal structure of a circuit arrangement including compensation.

If $\overline{H}$ has only an axial component and is independent of the axial coordinate z and angle ϕ, this equation simplifies in cylindrical coordinates to the equation

$$\frac{1}{r}\frac{\partial}{\partial r}\frac{r}{\sigma}\frac{\partial H_z}{\partial r} = \mu\frac{\partial H_z}{\partial t} \qquad (2)$$

For this differential equation the two following boundary conditions apply:

1) The magnetic field strength is the same on both sides of the surface boundary of the billet because of Ampere's law

$$H_z(R) = H_z^*(R) = IN_p/\ell \qquad (3)$$

where I is the current in the primary coil, N_p/ℓ the number of windings per length, and the asterisk indicates the region $R \leq r \leq {}^1R$ between the billet and the coil.

2) Along the axis the inner boundary condition holds that the angular component of the electric current has to vanish and therefore also the derivative of the magnetic field with respect to radius

$$\lim_{r\to 0} j_\phi = \lim_{r\to 0}(\operatorname{curl}\overline{H})_\phi = \lim_{r\to 0}\left(-\frac{\partial H_z}{\partial r}\right) = 0 \qquad (4)$$

For the electromotive force u in the secondary coil, it follows that

$$u = N_s\left[\int_0^R \mu\frac{\partial H_z}{\partial t}2\pi r\,dr + \int_R^{{}^1R}\mu\frac{\partial H_z^*}{\partial t}2\pi r\,dr\right]$$

$$= N_s\left[\lim_{r\to R}\frac{2\pi}{\sigma}r\frac{\partial H_z}{\partial r} + \pi({}^1R^2 - R^2)\mu^*\frac{\partial H_z^*}{\partial t}\right] \qquad (5)$$

where N_s is the number of windings of the secondary coil.

The determination of the conductivity σ as a function of the radius of the billet from the differential equation (2), the boundary conditions Eqs. (3) and (4), and the measured electromotive force in the secondary coil [Eq. (5)] leads to an unusual mathematical problem, because the differential equation (2) for the magnetic field H_z cannot be generally integrated if the conductivity is an unknown function of the radius. Three methods were developed for solving this problem.

In the first method the conductivity distribution is replaced by a series of suitable step functions with assumed parameters. For constant conductivity and harmonic time dependence of the primary field, Eq. (2) reduces to the zero-order Bessel equation. The solution is given by a linear combination of the Bessel function J_0 and the Neumann function N_0. The boundary conditions between adjacent layers are derived from the continuity of the tangential components of the magnetic and the electric fields. If values for the conductivity and the radii of the shells are assumed, the magnitude and phase of the electromotive force in the secondary coil can be calculated in general for any frequency: $^i r$, $^i \sigma \rightarrow U(f_n)$.

These solutions are compared with the electromotive force measured using primary currents of different frequencies. If the billet is divided into N shells, measurements at N different frequencies are required.

Figure 4 shows as an example the calculated voltage difference in appropriate units for a cylindrical 85 mm radius billet, in which the liquid phase has a conductivity of $^1\sigma$ = 725 kS/m and the conductivity of the solidified skin is assumed to have an average value of $^2\sigma$ = 806 kS/m. It depicts the voltage for solidified skin thicknesses of 9 and 10 mm. If the skin thickness is small compared to the radius of the billet and if the radius is small compared to the length of the primary coil, then experimental and theoretical results correspond quite closely.

The second method is to measure in the same manner as in the first method at different frequencies and to calculate the shell radii and their conductivities directly from the measurements

$$U(f_n) \rightarrow {}^i r, {}^i \sigma$$

Because the linear system of the boundary equations contains the unknown radii and conductivities within the Bessel and Neumann functions, it is extremely laborious to solve this transcendental system. This calculation can be performed by applying a technique by Booth (1949).

The third method is applicable if the conductivity varies only to a small degree in the billet, which is normally true. In this case the resistivity is written as

$$\rho = \rho_0 + \rho_1 \qquad (6)$$

where ρ_0 is a mean value of ρ taken as constant and ρ_1 describes the change of ρ.

The magnetic field strength can be composed correspondingly as

$$H = H_0 + H_1 \qquad (7)$$

For small changes in ρ, a ratio $H_1/H_0 \ll 1$ can be expected.

Neglecting terms of second-order magnitude in ρ_1 and H_1, the fundamental Eq. (2) results in a linear differential equation for H_1. The unknown function ρ_1 appears only in the inhomogeneous part. Therefore, it is possible to find the total solution by superposition of the homogeneous solution (which has to be zero in this case) and the particular solution which may be determined by using Green's method.

For the description of the correction ρ_1 as a function of the radius, suitable functions with unknown parameters are used and the parameters are determined by the least square method.

The first method for calculating conductivity distribution is the simplest way of finding the relationship between conductivity distribution and measured voltages. However, this method is too time-consuming for continuous measurements if the skin thickness is not restricted to a small range of values. This restriction does not apply for the second method of calculating the conductivity distribution. On the other hand, this method requires an extensive computing facility.

If the conductivity changes over the billet radius are small with respect to the average conductivity, the third method is preferable.

A prototype for factory operation is in development in cooperation with Betriebsforschungsinstitut des Vereins Deutscher Eisenhüttenleute.

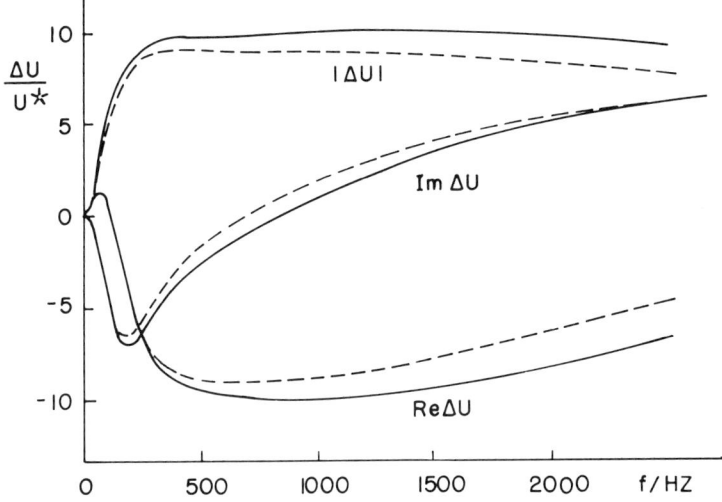

Fig. 4 Magnitude and real and imaginary parts of the induced voltage difference as a function of frequency for two different skin.

References

Booth A.D. (1949) "An application of the method of steepest descents to the solution of systems of nonlinear simultaneous equations," Q.J. Mech. Appl. Math., Vol. 2, pp. 460-468.

Landolt-Börnstein (1972) Stoffwerte von Eisen, Springer-Verlag, New York, Vol. 4 (2a), p. 228.

Pauli W. (1963) Theorie of Relativity, Pergamon Press, London.

Author Index for Volume 84

Alemany, A. 402
Alexion, C.C. 263,266
Baker, R.C. 225
Block, F.R. 445
Branover, H. 83,160,176
Brown, P. 113
Ciarelli, P.A. 263
Claesson, S. 83,160
El-Boher, A. 160
Etay, J. 422
Fabris, G. 216
Fautrelle, Y. 374
Fujii-e, Y. 53,323,339
Garnier, M. 422,433
Herman, H. 127,138
Hide, R. 90
Hughes, W.F. 287,313
Hunt, J.C.R. 359
Inoue, S. 53
Itoh, Y. 323,339
Jackson, W.D. 193
Kanagawa, T. 323,339
Keeton, A.R. 263

Lederman, S. 113
Marty, Ph. 402
McNab, I.R. 263,266,287
Mestel, A.J. 442
Miyazaki, K. 53,323,339
Moore, D.J. 359
Moreau, R. 20
Naot, D. 98
Narasimha, R. 30
Petrick, M. 127
Pierson, E.S. 127,138
Posillico, T. 113
Ricou, R. 387,402
Rodi, W. 98
Shercliff, J.A. 414
Timnat, Y.M. 94
Umezawa, S. 339
Vives, Ch. 387,402
Walker, J.S. 3
Winkleblack, R.K. 266
Winowich, N.S. 313
Yakhot, A. 160,176
Yamaoka, N. 53,323